Nanoscale Assembly

Chemical Techniques

Nanostructure Science and Technology

Series Editor: David J. Lockwood, FRSC
National Research Council of Canada
Ottawa, Ontario, Canada

Current volumes in this series:

Alternative Lithography: Unleashing the Potentials of Nanotechnology
Edited by Clivia M. Sotomayor Torres

Interfacial Nanochemistry: Molecular Science and Engineering at Liquid–Liquid Interfaces
Edited by Hiroshi Watarai, Norio Teramae, and Tsuguo Sawada

Nanoparticles: Building Blocks for Nanotechnology
Edited by Vincent Rotello

Nanoscale Assembly: Chemical Techniques
Edited by Wilhelm T.S. Huck

Nanostructured Catalysts
Edited by Susannah L. Scott, Cathleen M. Crudden, and Christopher W. Jones

Nanotechnology in Catalysis, Volumes 1 and 2
Edited by Bing Zhou, Sophie Hermans, and Gabor A. Somorjai

Ordered Porous Nanostructures and Applications
Edited by Ralf Wehrspohn

Polyoxometalate Chemistry for Nano-Composite Design
Edited by Toshihiro Yamase and Michael T. Pope

Self-Assembled Nanostructures
Jin Z. Zhang, Zhong-lin Wang, Jun Liu, Shaowei Chen, and Gang-yu Liu

Semiconductor Nanocrystals: From Basic Principles to Applications
Edited by Alexander L. Efros, David J. Lockwood, and Leonid Tsybeskov

Surface Effects in Magnetic Nanoparticles
Edited by Dino Fiorani

Nanoscale Assembly

Chemical Techniques

Edited by

Wilhelm T.S. Huck
University of Cambridge, Cambridge
United Kingdom

Wilhelm T.S. Huck
Department of Chemistry
Melville Laboratory
University of Cambridge
Lensfield Road
CB2 IEW, Cambridge, UK

Series Editor:
David J. Lockwood
National Research Council of Canada
Ottawa, Ontario
Canada

Cover illustration: Top left, top right and bottom left image by Kristen Frieda and bottom right by W.T.S. Huck

ISBN-10: 1-4614-9856-2
ISBN-13: 978-1-4614-9856-8
ISBN 0-387-25656-3 (eBook)
Printed on acid-free paper.

Softcover re-print of the Hardcover 1st edition 2005

Printed in the United States of America. (TB/EB)

9 8 7 6 5 4 3 2 1

springeronline.com

Preface

Nanotechnology has received tremendous interest over the last decade, not only from the scientific community but also from a business perspective and from the general public. Although nanotechnology is still at the largely unexplored frontier of science, it has the potential for extremely exciting technological innovations that will have an enormous impact on areas as diverse as information technology, medicine, energy supply and probably many others. The miniaturization of devices and structures will impact the speed of devices and information storage capacity. More importantly, though, nanotechnology should lead to completely new functional devices as nanostructures have fundamentally different physical properties that are governed by quantum effects. When nanometer sized features are fabricated in materials that are currently used in electronic, magnetic, and optical applications, quantum behavior will lead to a set of unprecedented properties. The interactions of nanostructures with biological materials are largely unexplored. Future work in this direction should yield enabling technologies that allows the study and direct manipulation of biological processes at the (sub) cellular level.

Nanotechnology has made considerable progress due to the development of new tools making the characterization and manipulation of nanostructures available to researchers around the world. Scanning probe technologies such as STM and AFM (and a range of modifications) allow the imaging and manipulation of individual nanoparticles or even individual molecules. At the same time, the development of extreme lithographic techniques such as e-beam, focused ion beam and extreme UV, now allow the fabrication of metal and polymer colloids with nanometer dimensions. Still, the fabrication of nanoscale building blocks is not a trivial task, especially when large numbers of identical nanostructures are required. For example, fascinating structures and devices can be made from nanosized GaAs islands grown on surfaces via nucleation and growth strategies. One of the inherent problems associated with such strategies is the variation of structures within the system. Even colloidal metals that are grown in solution like gold or CdSe quantum dots are not identical. There is reason to believe that entirely new manufacturing processes need to be invented to deliver these structures for economically viable processes. At the same time, new device layouts need to be developed that can tolerate a specific uncertainty in its building blocks.

Fabrication is difficult, but the large-scale assembly of nanoscale building blocks into either devices (e.g. molecular electronic, or optoelectronic devices), nanostructured materials, or biomedical structures (artificial tissue, nerve-connectors, or drug delivery devices) is an even more daunting and complex problem. There are currently no satisfactory strategies that allow the reproducible assembly of large numbers of nanostructures into large numbers of functional assemblies. It is unlikely that a robotic system could assemble nanoscale devices. A key issue will be the development of tools to integrate nanostructures into functional assemblies. Scanning probe lithographies such as AFM and STM that allow the manipulation of single molecules or nanoparticles could certainly provide a route towards functional structures and prototype devices. Recent examples such as the Millipede project of IBM have shown that 1000's of AFM tips that are individually addressable can be fabricated. However, such strategies require immense engineering efforts and are not generically applicable to a wide range of materials or structures. Furthermore, scanning probe techniques are essentially 2D and the fabrication of 3D nanostructures materials would present a significant hurdle. It is therefore very likely that any economically feasible assembly route will incorporate to a certain extent the principles of self-assembly and self-organization. After all, many inspirations for nanotechnology come from Nature where precisely these processes control the very fabric of life itself: The chemical recognition and self-assembly of complementary DNA strands into a double helix.

Chemists are beginning to master self-assembly as a tool to mimic biological processes using non-natural molecules or even nanoparticles. At the same time, our increased understanding of molecular biology should enable us to exploit biological "machinery" directly for the fabrication of synthetic nanostructures. Self-assembly is the spontaneous formation of ordered structures via non-covalent (or reversible) interaction between two objects (molecules, proteins, nanoparticles, or microstructures) can lead to a well-defined assembly. Directionality can be introduced through the type of interaction or via the shape of the object. Self-assembly is a spontaneous, energetically favorable process and leads, in principle, to perfect structures, if allowed to reach its lowest energy level. No nanoassemblers or nanorobots are required to physically manipulate objects. All information required for the assembly of a well-defined superstructure is present in the building blocks that are to be incorporated in the assembly. In practice, defect-free structures are difficult to obtain as it can take very long to reach equilibrium. Furthermore, all structures that are formed are dynamic, i.e. changing over time, as they are not covalently bound. It will hence be necessary to design device layouts with built-in defect tolerance.

In this book we will take a closer look at a great variety of different strategies that are pursued to assemble and organize nanostructures into larger assemblies and even into functional devices or materials.

Contents

1

Structure Formation in Polymer Films

From Micrometer to the sub-100 nm Length Scales

Ullrich Steiner

INTRODUCTION

Applications ranging from state-of-the-art lithography in the semiconductor industry to molecular electronics require the control of polymer structures on length scales down to individual molecules. Structures on nanometer length scales can be achieved by employing a "bottom-up" approach, in which individual molecules are assembled to form a structural entity [1]. By using bottom-up technologies it is, however, by no means trivial to interface the macroscopic world. Technologies that are applied in practice usually require the modification and control of structures extending from the smallest units to the millimeter length scale. Traditionally, this is achieved by a "top-down" approach that has miniaturized the originally 1 centimeter-sized transistor down to the 100 nm structures found on a Pentium® chip [2].

Neither bottom-up nor top-down technologies will by themselves achieve structural control on a molecular level combined with macroscopic addressability. In terms of the top-down approach, the challenge lies in the drive for ever decreasing structure sizes. A second aspect is, how existing top-down technologies can be extended to interface with structures made using a bottom-up method. The top-down approach is pursued by the semiconductor industry, with the aim to implement optical lithography down to length scales of several tens of nanometers [3]. Alternatively, new top-down methods have demonstrated the transfer of structures down to 100 nm (in some instances down to 10 nm). This includes the various "soft lithography" techniques (micro-contact printing, micro-molding, etc.) [4], but also the

Cavendish Laboratory, Department of Physics Madingley Road, Cambridge CB3 OHE, UK. u.steiner@phy.cam.ac.uk creation of surface patterns by embossing [5], injection molding [6], or various scanning probe techniques.

In addition, patterns created by surface instabilities can be used to pattern polymer films with a lateral resolution down to 100 nm [7]. Here, I summarize various possible approaches that show how instabilities that may take place during the manufacture of thin films can be harnessed to replicate surface patterns in a controlled fashion. Two different approaches are reviewed, together with possible applications: (a) patterns that are formed by the demixing of a multi-component blend and (b) pattern formation by capillary instabilities.

1.1. PATTERN FORMATION BY DEMIXING

Most chemically different polymers are immiscible due to their much reduced entropy of mixing compared to their low molecular weight analogs [8]. The control of the bulk phase morphology of multicomponent polymer blends is therefore an important topic in materials science and engineering. In thin films, the phase separation process is strongly influenced by the confining surfaces both thermodynamically [9], and by kinetic effects that take place during the preparation of the film [10]. This sensitive dependence of the polymer phase morphology on the boundary conditions provides a possibility to steer the phase separation process. Using suitably chosen processing parameters, a simple film deposition process can be harnessed for micrometer and sub-micrometer pattern replication. We limit ourselves here to structure formation processes caused by the demixing of homopolymer blends, but note that there are various similar attempts involving block-copolymer systems [1].

1.1.1. Demixing in Binary Blends

A weakly incompatible polymer blend quenched to a temperature belows its critical point of demixing develops a phase morphology exhibiting a single characteristic length scale [11]. Initially, a well defined spinodal pattern evolves which coarsens with increasing times. Most practically relevant polymer blends are, however, strongly incompatible. They cannot be blended into a homogeneous phase and their phase morphology is therefore determined by the sample preparation procedure. Thin polymer films are typically made by a solvent casting procedure, often by spin-coating (Fig. 1.1). When using a polymer blend, the polymers and the solvent form initially a homogeneous mixture. Solvent evaporation during spin-coating causes an increase in the polymer concentration that eventually leads to polymer-polymer demixing [12]. Films made this way exhibit a characteristic phase morphology, as shown in Fig. 1.2 [13].

The lateral morphology in Fig. 1.2 seems similar to the morphologies observed in bulk demixing [11]. It is therefore tempting to compare this phase separation process with the well understood demixing in a solvent-free weakly incompatible blend. This may, however, not be appropriate, for several reasons. Due to the high viscosity of polymer blends, hydrodynamic effects are strongly suppressed in weakly incompatible melts, while they are by no means negligible in solvent containing mixtures. Secondly, the presence of

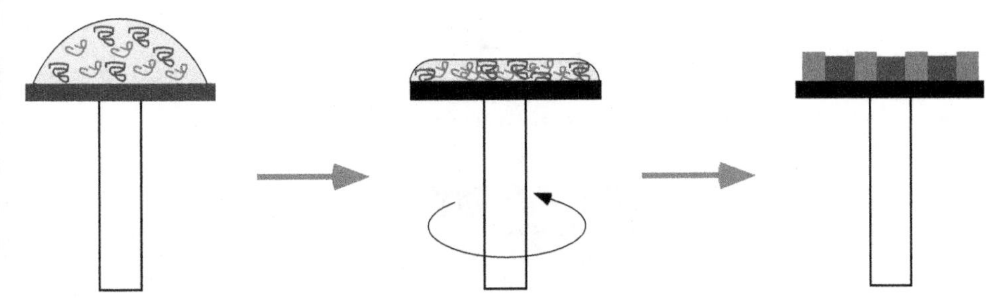

FIGURE 1.1. Schematic representation of a spin-coating experiment. Initially, the two polymers and the solvent are mixed. As the solvent evaporates during film formation, phase separation sets in resulting in a characteristic phase morphology in the final film (from [7]).

the two confining surfaces in thin films modify the demixing process [10], and thirdly, the rapid film formation by spin-coating is a non-equilibrium process, as opposed to the quasi-static nature of phase formation in the melt. In particular, the rapid solvent evaporation gives rise to polymer concentration gradients in the solution and to evaporative cooling of the film surface. Both effects may be the origin of convective instabilities [14].

Preliminary studies have identified a likely scenario that gives rise to the lateral morphologies observed in Fig. 1.2. This is illustrated in Fig. 1.3 [15]. The continuous increase in polymer concentration during spin-coating initiates the formation of two phases, each rich in one of the two polymers. Since both phases still contain a large concentration of solvent ($\sim$90%), the interfacial tension of the interface that separates the two phases is much smaller compared to the film boundaries. The film therefore prefers a layered over a laterally structured morphology. As more solvent evaporates, two scenarios can be distinguished. Either the layered configuration is stable once all the solvent has evaporated (as, for example in Fig. 1.2c), or a transition to lateral morphology takes place.

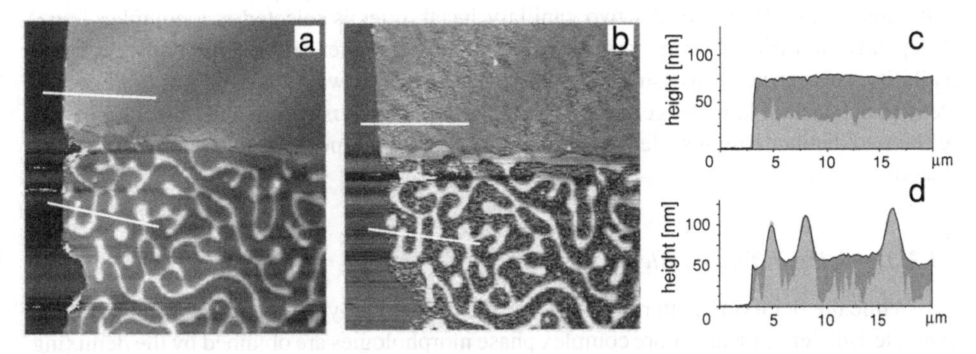

FIGURE 1.2. Atomic force microscopy (AFM) topography images showing the demixing of a polystyrene/poly(2-vinylpyridine) (PS/PVP) blend spin-cast from tetrahydrofuran (THF) onto a gold surface. The lower part of (a) was covered by a self-assembled monolayer (SAM) prior to spin-coating. (b) Scan of the same area as (a), after removal of the PS by washing in cyclohexane. The superposition of the cross sections (indicated by lines in (a) and (b) reveal a layered phase morphology on the Au surface (c) and a lateral arrangement of the PS and PVP phases on the SAM surface (d). Adapted from [13].

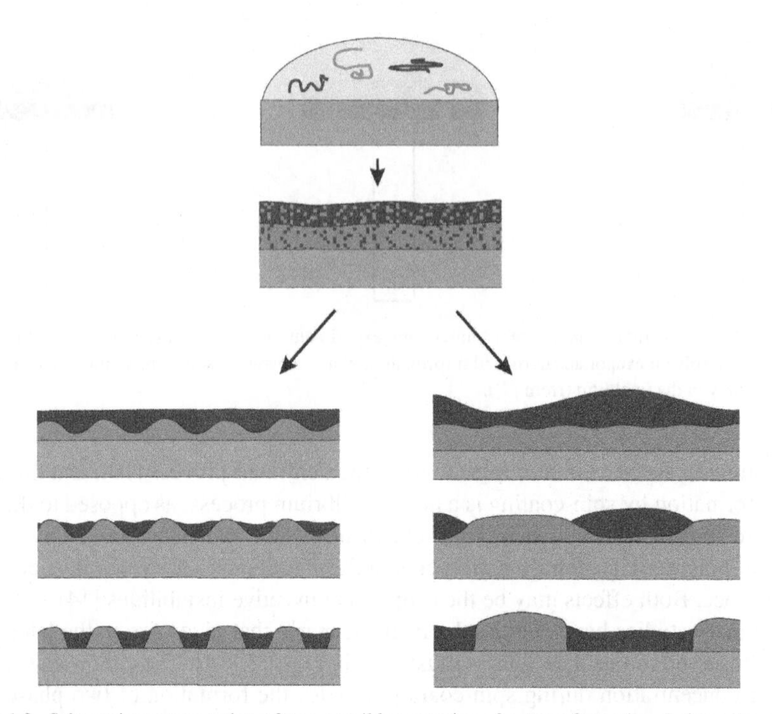

FIGURE 1.3. Schematic representation of two possible scenarios of pattern formation during spin-coating. In the initial stage, phase separation results in a layered morphology of the two solvent swollen phases. As more solvent evaporates, this double layer is destabilized in two ways: either by a capillary instability of the liquid-liquid interface (left) or by a surface instability (right), which, most likely, has a hydrodynamic origin (from [15]). Note the difference in morphological length scales resulting from each mechanism.

This occurs by an instability of one of the free interfaces: the polymer-polymer interface, the film surface, or a combination of the two, each of which gives rise to a distinct lateral length scale. Which of the two capillary instabilities is selected is a complex issue. It depends on various parameters, such as polymer-polymer and polymer-solvent compatibility, solvent volatility, substrate properties, etc. in a way which is not understood. Despite this lack of knowledge, playing with these parameters permits the selection of one of the two distinct length scales associated with these two mechanisms, or a combination thereof.

1.1.2. Demixing in Ternary Blends

While the demixing patterns in Fig. 1.2 are conceptually simple and exhibit only one characteristic length scale, more complex phase morphologies are obtained by the demixing of a multi-component blend [16]. With more than two polymers in a film, the pattern formation is (in addition to the factors discussed in the previous section) governed by the mutual wetting behavior of the components. Two different scenarios are shown in Fig. 1.4 [17]. While both films in Fig. 1.4(a) and (b) consist of the same three polymers, their mutual interaction was modulated by preparing the films under different humidity conditions [15].

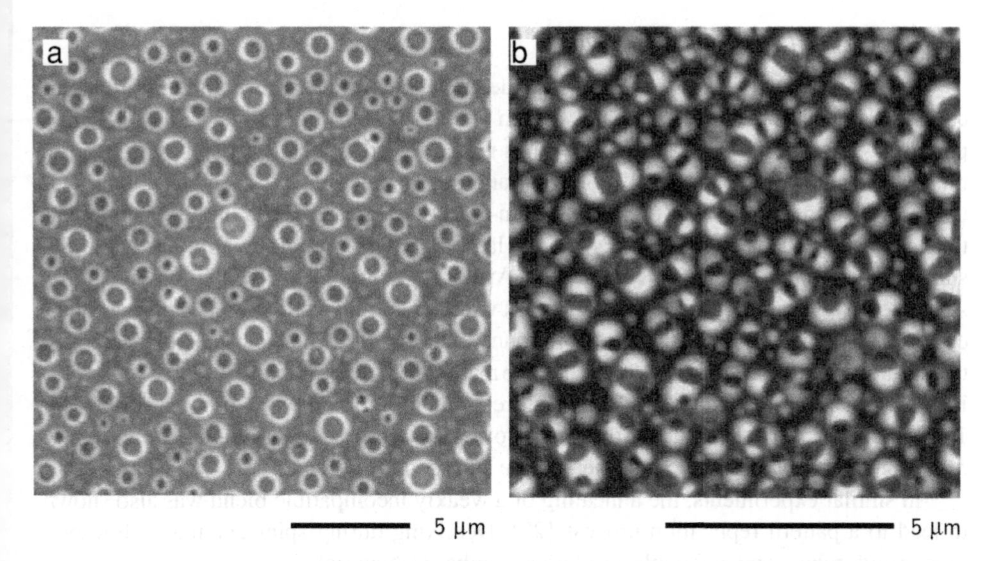

FIGURE 1.4. AFM images of ternary polystyrene/polymethylmethacrylate/poly(2-vinylpyridine) (PS/PMMA/PVP) blends cast from THF onto apolar (SAM covered) Au surfaces. Spin-casting at high humidities results in PMMA rings, which are characteristic for the complete wetting of PMMA at the PS/PVP interface (a), while a lowering of the humidity gives rise to three phases that show mutual partial wetting (b) (from [17]).

The differing water uptake of the three polymers during spin-coating results in a variation of the polymer-polymer interaction parameters and thereby in a change in their wetting behavior. In Fig. 1.4(a), the polystyrene (PS) – poly(2-vinylpyridine) (PVP) interface is completely wetted by an intercalating polymethylmethacrylate (PMMA) phase. This is contrasted by a partial wetting of the PS–PVP interface by PMMA in Fig. 1.4. While the interaction of the phase morphology with the vapor phase gives a certain amount of structural control, a richer variety of patterns can be achieved by changing the relative composition of the film (Fig. 1.5) [16].

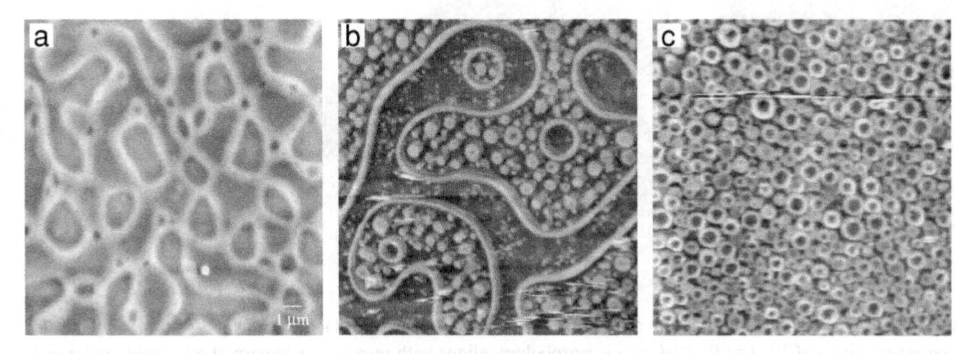

FIGURE 1.5. Same system as in Fig. 1.4a. A change in the relative PS: PMMA: PVP composition results in a variation of the lateral phase morphology. Polymer compositions: a:1:1:1; b:2:1:2; c:3:1:1. Adapted from [16].

1.1.3. Pattern Replication by Demixing

Figure 1.2 illustrates a strong substrate dependence of pattern formation during spin-coating. This observation can be harnessed in a pattern replication strategy. To this end, a pattern in surface energy of the substrate has to be created. While this can be achieved in many ways, it is most conveniently done by stamping a patterned self-assembled monolayer using micro-contact printing (μCP) [18]. Spin-casting a polymer blend onto such a prepatterned substrate leads to an alignment of the lateral phase morphology with respect to the substrate pattern, as shown in Fig. 1.6 [13]. After dissolving one of the two polymers in a selective solvent, a lithographic polymer mask with remarkably vertical side walls and sharp corners is obtained. As opposed to a more rounded morphology that is usually expected for two liquids in equilibrium at a surface [19], the rectangular cross-section observed in Fig. 1.6 is a consequence of the non-equilibrium nature of the film formation process, shown in Fig. 1.6c. The vertical side walls and the sharp corners are a direct consequence of a slightly differing solubility of the two polymers in the spin-coating solvent [12].

In similar experiments, the annealing of a weakly incompatible blend was also shown to lead to a pattern replication process [20]. Demixing during spin- coating is, however more rapid, robust and amenable to a larger number of materials.

The surface-directed process leading to the replication technique illustrated in Fig. 1.6 is also its main limitation. The pattern formation process is governed by two length scales: (i) the characteristic length scale that forms spontaneously during demixing (e.g. Fig. 1.2), and (ii) the length scale that is imposed by the prestructured surface. Since these two length-scales must be approximately matched, a reduction in lateral feature size entails a reduction of both length scales, which is a considerable challenge if sub-100 nm structures are required.

A second limiting issue is the substrate oriented nature of this process. Since the pattern replication is essentially driven by a difference in wettability of the two components on the modified substrate, the aspect ratio (height/width) of the polymer structures is smaller than 1. It is unlikely that high aspect ratio polymer patterns can be made this way.

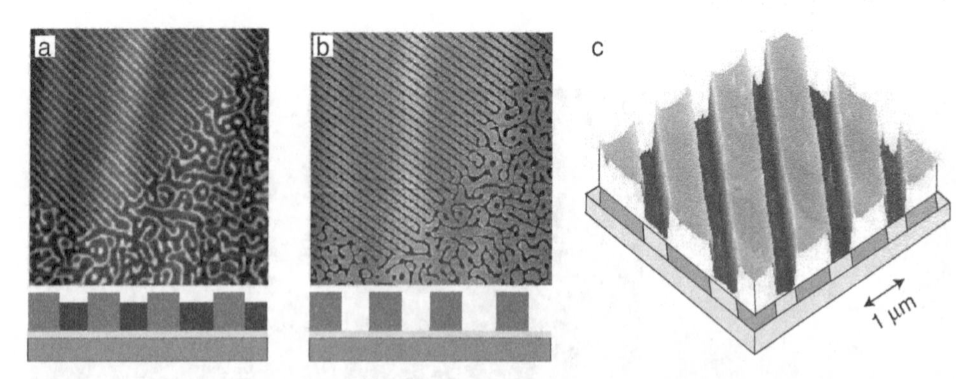

FIGURE 1.6. Same polymer mixture as in Fig. 1.2 spin-cast onto a Au surface that was pre-patterned by micro-contact printing (μCP). The PS/PVP phase morphology aligns with respect to a pattern of alternating polar and apolar lines (a), top-left), as opposed to the phase morphology on the unpatterned SAM layer (a), bottom right). After removal of the PVP phase by washing in ethanol (b), PS lines with nearly rectangular cross-sections are revealed (c). Adapted from [13].

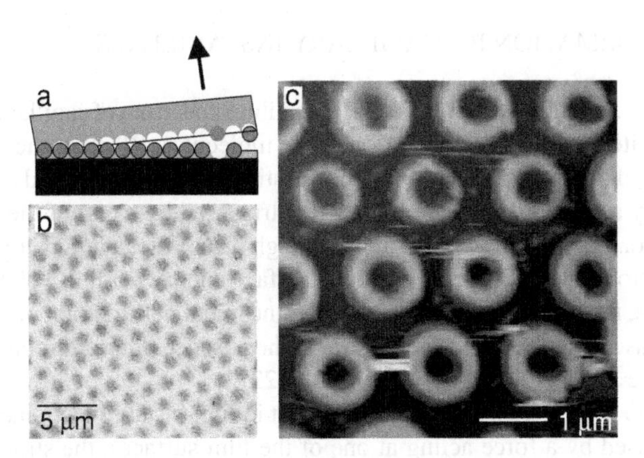

FIGURE 1.7. Alignment of the phase morphology in Fig. 1.4a, with respect to a pre-patterned substrate (see Fig. 1.6). The PS/PVMMA/PVP solution was spin cast onto a substrate, which consisted of hexagonally ordered polar dots in a SAM covered matrix (b), made by a μCP technique that employs a packed layer of colloidal spheres, schematically shown in (a): polydimethylsiloxane is cast onto a self-assembled monolayer of colloidal spheres and is cured to form a rubber stamp that mirrors the hexagonal symmetry of the colloidal layer. The PMMA rings that were obtained after dissolving PS and PVP mirror the hexagonal symmetry of the surface in (b). Adapted from [17].

Ternary blends One way to overcome these limitations is the use of ternary polymer blends. This approach makes use of the principle described in section 1.1.2, in which one of the polymer components wets the interface of the other two. By providing a pre-patterned substrate with surface regions, to which these two polymer segregate, it is possible to form structures in the intercalated polymer with dimensions that are not directly connected to the substrate pattern.

This principle is illustrated in Fig. 1.7 [17], making use of the blend that led to the PMMA rings in Fig. 1.4. To control the arrangement and size of the rings, the solution used in Fig. 1.4 was cast onto a substrate with a hexagonal pattern of polar dots in an apolar matrix, made by a colloidal stamp (Fig. 1.7a,b) [21]. The comparison of Figs. 1.4 and 1.7 shows the effect the substrate pattern has on the ternary morphology. The polydisperse distribution of PMMA ring sizes (initially located at the PS/PVP interface) was replaced by monodisperse rings, all in register with the substrate pattern. The wall size of $\approx$200 nm was one order of magnitude smaller compared to the lattice periodicity of 1.7 μm [17].

The main advantage of using a ternary blend (as opposed to the direct replication of Fig. 1.6, where the width of the polymer structures was directly imposed by the substrate pattern), is the relative independence of the structure parameters (width, aspect ratio) with respect to the substrate pattern. The width (and thereby the aspect ratio) of the PMMA rings in Fig. 1.7 is controlled by the relative amount of PMMA in the PS/PMMA/PVP blend. While the lateral periodicity of the polymer structures is determined by the substrate, the structure size is controllable by the relative amount of PMMA in the blend. Similar to the replication technique using two polymers, pattern replication by demixing of ternary blends should be expandable to other polymer system, with the main requirement that one of the components wets the interface of the other two.

1.2. PATTERN FORMATION BY CAPILLARY INSTABILITIES

While macroscopically flat, liquid surfaces exhibit a spectrum of capillary waves that are continuously excited by the thermal motion of the molecules. Whether these perturbations cause a break-up of the surface depends on the question, whether the liquid can minimize its surface energy by a change in morphology that is triggered by a part of the capillary wave spectrum [22]. For example in the case of a Rayleigh instability, a liquid column breaks-up spontaneously into drops, reducing the overall surface energy per unit volume. In contrast to liquid columns, flat surfaces are stable, since a sinusoidal perturbation of any wavelength leads to an increase in surface area. Therefore, in the absence of an additional destabilizing force acting at the surface, liquid films are stable [22].

There are two objectives triggering the interest in film instabilities. Since film instabilities must be caused by a force acting at one of the film surfaces, the structure formation process mirrors these forces. The observation of film instabilities can therefore be used as a sensitive measurement device to detect interfacial forces. The knowledge of these forces enables us, on the other hand, to control the morphology that is formed by the film break-up.

1.2.1. Capillary Instabilities

The theoretical framework, within which the existence of surface instabilities created by capillary waves can be predicted is the linear stability analysis [23, 24]. This model assumes a spectrum of capillary waves with wave vectors q and time constant τ (Fig. 1.8a).

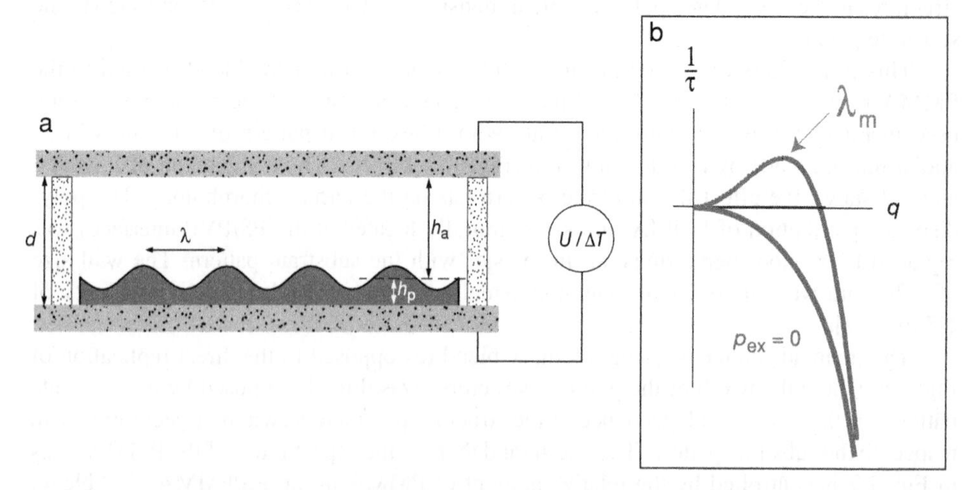

FIGURE 1.8. (a) Schematic representation of the device used to study capillary surface instabilities. A polymer-air bilayer of thicknesses h_p and h_a, respectively, is formed by two planar silicon wafer held at a separation d by spacers. A capillary instability with wavelength $\lambda = 2\pi/q$ is observed upon applying a voltage U or a temperature difference ΔT. (b) Dispersion relation (prediction of Eq. (1.6)). While all modes are damped ($\tau < 0$) in the absence of an interfacial pressure p_{el}, the application of an interfacial force gradient leads to the amplification of a range of λ-values, with λ_m the maximally amplified mode.

In one dimension (with lateral coordinate x), we have for the local height of the film surface

$$h(x,t) = h_{\rm p} + \zeta \exp(iqx + t/\tau). \tag{1.1}$$

ζ is the amplitude of the capillary wave and $h_{\rm p}$ is the position of the planar surface ($\zeta = 0$). For negative values of τ, the mode with wave vector q is damped. For positive τ the surface is destabilized by an exponential growth of this mode.

The formation of a surface wave in Fig. 1.8a requires the lateral displacement of liquid. Assuming a non-slip boundary condition at the substrate surface (lateral velocity $v(z) = 0$ at the surface ($z = 0$)), and the absence of normal stresses at the liquid surface, this implies a parabolic velocity profile (half-Poiseuille profile) in the film

$$v(x,z) = \frac{1}{2\eta} z(z - 2h)\partial_x p \tag{1.2}$$

with η the viscosity of liquid and ∂_i represents the partial derivative with respect to i. $\partial_x p$ is the lateral pressure gradient that drives the liquid flow in the film. In the one dimensional case considered here, the lateral flow causes an averaged flux $\bar{j} = h\bar{v}$ through the film cross section h, given by

$$j = -\frac{h^3}{3\eta}\partial_x p. \tag{1.3}$$

The third necessary ingredient for the model is a continuity equation for the non-volatile polymer melt

$$\partial_t h + \partial_x j = 0. \tag{1.4}$$

Inserting Eq. (1.3) into Eq. (1.4) yields the equation of motion

$$\partial_t h = \frac{1}{3\eta}\partial_x \left(l^3 \partial_x p\right). \tag{1.5}$$

Together with the ansatz Eq. (1.1), Eq. (1.5) describes the response of a liquid film to an applied pressure p. The resulting differential equation is usually solved in the limit of small amplitudes $\zeta \ll h \approx h_{\rm p}$ and only terms linear in ζ are kept ("linear stability analysis"). This greatly simplifies the differential equation. The pressure inside the film $p = p_{\rm L} + p_{\rm ex}$ consists of the Laplace pressure $p_{\rm L} = -\gamma \partial_{xx} h$, minimizing the surface area of the film, and an applied destabilizing pressure $p_{\rm ex}$, which does not have to be specified at this point. This leads to the dispersion relation

$$\frac{1}{\tau} = -\frac{h_{\rm p}^3}{3\eta}\left(\gamma q^4 + q^2 \partial_l p_{\rm ex}\right). \tag{1.6}$$

The predictions of Eq. (1.6) are schematically shown in Fig. 1.8b. For $p_{\rm ex} = 0$, $\tau < 0$ for all values of q. This confirms that films are stable in the absence of a destabilizing pressure. If a (possibly externally imposed) force is switched on, so that $\partial_h p_{\rm ex} < 0$, $\tau > 0$ if q is smaller than a critical value qc and has a maximum for $0 < q_{\rm m} < q_{\rm c}$.

Qualitatively, modes with a large wave vector q corresponding to surface undulations with short wavelengths $\lambda = 2\pi/q$ are suppressed ($\tau < 0$), since the amplification of such waves involves a large increase in liquid-air surface area. On the opposite end of the spectrum, long wavelength (small q) modes, while allowed, amplify slowly due to the large lateral transport of material involved in this process. As a consequence the mode with the highest positive value of $\tau_{\rm m}$ is maximally amplified

$$\lambda = \frac{2\pi}{q_{\rm m}} = 2\pi\sqrt{-\frac{2\gamma}{\partial_h p_{\rm ex}}} \tag{1.7}$$

and

$$\tau_{\rm m} = \frac{3\eta}{\gamma h_{\rm p}^3} q_{\rm m}^{-4}. \tag{1.8}$$

Eq. (1.7) is a generic equation describing film instabilities in the presence of an applied pressure. It is the basis for the film instabilities driven by van der Waals forces, or forces caused by electrostatic or temperature gradient effects discussed below. Eq. (1.7) also illustrates that film instabilities mirror the forces that cause them. The systematic study of film instabilities can therefore be used to quantitatively measure surface forces.

Van der Waals forces The case $p_{\rm ex} = 0$ is purely academic, since van der Waals interactions are omnipresent and are known to affect the stability of thin films. In the non-retarded case, the van der Waals disjoining pressure is

$$p_{\rm vdW} = \frac{A}{6\pi h^3} \tag{1.9}$$

where A is the effective Hamaker constant for the liquid film sandwiched between the substrate and a third medium (usually air). Depending on the sign of A, $p_{\rm vdW}$ can have either a stabilizing ($A < 0$) or a destabilizing ($A > 0$) effect. Eqs. (1.7) and (1.9) yield the well known dewetting equations [24]

$$\lambda = 4\pi\sqrt{\frac{\pi\gamma}{A}}h^2 \tag{1.10}$$

and

$$\tau_{\rm m} = 48\pi^2\frac{\gamma\eta}{A^2}h^5. \tag{1.11}$$

Dewetting driven by van der Waals forces has been observed in many instances [25]. It is characterized by a wave pattern, as opposed to heterogeneously nucleated film break-up caused by imperfections in the film, leading to the formation of isolated holes that cause the dewetting of the film [26, 27].

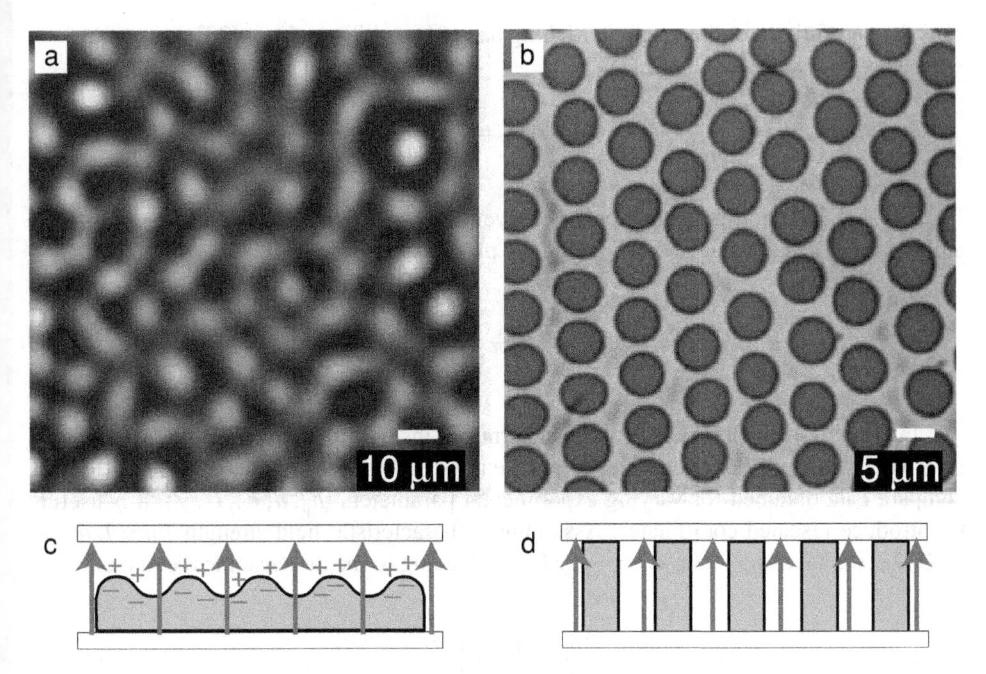

FIGURE 1.9. Electrohydrodynamic instability of a polymer film. Applying a voltage at the capacitor in Fig. 1.8a results in the amplification of a surface undulation with a characteristic wavelength λ (a). This leads to the formation of hexagonally ordered columns (b). The origin of the destabilizing pressure p_{el} is schematically shown in (c): the electric field causes the energetically unfavorable build-up of displacement charges at the dielectric interface. (d) The alignment of the dielectric interface parallel to the electric field lines lowers the electrostatic energy. Adapted from [30].

Electrostatic forces Films are also destabilized by an electric field applied perpendicular to the film surface. This is done by assembling a capacitor device that sandwiches a liquid-air (or liquid-liquid bilayer [28, 29]). After liquefying the film and applying an electric field, the film develops first an undulatory instability (Fig. 1.9a). With time, the wave pattern is amplified, until the wave maxima make contact to the upper plate, leading to an hexagonally ordered array of columns (Fig. 1.9a) [30].

The destabilizing effect arises from the fact that the electrostatic energy of the capacitor device is lowered for a liquid conformation that spans the two electrodes (Fig. 1.9d) compared to a layered conformation (Fig. 1.9c) [31]. The corresponding electrostatic pressure p_{el} is obtained by the minimization of the energy stored in the capacitor (constant voltage boundary condition) $F_{el} = QU = \frac{1}{2}CU^2$, with the capacitor charge Q and the applied voltage U. The capacitance C is given in terms of a series of two capacitances. This leads to a destabilizing pressure

$$p_{el} = -\epsilon_0(\epsilon_2 - \epsilon_1)E_1 E_2 \tag{1.12}$$

with ϵ_1, ϵ_2 the dielectric constants of the two media and the corresponding electric fields

$$E_i = \frac{\epsilon_j}{\epsilon_1 h_2 + \epsilon_2 h_1} \quad (i, j = 1, 2; i \neq j). \tag{1.13}$$

ϵ_0 is the permittivity of the vacuum. Making use of Eq. (1.7), we have [28]

$$\lambda_{el} = 2\pi \sqrt{\gamma U \frac{\sqrt{\epsilon_1 \epsilon_2}}{\epsilon_0(\epsilon_2 - \epsilon_1)^2} (E_1 E_2)^{-\frac{3}{4}}} = 2\pi \sqrt{\frac{\gamma(\epsilon_1 h_2 + \epsilon_2 h_1)^3}{\epsilon_0 \epsilon_1 \epsilon_2 (\epsilon_2 - \epsilon_1)^2 U^2}}. \tag{1.14}$$

For a double layer consisting of a polymer layer ($\epsilon_2 = \epsilon_p$) with film thickness $h_2 = h_p$ and an air gap ($\epsilon_1 = 1$), we have (introducing the plate spacing $d = h_1 + h_2$)

$$\lambda_{el} = 2\pi \sqrt{\frac{\gamma U}{\epsilon_0 \epsilon_p (\epsilon_p - 1)^2} E_p^{-\frac{2}{3}}} = 2\pi \sqrt{\frac{\lambda(\epsilon_p d - (\epsilon_p - 1)h_p)^3}{\epsilon_0 \epsilon_p (\epsilon_p - 1)^2 U^2}}. \tag{1.15}$$

In Fig. 1.10 [31], the experimentally determined instability wavelength is plotted versus d (at a constant applied voltage), reflecting the non-linear scaling predicted by Eq. 1.15. To compare data obtained for varying experimental parameters ($h_p, d, \epsilon_p, U, \gamma$), it is useful to introduce rescaled coordinates. Assuming a characteristic field strength $E_0 = U q_0 = 2\pi U / \lambda_0$, we have $\lambda_0 = 2\pi \epsilon_0 \epsilon_p (\epsilon_p - 1)^2 U^2 / \gamma$, leading to the dimensionless equation

$$\frac{\lambda}{\lambda_0} = \left(\frac{E_p}{E_0}\right)^{-\frac{3}{2}}. \tag{1.16}$$

The result of the rescaled equation is shown in Fig. 1.10b. The experimental data for a number of experiments corresponding to a range of experimental parameters collapse to a master curve. The line is the prediction of Eq. (1.16). It not only correctly predicts the $-3/2$ power-law, but quantitatively fits the data in the absence of adjustable parameters [31].

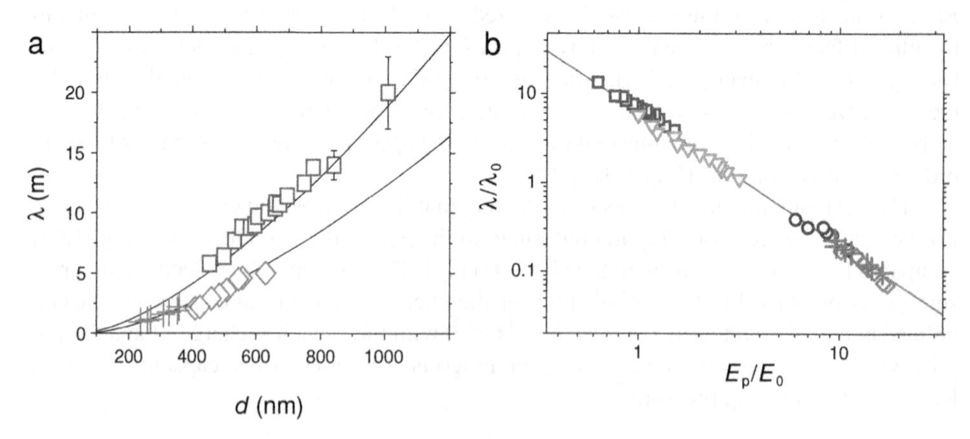

FIGURE 1.10. (a) Variation of λ versus d for electrostatically destabilized polymer films (□: PS, $h_0 = 93$ nm, $U = 30$ V, ◇: brominated PS, $h_0 = 125$ nm, $U = 30$ V). The crosses correspond to a 100 nm thick PMMA film that was destabilized by a alternating voltage of $U = 37$ V (rectangular wave with a frequency of 1 kHz). The lines correspond to the prediction of Eq. (1.15). (b) The data from (a) and additional data sets (▽: PS, $h_0 = 120$ nm, $U = 50$ V, ○: PMMA, $h_0 = 100$ nm, $U = 30$ V) plotted in dimensionless coordinates (see text) form a master-curve described by Eq. (1.16) (solid line). Adapted from [31].

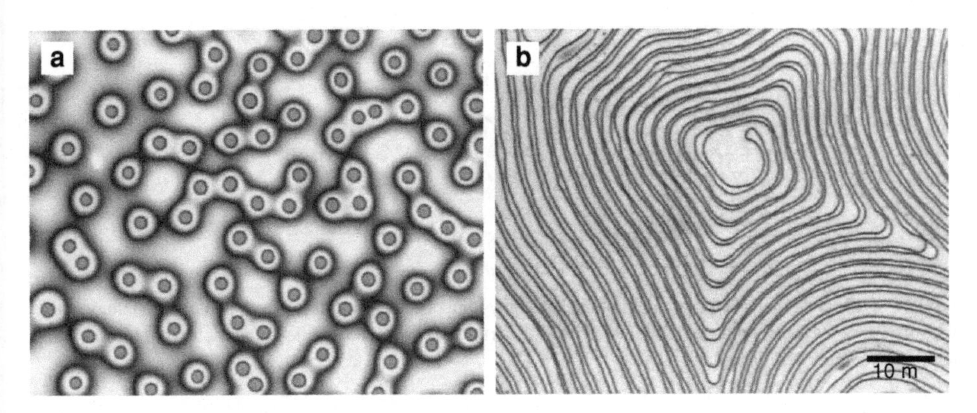

FIGURE 1.11. Pattern formation in a temperature gradient, using the set-up from Fig. 1.8, where the lower plate was set to a temperature T_1 and the upper plate to $T_2 = T_1 + \Delta T$. The transition of the film to columns (a) and stripes (b) was observed, often on the same sample. Adapted from [32].

The electric field experiment shown here can be considered as a test case for the quantitative nature of capillary instability experiments. It shows the precision, with which the capillary wave pattern reflects the underlying destabilizing force. In the case of electric fields, this force is well understood. Therefore, the good fit in Fig. 1.10b demonstrates the use of film instability experiments as a quantitative tool to measure interfacial forces. The application of this technique to forces that are much less well understood is described in the following section.

Temperature gradients In these experiments, the same sample set-up as in Fig. 1.9 is used, but instead of a voltage difference, a difference in temperatures is applied to the two plates (i.e. the two plates are set to two different temperatures T_1 and T_2 and they are additionally electrically short circuited to prevent the build-up of a electrical potential difference). Experimentally, structures similar to those caused by an applied electric field (Fig. 1.9) are observed. Figure 1.11 shows a transition from a layered morphology (polymer-air bilayer, not shown) to columns or lines spanning the two plates [32]. Since films are intrinsically stable, it is interesting to investigate the mechanisms that lie at the origin of this film instability. In particular, Eq. (1.7) requires a force at the interface that destabilizes the film.

Superficially considered, this morphological transition seems hardly surprising. Temperature gradients are known to cause instabilities in liquids either by convection or by surface tension effects [33, 34]. Convection is, however, ruled out in our experiments, since the liquid layer is extremely thin and highly viscous. In terms of surface tension, one has to consider whether the the creation of a surface wave lowers the overall surface free energy. This is not the case for the boundary conditions of this experiment (planar boundaries that are held at constant temperature). Therefore, neither of the known mechanisms account for the film instability. An additional complication arises from the fact, that the morphological transition in Fig. 1.11 cannot be described in terms of the minimization of a Gibbs free energy [35]. Since heat flows through the system, the morphological change in Fig. 1.11 is a transition between two non-equilibrium steady states, rather than the (slow) relaxation of an unstable towards a stable state (as in the case of an applied electric field).

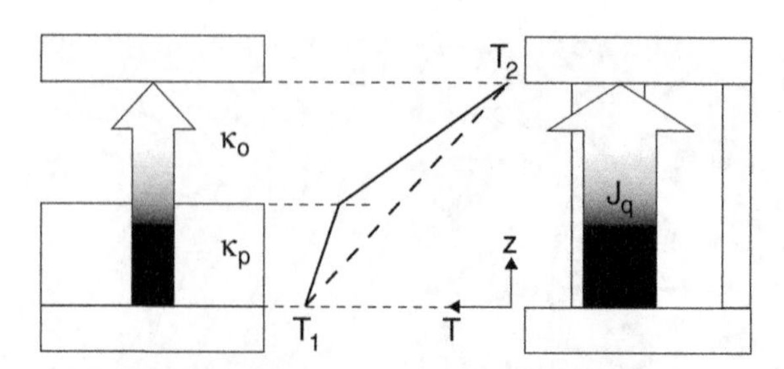

FIGURE 1.12. Schematic representation of the heat-flow for a polymer-air bilayer (left) and a morphology where the polymer spans the two plates (right), which maximizes the heat flow. The middle frame shows the corresponding temperature gradients. From [36].

Despite the intrinsic non-equilibrium nature of the phenomenon, it is possible to gain insight from a qualitative argument. Rearranging the polymer from a bilayer to a conformation spanning the two plates increases the heat flux between the two plates by forming "bridges" of the material with the higher heat conductivity (Fig. 1.12) [36]. While the maximization of the heat-flow (and thereby a maximization of the rate of entropy increase) is not a sufficient condition for the morphology change, it is a principle that is often observed [35].

Instead of a thermodynamic argument, we resort to a description that is based on the microscopic mechanisms that transport the heat [32, 36]. In the absence of convection and radiative transfer of heat (which is significant only at very high temperatures), heat is transported by diffusion. In the present bilayer system there are two differing diffusive mechanisms. In the air layer, heat diffusion takes place by the center of mass diffusion of gas molecules. In the polymer layer, on the other hand, heat is transported by high-frequency molecular excitations (phonons). Due to the high molecular weight and the entangled nature of the polymer melt, the contribution of center of mass-diffusion of polymer molecules to the heat transport is negligible.

We have previously reported that the destabilizing force is a consequence of the heat diffusion mechanism (for details see ref. [36]). The diffusive heat flux across a medium with thermal conductivity κ is given by Fourier's law.

$$J_{\mathrm{q}} = -\kappa \partial_z T. \tag{1.17}$$

For a bilayer with differing heat capacities κ_{p} and κ_{a}, we have

$$J_{\mathrm{q}} = \frac{\kappa_{\mathrm{a}}\kappa_{\mathrm{p}}(T_1 - T_2)}{\kappa_{\mathrm{a}} h_{\mathrm{p}} + \kappa_{\mathrm{p}} h_{\mathrm{a}}}. \tag{1.18}$$

We focus on the polymer film. Since heat diffusion is propagated by segmental thermal excitations, it corresponds to the propagation of longitudinal phonons from the hot substrate-polymer interface to the colder polymer-air surface. Associated with the heat flux is a momentum flux (or rather, a flux in quasi-momentum [36])

$$J_{\mathrm{p}} = \frac{J_{\mathrm{q}}}{u} \tag{1.19}$$

where u is the velocity of sound in the polymer. Phonons impinging onto an interface between two media of different acoustic impedances cause a radiation pressure

$$p = -2R\frac{J_q}{u} \tag{1.20}$$

with the reflectivity coefficient R. This pressure can, in principle destabilize the polymer film.

Equation (1.20) is, however only valid for the coherently propagating phonons, i.e. phonons with a mean free path length larger than the polymer film thickness. The propagation behavior of phonons depends on their frequency. In polymer melts, 100 GHz phonons (corresponding to phonon wavelengths comparable to the film thickness) have a mean-free path length of several micrometers [37], while phonons close to the Debye limit (several THz) scatter after very short (Å) distances and propagate therefore predominantly diffusively. Only low frequency phonons exert a destabilizing radiation given by Eq. (1.20).

The frequency dependent derivation of J_q and p is somewhat lengthy and is therefore discussed here only qualitatively (see [36] for a full discussion). Essentially, one has to write the heat flux and the pressure at the polymer-air interface in terms of reflectivities and transmittances of all three interfaces (all of which are a function of the phonon frequency). The total heat-flux and interfacial pressure are then obtained in a self-consistent way by an integration over the Debye density of states [36].

This leads to a rather simple scaling form of the interfacial pressure

$$p = \frac{2\bar{Q}}{u}J_q. \tag{1.21}$$

$\bar{Q}$ is the acoustic quality factor of the film. It depends on all interfacial transmission and reflection coefficients, and therefore contains all the complexity indicated above. On the level of this review, we regard $\bar{Q}$ as a scaling coefficient, but note that it can be calculated in detail [36].

Using Eq. (1.21), (1.18) and (1.7), we can analyze the instability of a polymer-air bilayer exposed to a temperature gradient

$$\lambda = 2\pi\sqrt{\frac{\gamma u(T_1 - T_2)}{\bar{Q}}\frac{\kappa_a\kappa_p}{\kappa_p - \kappa_a}\frac{1}{J_q}}. \tag{1.22}$$

In Fig. 1.13a the experimentally determined instability wavelength λ (e.g. determined from Fig. 1.11) is plotted versus the total heat flux J_q. The linear $1/J_q$ dependence of Eq. (1.22) describes well the experimental data. A second verification of the experimental model stems from the value of $\bar{Q}$ that is determined by a fit to the data. Rather than a different value of $\bar{Q}$ for each data-set, we find a universal value of $\bar{Q}$ that depends only on the materials used (substrate, polymer), but not on any of the other experimental parameters (sample geometry, temperature difference). A value of $\bar{Q} = 6.2$ described all data sets for PS on silicon in Fig. 1.13a, with a value of $\bar{Q} = 83$ for PS on gold. This allows us, in similarity to the electric field experiments in the previous section to introduce dimensionless

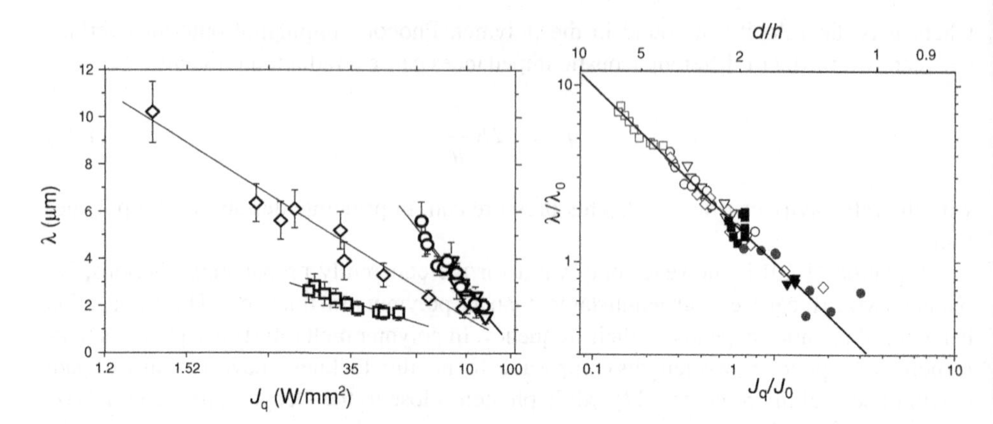

FIGURE 1.13. (a) λ versus J_q for PS films of various thicknesses and values of ΔT [32]. (b) When plotted in a dimensionless representation, the data from (a) (plus additional data [36]) collapes to a single master curve described by Eq. (1.23). Adapted from [32] and [36].

parameters $J_0 = \kappa_a \kappa_p (T_1 - T_2)/(\kappa_p - \kappa_a) h_p$ and $\lambda_0 = 2\pi\sqrt{\gamma u h_p / \bar{Q} J_0}$. Eq. (1.22) is then written as

$$\frac{\lambda}{\lambda_0} = \left(\frac{J_q}{J_0}\right)^{-1}. \tag{1.23}$$

In this representation all data collapses onto a single master curve. The $1/J_q$ scaling of λ, on one hand, and the master curve in Fig. 1.13b, on the other hand, are strong evidence for the model, which assumes the radiation pressure of propagating acoustic phonons as the main cause for the film instability.

1.3. PATTERN REPLICATION BY CAPILLARY INSTABILITIES

The previous section described pattern formation processes triggered by homogeneous forces acting at a film surface. While this leads to the formation of patterns exhibiting a characteristic length scale, these pattern are laterally random. By introducing a lateral variation into the force field, the pattern formation process can be guided to form a well defined structure. While such a lateral modulation of the destabilizing interfacial forces can, in principle, be achieved by several means, perhaps the most simple approach is the replacement of one of the planar bounding plates by a topographically structured master, schematically shown in Fig. 1.14.

Electric fields A patterned top electrode generates a laterally inhomogeneous electric field [30]. The replication of the electrode pattern is due to two effects. Since the time constant for the amplification of the surface instability scales with the fourth power of the plate spacing (Eq. (1.8)), the film becomes unstable first at locations where the electrode topography protrudes downward towards the polymer film. In a secondary process, the

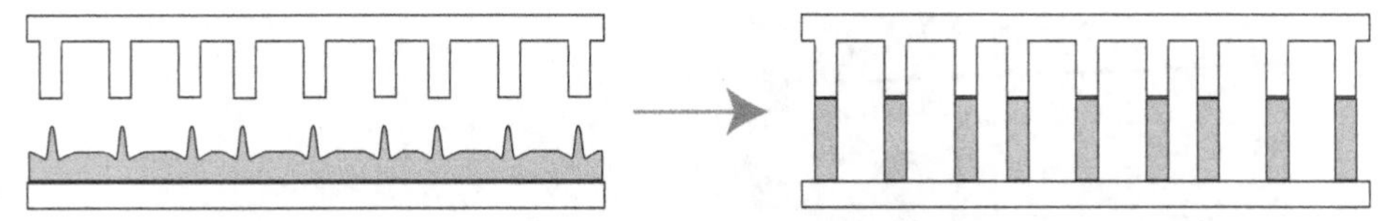

FIGURE 1.14. Schematic representation of the pattern replication process. The topography of the top plate induced a lateral force gradient that focuses the instability towards the downward pointing protrusions of the master plate.

polymer is drawn towards the locations of highest electric field, i.e. in the direction of these protrusions. This leads to the faithful replication of the electrode pattern shown in Fig. 1.15. Patterns with lateral dimensions down to 100 nm were replicated [30].

Interestingly, the patterns generated by the applied electric field are not stable in its absence. The change in morphology (from a flat film to stripes) significantly increases the polymer-air surface area. The vertical side walls of these line structures are, however, stabilized by the high electric field ($\sim 10^8$ V/m). If the polymer is cooled below the glass transition temperature before removing the electric field, as was done in our experiments, it is nevertheless possible to preserve the polymer pattern in the absence of an applied voltage.

Temperature gradients The same principle as in the case of the electric fields applies for an applied temperature gradient. Since the destabilizing pressure depends linearly on J_q (Eq. (1.21)), which scales inversely with the plate spacing (Eq. (1.18)), there is also a strong dependence of the corresponding time constant with d. Therefore, the same arguments as above apply here: the instability is generated first at locations where d is smallest and the liquid material is drawn toward regions where the temperature gradient is maximal. This

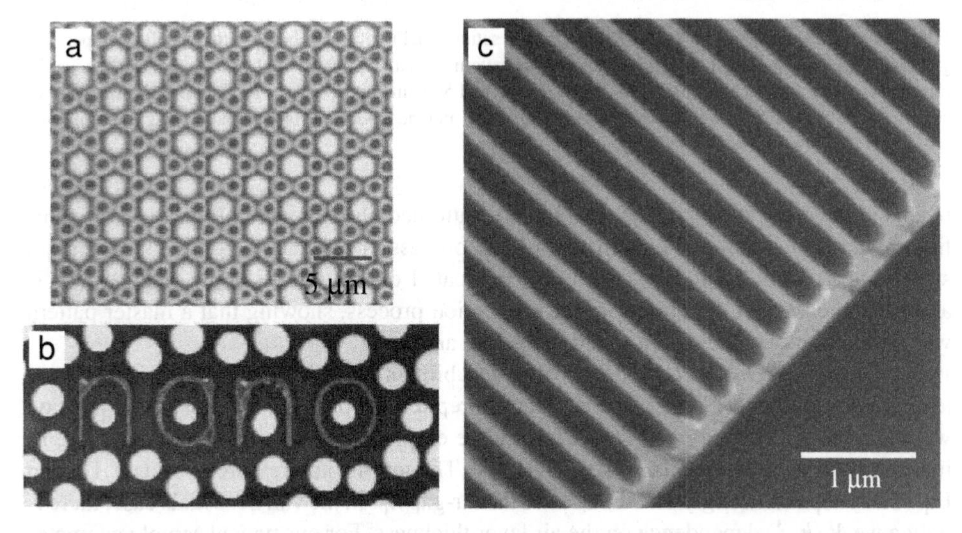

FIGURE 1.15. Electrohydrodynamic pattern replication. (a): double-hexagonal pattern, (b): the word "nano", (c): 140 nm wide and 140 nm high lines. In (b) the line width was ≈300 nm. The larger columns stem from a secondary (much slower) instability of the homogeneous (not structured) film. Adapted from [30] and [38].

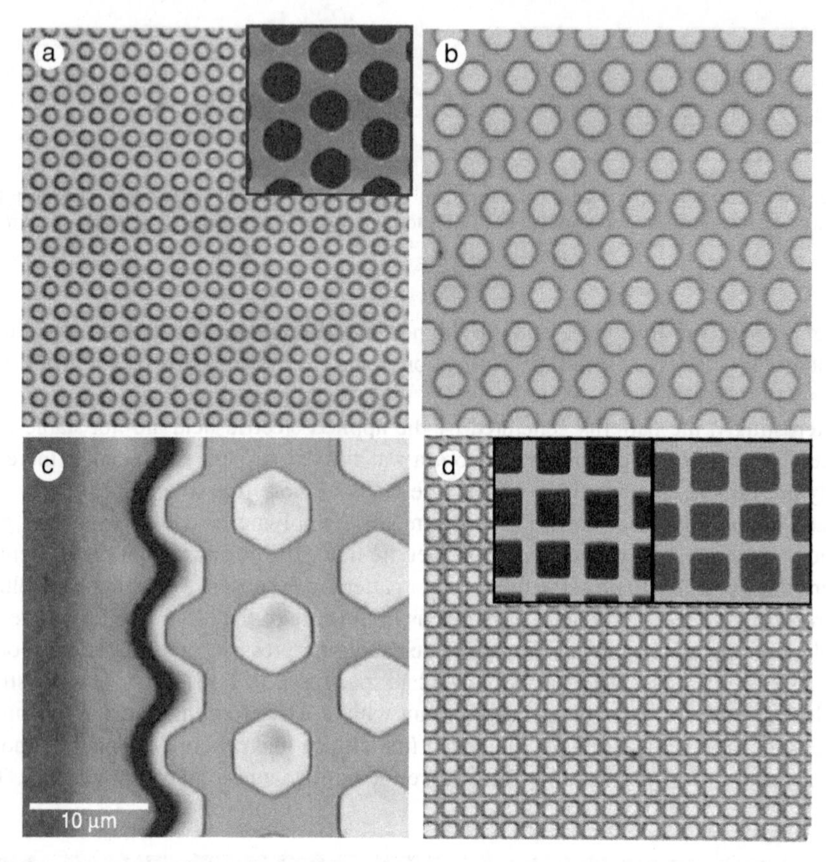

FIGURE 1.16. Pattern replication in a PS film using the setup from Fig. 1.14 combined with an applied temperature gradient. (a)–(c) show optical micrographs of various pattern sizes all replicated on the same sample, showing the robustness of this approach. In (d) a cross-hatched pattern of 500 nm wide lined was replicated. The inset shows a comparison of AFM images of the master plate pattern (left) and the replicated polymer pattern (right). From [39].

results in the replication of a topographically patterned plate shown in Fig. 1.16 [39]. This figure shows also that the pattern replication process is reasonably robust, with pattern sizes ranging from 5 μm down to 1 μm replicated on the same sample. Figure 1.17 is another indication of the quality of the replication process, showing that a master pattern was perfectly replicated over a 4 mm^2 substrate area.

van der Waals driven dewetting Since film instabilities triggered both by electric fields and temperature gradients can be used in a pattern replication process, this raises the question whether van der Waals driven dewetting can be employed to the same end. While seemingly similar, there is a fundamental difference. The interfacial pressures in Eq. (1.12) and Eq. (1.20) depend strongly on the width of the air-gap h_p. In the van der Waals case, there is only a weak (h_a^{-3}) dependence on the air layer thickness. For our typical sample geometry with $h_\mathrm{p} < h_\mathrm{a}$, the reduction in the instability time constant due to the presence of a top plate is negligible considering only van der Waals forces. In addition, a force gradient towards the top-plate protrusions becomes significant only in the close vicinity of the upper surface.

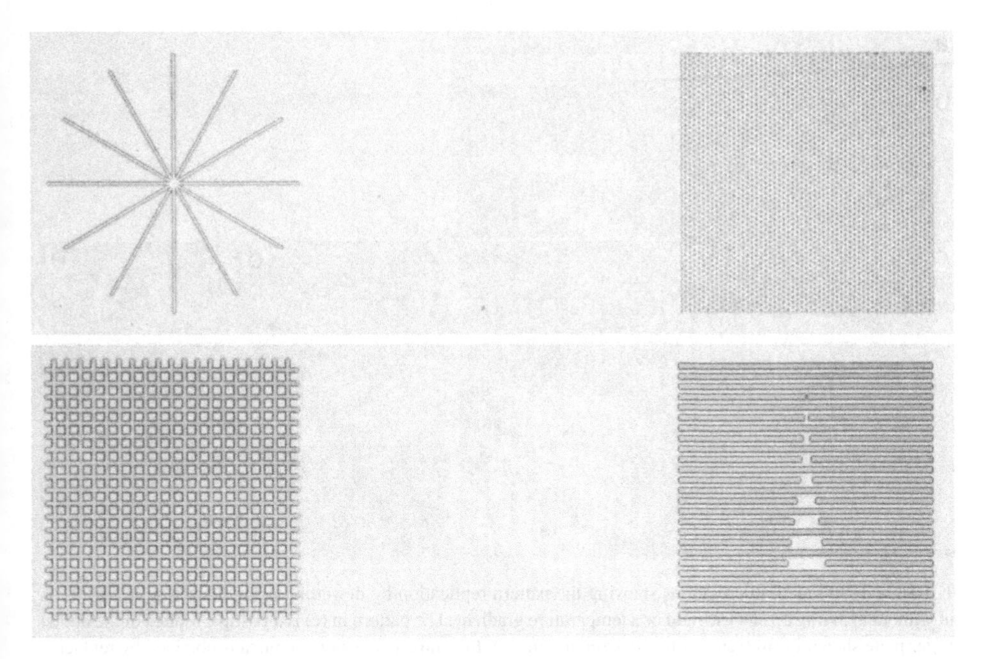

FIGURE 1.17. Large area images of patterns replicated with the help of a temperature gradient. The four patterned areas are $200 \times 200\ \mu m^2$. The top-left and bottom right area lie at a lateral distance of 2.7 mm on the sample, illustrating the reliability of this replication technique. From [39].

Therefore, the strong driving forces that are responsible for the pattern replication process in Figs. 1.15 and 1.16 are absent in the van der Waals driven case.

Nevertheless, if the parameters are properly chosen, pattern replication is observed for the set-up of Fig. 1.14 without applied voltages or temperature differences. This is shown in Fig. 1.18a–c [40]. Here, the dewetting of a PS—polyvinylmethylether (PVME) blend produced the positive replication of a topographically structured master. A PS-PVME blend was chosen, since the polymers and substrates used in the experiments involving electric fields and temperature gradients do not exhibit a capillary instability triggered by van der Waals forces [27]. As opposed to Figs. 1.15 and 1.16 the instability is not guided to the master protrusions, but the capillary instability is expected to develop randomly (in an identical fashion as in an unconfined film). As the undulation amplitudes increase, they touch the downward pointing protrusions and are pinned there. The minimization of surface energy causes the liquid to spread on the surfaces of the protrusions facing the substrate leading to a straightening of the vertical polymer-air surfaces.

Whether this leads to a positive or negative replication of the master pattern depends on the surface energy of the upper surface. In the case of a high energy master surface, it is completely wetted by the polymer, which is drawn into all its cavities, expelling the air. This is the well known capillary molding technique [4], shown in Fig. 1.18d, where a negative replication of the master reveals the nature of the lithographic process. In contrast, a low energy patterned surface is not wetted by the polymer. In this case the liquid bridges that are formed by the capillary instability do not spread into the concave parts of the master structure, leading to the positive replica shown in Fig. 1.18c.

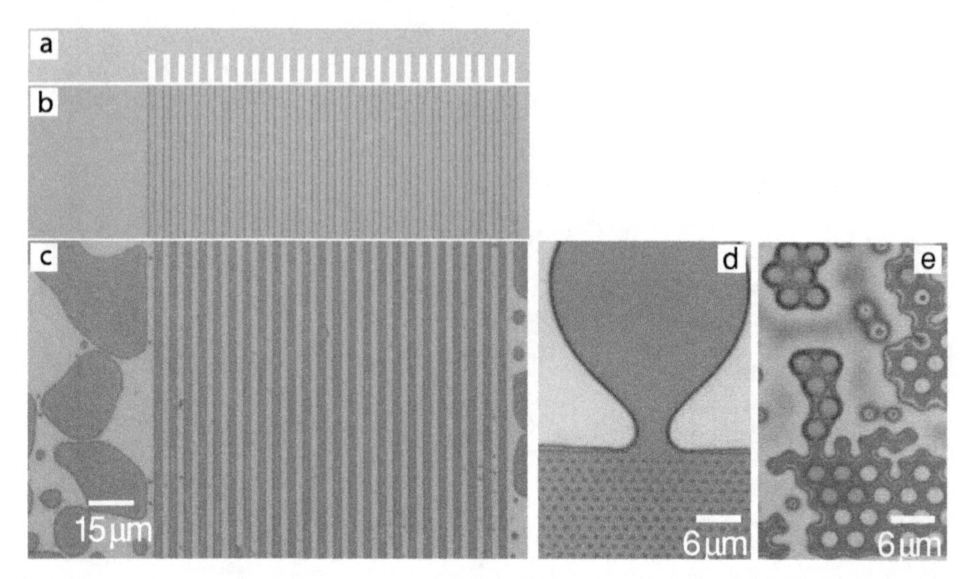

FIGURE 1.18. Optical micrographs showing the pattern replication by dewetting using the set-up in Fig. 1.14, but without applying an electric field or a temperature gradient. The pattern in (c) is a positive replica of the apolar master plate shown in (b) (schematic cross-section in (a)). In contrast, a negative replica is obtained by replacing the apolar master surface by a polar patterned surface. This is shown in (d), where the drop-shaped plug is drawn into the patterned slit-pore. The image in (e) shows a filling transition, which reduces the polymer-air vertical surface area. Adapted from [40].

The polymer structure in Fig. 1.18a is, however, not the thermodynamic equilibrium, if the aspect ratio (height/width) of the polymer structures is larger than 1. The formation of polymer bridges involves the creation of a large amount of polymer-air surface in this case. The overall free energy can therefore be lowered by the coalescence of these bridges to form a plug. The transition from bridges to a plug (filling transition) can be seen in Fig. 1.18e, which has taken place at the smallest of the replicated areas of an imperfectly replicated master structure [40].

A similar type of filling transition should, in principle also occur in the case of an applied electric field or a temperature gradient. The presence of additional forces which act to stabilize the vertical sidewalls of the replicated pattern imposes, however, a much higher energy barrier between the bridge and the plug conformation, significantly reducing the probability for such a transition. Indeed, the filling transition was only observed for films destabilized by van der Waals forces (Fig. 1.18e), but not for the other two cases.

1.3.1. Hierarchical Pattern Formation and Replication

All the examples of pattern formation and replication by capillary instabilities discussed so far rely on the amplification of a single very narrow band of instability wavelength. Pattern replication succeeds only if (within certain bounds—see for example Fig. 1.16) the length scale of the master pattern matches the instability wavelength. For many practical applications, the simultaneous replication of more than one length scale and more than one material is required.

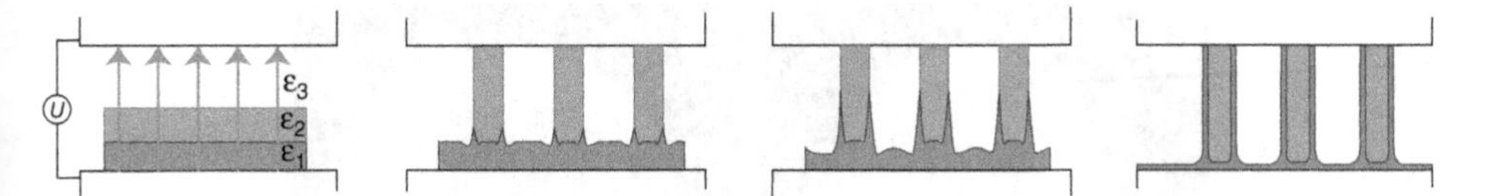

FIGURE 1.19. Schematic representation of a hierarchic pattern formation in by an electric field. First, the top polymer layer is destabilized, in similarity to Fig. 1.9, leaving the lower layer essentially undisturbed. In a secondary process, the polymer of the lower layer is drawn upward along the outside of the primary polymer structure, leading to the final morphology, in which the the polymer from the lower layer has formed a mantle around the initial polymer structure. From [41].

It is possible to extend the pattern replication processes by capillary instabilities to produce a hierarchical range of length scales [41]. This is illustrated schematically in Fig. 1.19. A polymer bilayer in a capacitor is destabilized by an applied electric field. While both the polymer-polymer and the polymer-air interface experience a destabilizing pressure, the polymer-air interface destabilizes first for hydrodynamic reasons [41, 42]. Since the two polymer layers have no strong hydrodynamic coupling, this leads to the formation of a pillar structure on the upper polymer layer on top of the essentially undisturbed lower polymer layer, in analogy to the single polymer layer case. The formation of this structure involves, however, the creation of a retracting polymer-polymer-air contact line. The hydrodynamic stresses at a dynamic contact line are very large, leading to a local deformation of the lower polymer layer [41–43].

The completion of the primary structure formation process entails a change in boundary condition for the lower polymer surface. The changed hydrodynamics of a polymer-air surface (as compared to the initial polymer-polymer surface) significantly reduces the time constant for an instability of this surface. This secondary instability is nucleated at the locations of highest film thickness (or lowest air gap), i.e. at the locations of the contact line. The polymer is drawn upward along the outside of the initially formed polymer structure. This secondary coating of the initial structure is facilitated by a reduced polymer-polymer surface energy, compared to a polymer-air surface of a pattern replication process that occurs independent of the primary structure. The results of such experiments are shown in Fig. 1.20, both for the formation of columns in a laterally homogeneous electric field, as well as for the lithographic replication of lines. In Fig. 1.20b and d, the polymer that formed the primary instability was removed by dissolution in a selective solvent, revealing the secondary structure.

As opposed to the primary instability, where the structure size is essentially determined by the instability wavelength, the width of the secondary structure is determined by two factors that can be independently controlled and adjusted. Since the secondary process is comparably slow, the width of the mantle that forms around the primary structure can be controlled by the exposure time to the electric field. Secondly, the structure width after full equilibration depends only on the thickness of the lower polymer film, a second parameter that can be adjusted independent of the wavelength of the primary instability. This allows us the generate structure widths and aspect ratios that are significantly smaller than the initial instability wavelength, e.g. 100 nm in Fig. 1.20 [41]. Apart from this reduction in structure sizes, a further advantage is the structuring of two different materials in a single processing step. This procedure should be extendable to to three or more layers for the independent

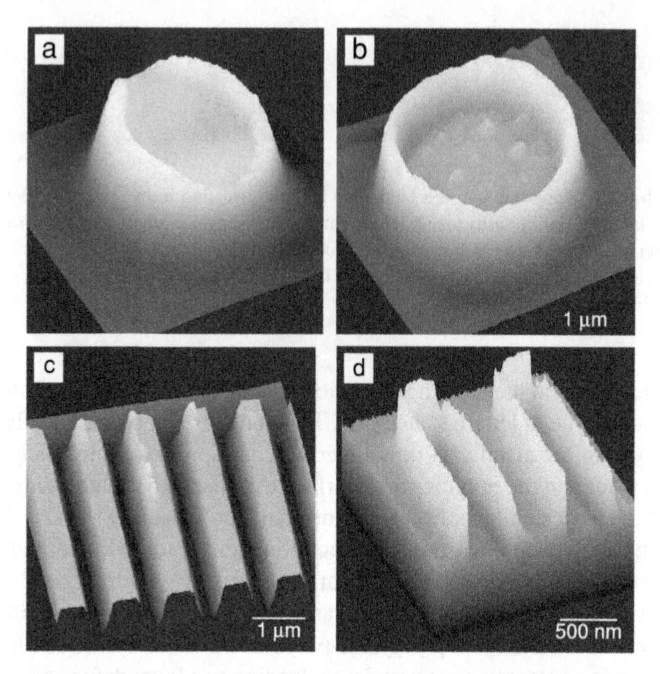

FIGURE 1.20. Hierarchical structure formation of a PMMA/PS bilayer. (a) composite PMMMA/PS column. (b) Same column as in (a) after removal of PS. (c) Composite PMMA/PS line pattern. (d) After PS removal. The PMMA structures have a wall width of ~100 nm. Adapted from [41].

replication of a larger number of structure widths and the simultaneous replication of more than two different materials.

1.4. CONCLUSIONS

In recent years, an ever increasing number of lithographic techniques have emerged to complement optical lithography, which is still the work horse for practically all pattern replication processes. These developments are driven not only by the need for methods for the replication of sub-100 nm patterns (where conventional lithography is expected to meet its limits), but also by the invention of new high performance, low cost technologies, for example all-polymer based electronics, displays or photovoltaic devices.

The formation and replication of patterns into polymer films using instabilities is a new contribution in the field of soft lithography, which typically requires the mechanical contact between a patterned master and the resist. Two classes of instabilities were discussed. The demixing of two incompatible polymers leads to a well known spinodal pattern. In thin films, this structure formation process can be guided by a pattern in surface energy.

A second approach makes use of capillary surface instabilities that occur in the presence of a destabilizing surface force. Since such a instability mirrors the details of the destabilizing force field, it can be employed as a sensitive tool to study and explore forces that act at the surface of liquid films. A controlled lateral variation of the force field provides, on the other hand a novel technique for the lithographic replication of structure down to 100 nm, and possibly below.

REFERENCES

[1] Crego-Calama, M.; Reinhoudt, D.N.; Garciá-López, J.J.; Kerckhoffs, J.M.C.A. Chapter 4.
[2] Zimelis, K. *Nature* **2000**, *406*, 1021.
[3] *The National Technology Roadmap for Semiconductors*; Semiconductor Industry Association Report, 1997.
[4] Xia, Y.; Rogers, J. A.; Paul, K. E.; Whitesides, G. M. *Chem. Rev.* **1999**, *99*, 1823.
[5] Chou, S. Y.; Krauss, P. R.; Renstrom, P. J. *Science* **1996**, *272*, 85.
[6] Schift, H. *et al. Microelectron. Eng.* **2000**, *53*, 171.
[7] Walheim, S.; Schäffer, E.; Mlynek, J.; Steiner, U. *Science* **1999**, *283*, 520.
[8] Flory, P. J. *Principles of Polymer Chemistry*; Cornell University Press: Ithaca, 1971.
[9] Schmidt, I.; Binder, K. *J. Phys. II(Paris)* **1985**, *46*, 1631.
[10] Jones, R. A. L.; Norton, L. J.; Kramer, E. J.; Bates, F. S.; Wiltzius, P. *Phys. Rev. Lett.* **1991**, *66*, 1326; Fischer, H.-P.; Maass, P.; Dieterich, W. *Phys. Rev. Lett.* **1997**, *79*, 893.
[11] Gunton, J. D.; San Miguel, M.; Sahni, P.S. In *Phase Transitions and Critical Phenomena*; Domb, C.; Lebovitz, J. L., Eds.; Academic Press: London, 1983; Vol 8, p. 267.
[12] Walheim, S.; Böltau, M.; Mlynek, J.; Krausch, G.; Steiner, U. *Macromolecules* **1997**, *30*, 4995.
[13] Böltau, M.; Walheim, S.; Mlynek, J.; Krausch, G. Steiner, U. *Nature* **1998**, *391*, 877.
[14] de Gennes, P.G. *Eur. Phys. J. E* **2001**, *6*, 421.
[15] Sprenger, M.; Walheim, S.; Budkowski, A.; Steiner, U. *Interf. Sci.* **2003**, *11*, 225.
[16] Walheim, S.; Ramstein, M.; Steiner, U. *Langmuir* 1999, 15, 4848.
[17] Sprenger, M.; Walheim, S.; Schäfle, C.; Steiner, U. *Adv. Mater.*, **2003**, *15*, 703.
[18] Xia, Y.; Zhao, X.-M.;Whitesides, G. M. *Microelectronic Engineering* **1996**, *32*, 255.
[19] Gau, H.; Herminghaus, S.; Lenz, P.; Lipowsky, R. *Science* **1999**, *283*, 46.
[20] Karim, A. *et al. Phys. Rev. E* **1998**, *57*, 273.
[21] Xia, Y.; Tien, J.; Qin, D.;Whitesides, G. M. *Langmuir* **1996**, *12*, 4033; Bechinger, C.; Muffer, H.; Schäfle, C.; Sundberg, O.; Leiderer, P. *Thin Solid Films* **2000**, *366*, 135.
[22] Langbein, D. *Capillary Surfaces*; Springer: Berlin, 2002.
[23] Vrij, A. *Discuss. Faraday Soc.* **1966**, *42*, 23.
[24] Brochard-Wyart, F.; Daillant, J. *Can. J. Phys.* **1990**, *68*, 1084.
[25] Seemann, R.; Herminghaus, S.; Jacobs, K. *J. Phys. Condes. Mat.* **2001**, *13*, 4925.
[26] Reiter, G. *Phys. Rev. Lett.* **1992**, *68*, 75.
[27] Seemann, R.; Herminghaus, S.; Jacobs, K. *Phys. Rev. Lett.* **2001**, *86*, 5534.
[28] Lin, Z.; Kerle, T.; Baker, S. M.; Hoagland, D. A.; Schäffer, E.; Steiner, U.; Russell, T. P. *J. Chem. Phys.* **2001**, *114*, 2377.
[29] Lin, Z.; Kerle, T.; Russell, T. P.; Schäffer, E.; Steiner, U. *Macromolecules* **2002**, *35*, 3971.
[30] Schäffer, E.; Thurn-Albrecht, T.; Russell, T. P.; Steiner, U. *Nature* **2000**, *403*, 874.
[31] Schäffer, E.; Thurn-Albrecht, T.; Russell, T.P.; Steiner, U. *Europhys. Lett.* **2001**, *53*, 518.
[32] Schäffer, E.; Harkema, S.; Blossey, R.; Steiner, U. *Europhys. Lett.* **2002**, *60*, 255.
[33] Cross, M. C.; Hohenberg, P. C. *Rev. Mod. Phys.* **1993**, *65*, 851.
[34] Li, M.; Xu, S.; Kumacheva, E. *Macromolecules* **2000**, *33*, 4972.
[35] Schmittmann, B.; Zia, R. In *Phase Transitions and Critical Phenomena*; Domb, C.; Lebovitz, J. L., Eds.; Academic Press: London, 1983; Vol 17.
[36] Schäffer, E.; Harkema, S.; Roerdink, M.; Blossey, R.; Steiner, U. *Macromolecules* **2003**, *36*, 1645.
[37] Sette, F.; Krisch, M. H.; Masciovecchico, C.; Ruocco, G.; Monaco, G. *Science* **1998**, *280*, 1550.
[38] Schäffer, E. Ph.D. thesis, University of Konstanz, Konstanz, Germany, 2001.

[39] Schäffer, E.; Harkema, S.; Roerdink, M.; Blossey, R.; Steiner, U. *Adv. Mater.* **2003**, *15*, 514.
[40] Harkema, S.; Schäffer, E.; Morariu, M. D.; Steiner, U. *Langmuir* **2003**, *19*, 9714.
[41] Morariu, M. D.; Voicu, N. E. Schäffer, E.; Lin, Z.; Russell, T. P.; Steiner, U. *Nature Materials* **2003**, *2*, 48.
[42] Lin, Z.; Kerle, T.; Russell, T. P.; Schäffer, E.; Steiner, U. *Macromolecules* **2002**, *35*, 6255.
[43] Lambooy, P.; Phelan, K. C.; Haugg, O.; Krausch, G. *Phys. Rev. Lett.* **1996**, *76*, 1110.

2

Functional Nanostructured Polymers

Incorporation of Nanometer Level Control in Device Design

Wilhelm T. S. Huck

2.1. INTRODUCTION

Macromolecules make up the fabric of life. Without the protein machinery and DNA/RNA as information carriers, no cell would be able to complete its life cycle and to remain in a non-equilibrium thermodynamical state. Looking beyond the chemical structure of proteins and DNA, it becomes clear that their intricate interactions with other macromolecules form the core of their functionality. In DNA, this is evident in the formation of the double helix through H-bonding, whereas in proteins, numerous examples of functional macromolecular assemblies exist. A particularly impressive example of such a macromolecular assembly is the photosystem I,[1] which is a trimeric complex forming a large disc (Figure 2.1). However, each complex is an assembly of a dozen proteins, bringing together and precisely positioning hundreds of co-factors (chlorophyll). An equally impressive example of cellular machinery based on macromolecules is the ribosome complex, where RNA read-out and protein synthesis take place (Figure 2.1).[2] It is beyond the scope of this chapter to discuss the exact mechanisms of assembly and function, but these examples do illustrate the tremendous potential of polymers in nanotechnology, if, at least, we learn to harness such systems in man-made devices. A first step towards harnessing the power of biological 'machines' has been demonstrated by the seminal work of Montemagno and co-workers.[3] By engineering a

Melville Laboratory for Polymer Synthesis, Department of Chemistry, University of Cambridge, Lensfield Road, Cambridge, CB2 1EW, UK.
The Nanoscience Centre, Interdisciplinary Research Collaboration in Nanotechnology, University of Cambridge, 11 J. J. Thomson Avenue, CB3 0FF, UK. wtsh2@cam.ac.uk

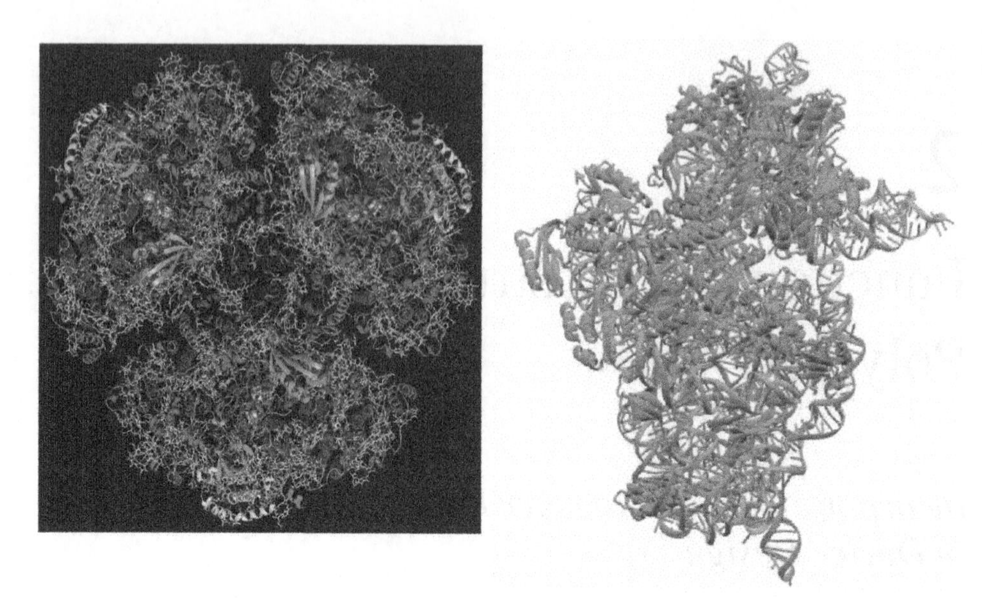

FIGURE 2.1. *Left*: Three-Dimensional Structure of Cyanobacterial Photosystem I at 2.5 Å resolution (reproduced with permission from reference).[1] *Right*: crystal structure of the 30S subunit from Thermus thermophilus, refined to 3 Å resolution (grey: RNA, blue: proteins) (reproduced with permission from reference).[2]

biomolecular nanomotor F1–adenosine triphosphate synthase (F1-ATPase) and integrating this biomolecule into an inorganic nanoscale system, they demonstrated the feasibility of building a nanomechanical device powered by a biomolecular motor.

Proteins have some obvious drawbacks. They cannot be designed *de novo*, the stability of the tertiairy structure and hence their functionality is strongly dependent on solvent, temperature, and salt concentration, and the large scale synthesis of complex proteins is not well-developed. Synthetic polymers should be able to overcome those problems, but it is impossible to emulate the complexity of proteins using synthetic polymers. The synthesis of polymers with a similar range of diversity in monomers is daunting, but the design of a folding synthetic structure and its interactions with other folded structures is far too complex for our current understanding of protein chemistry. In order to realize the potential of macromolecules and to introduce some of the complexity generated by biological systems into silicon devices, we aim to exploit the synthetic accessibility of 'everyday' polymers, and combine these with nanolithographic techniques as well as self-assembly and self-organization. Instead of synthesizing and assembling polymers in solution, where there are very few methods of producing anything but spherical nanostructures, surface chemistry and topography can be used to induce nanoscale assembly and organization into (functional) nanostructures.

2.2. PHASE SEPARATION OF POLYMER BLENDS IN LIGHT EMITTING DEVICES

Elsewhere in this book, Prof. Steiner describes the wealth of patterns arising from phase separation of polymer blends and different strategies to control this phase separation into

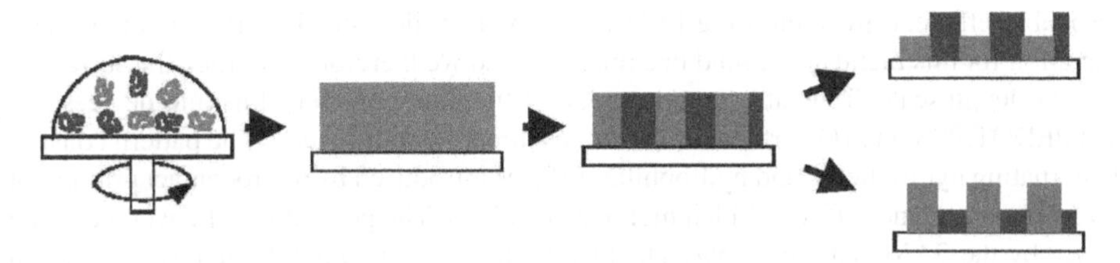

FIGURE 2.2. Phase separation of spincoated polymer blends on patterned surfaces.

ordered microstructures. This process is driven by the minimization of interfacial energy between the two polymers, the substrate surface and the air-polymer interface.[4] In short, polymer blends spin-coated onto surfaces patterned into areas of different surface energy, will phase separate following the underlying pattern (Figure 2.2).[5]

Patterns form efficiently if the difference in surface energy between the two regions is sufficient, so one component of the blend will preferentially migrate away from regions of higher surface energy.[6] This process is difficult to extend into the nanometer regime, because of the large wavelengths associated with the phase separation process. This is particularly the case for spinodal decomposition, but also in the nucleation-growth regime it is difficult to control size in the nanometer regime. However, the effect is interesting because of its potential use in Polymer Light Emitting Devices (PLEDs),[7] where it has been demonstrated that controlling the morphology of the phase separated structure inside blend devices can be used to improve device performance.[8,9,10] The polymers used in our study (for structures see Figure 2.3, below) consisted of poly(9,9-dioctylfluorene), F8, and poly(9,9-dioctylfluorene-alt-benzothiadiazole), F8BT, which are known to make

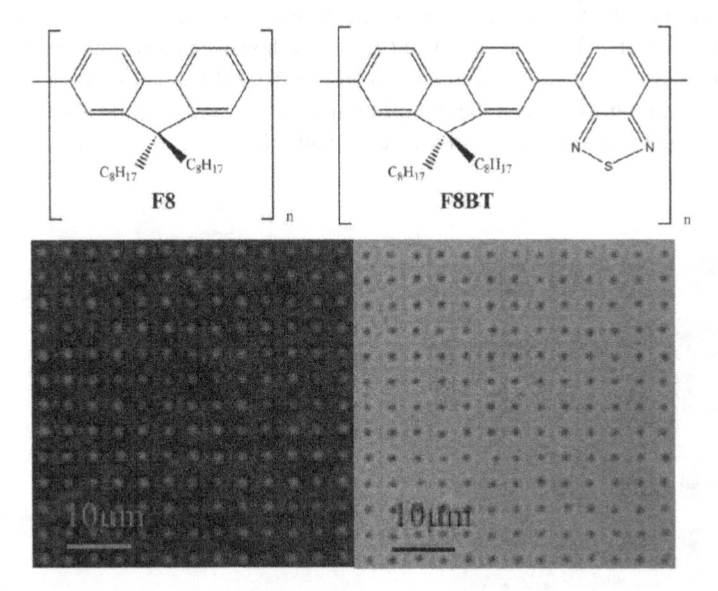

FIGURE 2.3. *Top*: Chemical structures of the semiconducting polymers F8 and F8BT. *Bottom*: Fluorescence microscopy images using different filters, showing F8 emission on the hydrophobic dots (*left*) and green emission from F8BT on the hydrophilic matrix.

reasonably efficient green emitting LEDs. The 'typical' domain sizes formed after phase separation for this blend are around one micron[11] and we therefore patterned the surface at a similar lengthscale. To enable incorporation of the final polymer films into devices, we used PEDOT:PSS on ITO surfaces as substrates throughout this study. The pattern consists of alternating hydrophobic and hydrophilic surfaces introduced by microcontact printing of a hydrophobic silane self-assembled monolayer (SAM). The percentage of coverage of the surface by the SAM is around 15%. The blend ratio of the two polyfluorenes was chosen to reflect the surface coverage of the hydrophobic and hydrophilic areas. F8BT is the more polar due to the benzothiadiazole group on the main chain and it is therefore expected to separate onto the hydrophilic plasma-treated PEDOT:PSS. The targeted morphology, *i.e.*, the same structure as patterned by the stamp, was obtained when the film was allowed to dry for 30 minutes in saturated atmosphere. Fluorescence micrographs of the patterned film (Figure 2.3) show that the blend morphology has replicated the underlying 2-D surface pattern and consists of well-defined blue-emitting F8-rich phases on the hydrophobic dots embedded in a green-emitting F8BT-rich matrix.

The films shown here are approximately 100 nm thick. Given the pattern is on the PEDOT:PSS layer and it is replicated on the surface of the blend layer, we conclude that the pattern extends *throughout* the bulk of the film, demonstrating the powerful effect of surface patterning. It should be noted that the components of the blend are quite similar and therefore the pattern will not consist of compositionally pure phases. The micrographs further show that the patterned morphology extends over the entire patterned area and is not confined to localised regions of the film. As discussed above, smaller patterns would not lead to smaller phase separated domains. However, the nanoscale order in these films only becomes apparent when investigating the surface topography of the 2-D patterned film using tapping-mode AFM (Figure 2.4). The image indicates that the PFO-rich domains are approximately 2 to 2.5 μm in diameter and have the same 4 μm periodicity in both lateral directions. The self-organized films are however more interesting when the surface is studied in more detail. In related work, Budkowski and co-workers[12] noted that after phase separation, significant height differences in the films can be observed, together with distinct curvature (concave and convex) inside domains. In our films, the periodic

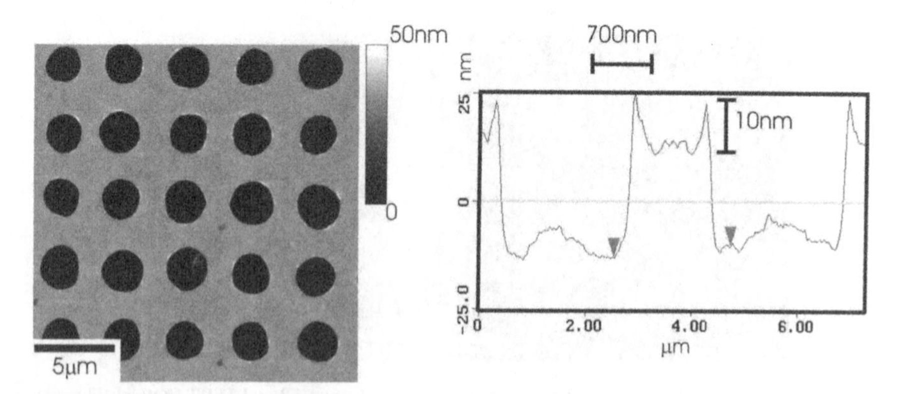

FIGURE 2.4. *Left*: 8 × 8 μm AFM image of 2-D patterned F8:F8BT blend film. *Right*: Line scan of part of the film shown left. The topographical variations at the edges of the patterned regions can clearly be seen.

PFO-rich domains are all ringed by "ripple" structures (Figure 2.4), which exhibit height variation on the order of 10 nm and a characteristic length-scale, λ, of approximately 500–800 nm. We speculate that the formation of these features is attributed to the differences in solubility of the two polymers in the common solvent: the less soluble component of the blend in the common solvent will first solidify and impose the domains architecture of the film. The other component still swollen at this time will then collapse as the evaporation continues.

Wetting phenomena and local surface tensions result in the curvature at the interface between the two polymers. Following this model we can deduce that PFO is less soluble in xylene as the F8BT-rich phase is roughly 20–30 nm higher on average than the PFO-rich regions. Similar 'ripple' features have been observed in unpatterned spincoated films, but the features are not as well defined and tend to be smeared out over longer wavelengths.

The resulting films should have interesting optical properties, because of their alternating regions of different refractive indices, in combination with small wavelength corrugations on the surface. Such structures are ideal photonic elements and we have investigated their effect on waveguided light in PLEDs. Waveguiding is an important loss mechanism in PLEDs and every strategy to minimize waveguiding and promote outcoupling will improve device performance. The patterned, hierarchically phase separated films were incorporated in devices by evaporating a 100 nm thick calcium cathode on top. Figure 2.5 presents the electroluminescence radiation pattern *i.e.*, the angular distribution of emitted photon flux

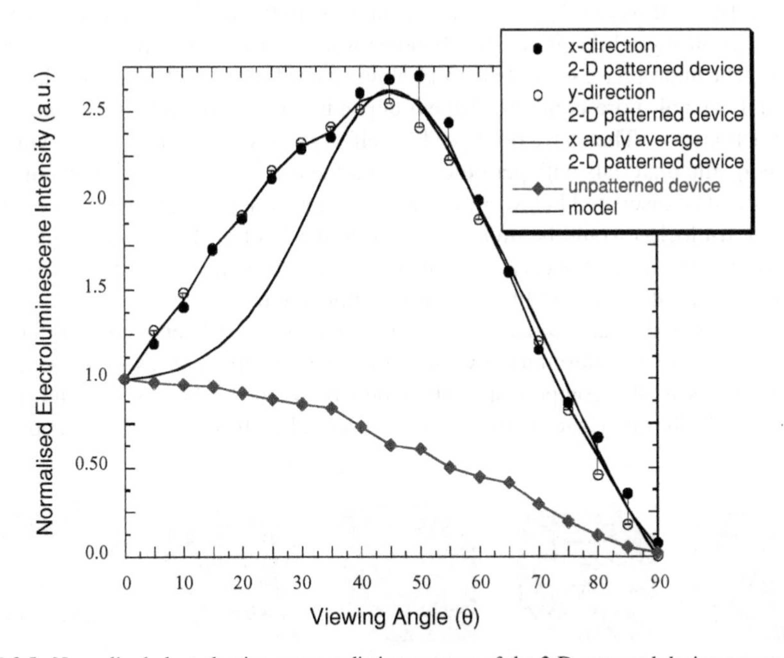

FIGURE 2.5. Normalised electroluminescence radiation patterns of the 2-D patterned device measured in the x-(filled circles) and y-directions (filled squares) are shown along with the average of the two (black line, no symbols). The radiation pattern of an unpatterned device is also shown (filled triangles) along with radiation pattern obtained from the lambertian emission (dashed line, no symbols).

per unit solid angle. Angles are described with respect to the normal to the glass surface. For unpatterned films, the surface emission through the surface of the glass substrate merely decreases in intensity geometrically with increasing viewing angle, *i.e.*, lambertian, which is to be expected as this device does not contain any structures that could alter the propagation angles of light waveguided in the polymer-ITO layers. For patterned devices, the electroluminescence intensity increases with the viewing angle until the peak at 50 degrees and decrease sharply towards 90 degrees. This non-lambertian behaviour of the emission pattern indicates the scattering of the waveguide mode out of the plane of the device. At the same time, neither the current density through the device, nor the switch-on voltages were significantly changed from unpatterned devices. The overall light output from the patterned devices was approximately 100% higher than unpatterned devices. Detailed model calculations (not discussed here) were done to confirm that this increase in device performance was indeed due to an outcoupling of waveguided light, as a result of interactions with the nanoscale surface corrugations and the micronscale phase separated bulk.

2.3. PHASE SEPARATION OF BLOCK COPOLYMERS

Polymer blends tend to phase separate into non-uniform micronscale features and only careful surface patterning can induce order in these features. In contrast, block copolymers comprised of two chemically incompatible and dissimilar blocks can microphase separate into a variety of morphologies with nanometer scale dimensions (typically in the 10–100 nm size range). This self-assembly process is driven by an unfavourable mixing enthalpy and a small mixing entropy, while the covalent bond between the two blocks prevents macrophase separation. The microphase separated morphology that is formed (spheres, lamellae, inverse spheres and several more complex shapes) depends on the polymers used and on their volume fractions.[13,14] They have been used as self-organized templates for the synthesis of various inorganic materials with periodic order on the nanometer scale,[15] and a number of these ideas will be discussed below. A full review of phase separation of block copolymers (diblocks or triblocks) would be impossible within the limit of this chapter. The reader is referred to several excellent overviews of the synthesis and properties of these polymers for a full appreciation of the possibilities for future applications.[16,17,18]

Block copolymer phase separation has first and foremost been studied in bulk. The mesoscale structure is determined by molecular parameters such as chain length (N), volume fractions of the components, interaction between the blocks (χ) and temperature (Figure 2.6). In this Chapter, we will be concerned mainly with diblock copolymers,

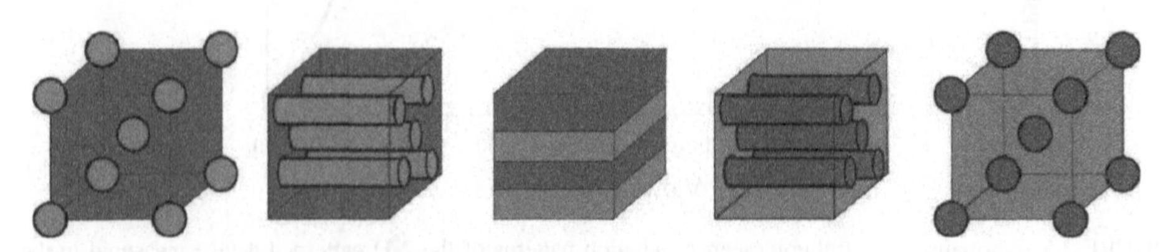

FIGURE 2.6. Block copolymer morphologies obtainable via phase separation. The exact structure will depend on the relative volume fractions of the two blocks, as well as χN (where χ is the interaction parameter and N the length of the polymer).

since these are relatively well-understood. Recent calculations on triblock copolymers show an extremely rich phase behaviour,[19,20] but only a few of these have been verified experimentally.[21]

Phase separation of polymers into lamellar structures has been used to generate 50–100 nm thick periodic layers with different refractive indices, which can be used as photonic crystals. Solution cast films of symmetrical polystyrene-polyisoprene (PS-PI) diblock copolymer films showed single, well-defined reflectivity peaks in the visible wavelength region.[22] The peak reflective wavelength could be tuned from 350–600 nm by adding homopolymers to the block copolymers, which increased the thickness of the lamellar layers.[23] By blending a triblock copolymer (poly(styrene)-b-poly(butadiene)-b-poly(t-butyl methacrylate) (PS-PB-PBA) with a PS-PBA diblock copolymer, non-centrosymmetric lamellar phases were obtained with three different alternating layers.[24] The simultaneous self-organization of block copolymers in the presence of ex-situ synthesized particles provides yet another approach to engineer 2D and 3D nanostructures that facilitates better control of the structural characteristics of the sequestered component, which becomes important when applications rely on size- or shape-related properties of nano-objects.[25] Kramer and co-workers exploited the lamellar assembly of symmetric poly(styrene-*b*-ethylene propylene) (PS-PEP) copolymer with a molecular weight of the respective blocks of 4×10^5 g/mol to organize gold or silicon nanoparticles in a well-defined way.[26] The diblocks form lamellae with domain spacings of 100 nm for the PS and 80 nm for the PEP domain. The block copolymer/nanocrystal composite films were subsequently obtained by casting a 5% polymer solution in toluene admixed with nanocrystals to result in a final amount of inorganic component in the composite of 2 vol%. TEM analysis of slices of these films showed particles either at the interface of the two polymers, or in the middle of one of the two components, depending on the type of particle (gold or silicon) and their diameter. This phenomenon had been theoretically predicted,[27,28] and clearly demonstrates that block copolymers will have a role to play in the design of new materials where nanoscale order will improve mechanic and electronic properties.

Thin films of block copolymers on surfaces are another important area of study, because of the extra constraints placed on the phase-separating system by the substrate-polymer and polymer-air interfaces. Spincoating block copolymers into thin films on the surfaces provides a number of possibilities to create very well-ordered nanostructures.[29] Parameters that can be investigated include film thickness, wetting energy of the surface and annealing conditions. Because the phase separation process is thermodynamically driven, it yields, in principle, highly ordered nanostructures over large areas. With perfect control, the phase separation of block copolymers could be a powerful tool for fabricating nanostructures without the use of costly lithography. An attractive feature of block copolymers containing either PMMA or poly(isoprene)/poly(butadiene) blocks is the ease with which these blocks can be removed. After microphase separation, a selective etch with UV/acetic acid or ozone, will render the microphase separated film nanoporous. Dense periodic arrays of holes and dots were fabricated from polystyrene-*b*-polybutadiene (PS-*b*-PB) or PS-*b*-PMMA diblock copolymer films.[30] In a typical example, 20 nm holes, separated by 40 nm and extending over very large areas, were obtained. The phase-separated structures can be used as an etch mask to transfer the features into an underlying silicon nitride film. Subsequent e-beam evaporation of metals and lift-off of the polyimide mask, resulted in dense arrays of metal dots.[31] The fabrication of ultra-high density storage devices via this

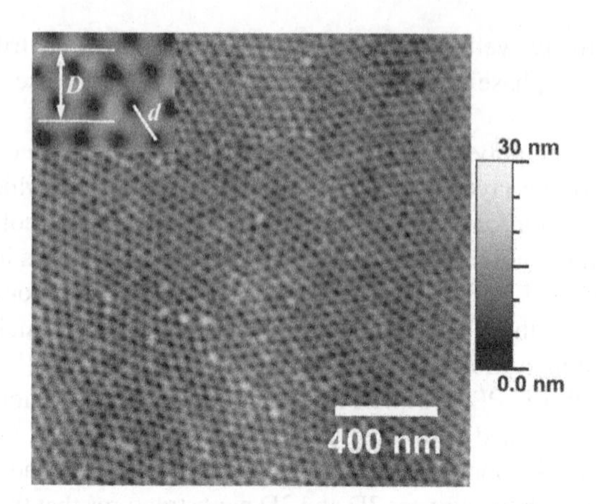

FIGURE 2.7. AFM image of PS matrix after removing PMMA columns on neutralized silicon surface with a total film thickness of 42 nm. The insert shows the centre-to-centre distance *d* and the distance *D* required to fit two rows of PMMA columns (42 nm and 72 nm respectively).

route is currently being explored. A prerequisite for the successful use of block copolymers as nanoscopic templates is the control over the orientation of the microdomains. In particular for cylindrical microdomains, an orientation normal to the substrate is required. One way of controlling the orientation of the nanoscale domains is by controlling the surface energy of the substrate, choosing the right solvent for drop or spin casting of the block copolymer and thermal annealing. On an unpatterned, 'neutral' surface, a spin-coated film of a PS-*b*-PMMA block copolymer (46.1 k PS and 21.0 k PMMA) shows the cylindrical nanodomains oriented normal to the surface, when the thickness of the film is around 40–45 nm and after annealing in vacuum. The 'neutral' surface is required to avoid preferential wetting by one of the blocks, which would result in a horizontal alignment of the films. Figure 2.7 shows an AFM image of a phase-separated film after removal of the PMMA block, illustrating that the cylinders are oriented perpendicular to the surface and showing short range hexagonal ordering.

As an alternative to surface energy matching, electrical poling can be used to orient the microdomains.[32,33] One of the advantages of this method is that the thickness of the film can be thicker, and the substrate can be varied.[34] When the block copolymer is allowed to phase separate on an electrode surface, the porous film can be filled via electroplating, leading to the formation of nanowires (~15 nm in diameter) in a polymer matrix.[35]

The same PS-*b*-PMMA diblocks were used as in the example above, with a volume fraction of styrene of 0.71 but a lower total molecular weight of 39.6 k The resulting 1 μm thick films consisted of 14 nm diameter PMMA columns with a 24 nm lattice constant. This again illustrates the flexibility by which the sizes can be varied and the small length-scales which are accessible without any lithography. After removal of the PMMA, the unfilled, nanoporous film on top of the electrode, acted as a nanoelectrode array, which at low scan rates behaves like a macroelectrode, but at high scan rates the nanoelectrodes act independently.[36] Alternatively, the holes can be filled with $SiCl_4$, and subsequent hydrolysis leads to an array of silicon oxide posts in an organic matrix. After etching away the organic material, the substrate has a very high surface area, which could be of use in sensor applications.[37] One of the main problems which precluded commercial applications is the lack of long range order in the phase separated domains. As illustrated in Figure 2.7, the PMMA columns aligned vertically and appear hexagonally close-packed, but the domains are generally small and not aligned (it should be noted that this is not an optimized example and much larger domains have been reported in the literature.[38] There is a strong driving force to align the columns, but there are no forces acting laterally on the film to order the domains. A topographically patterned substrate can be used to template a large, single crystalline grain of block copolymer spherical domains (graphoepitaxy).[39] Both the geometry of the substrate and the annealing temperature can be used to control the degree of order as well as the crystallographic orientation of the grain. The presence of a hard edge imparts translational order onto the hexatic, which has quasi long-range orientational order and short-range translational order, and onto the liquid which has short-ranged translational and orientational order. This ordering influence is seen to fade as the distance from the edge increases. Crystalline grains templated by a hard step wall also have superior translational order as compared to those templated by a vacuum edge. The adjacent step also plays a crucial role in determining the grain orientation of the crystalline array on top of a mesa.[40] Very recently,[41] Russell and co-workers have made important breakthroughs in this field by showing truly remarkable long range order in PS-*b*-PEO films (M_w 25,300; PS weight fraction 0.75) which were annealed for 48 hrs in benzene or benzene/water vapour (Figure 2.8). Virtually no defects were detected over areas spanning several hundreds of μm^2.

The mechanism by which this long range ordering occurs is unclear. It seems that the evaporation of the solvent from the top aligns the columns, while allowing sufficient mobility to form close-packed arrays. Solvent annealing clearly provides a route towards defect free nanoscopic ordered films. In the same paper, the authors also illustrate that graphoepitaxy can be used to create an additional level of control over the ordering and positioning of nanostructures. By spin-coating the same polymer onto a surface containing lithographically etched 875 nm wide, 325 nm deep channels, the phase separated domains lined up perfectly with the channel walls. Preferential wetting of these walls by one of the components of the block copolymer again induces long range order. This idea had been

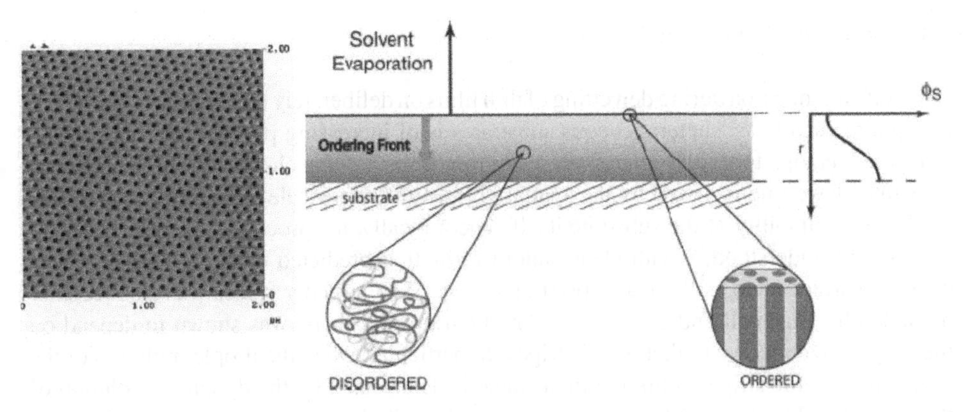

FIGURE 2.8. Long range ordering of PS-*b*-PEO blocks on planar substrates (reproduced with permission from reference [41], Copyright 2004 Wiley-VCH).

explored by other groups. Sibener and co-workers have shown that the same technique can be used to align cylinders of PS-*b*-PEP horizontally inside shallow trenches.[42] Kramer and Vancso have used ferrocenyldimethylsilane containing copolymers for alignment inside lithographically prepared trenches.[43,44] These block copolymers have the advantage that a simple plasma etch is sufficient to convert the polymer into nanoscale SiO_2 dots.[45] By varying the step width, integer numbers of columns can be fitted inside. The microphase separation appears remarkably robust in cases of 'mismatching' trench widths, where the domain sizes simply become compressed or stretched to fill the gap.

Nealey and co-workers[46] have taken surface-guided alignment of blocks to the extreme, by *chemically* patterning the surface at the 25 nm level to match the surface energy to the different blocks.[47] In this example of epitaxial assembly of block-copolymer films, molecular-level control over the precise size, shape and spacing of the ordered block copolymers is achieved on nanopatterned surfaces. Photoresist was patterned with alternating lines and spaces with periods between 45 and 55 nm using extreme ultraviolet interferometric lithography (EUV-IL), and the pattern in the photoresist was transferred to an underlying self-assembled monolayer (SAM) by chemical modification of the SAM. Following removal of the remaining photoresist, a 60-nm film of symmetric lamella-forming (PS-b-PMMA), molecular weight 104 k, lamellar period approximately 48 nm) was spin-coated and annealed on the chemically patterned surface. The PMMA block preferentially wets the SAMs that were chemically modified to contain polar groups, and the other regions exhibited neutral wetting behaviour by the block copolymer. The domain structure of the block copolymer film after annealing was imaged using scanning electron microscopy (SEM). When the period of the chemically patterned SAM, Ls, was equal to the period of the lamellar spacing of the block copolymer, Lo, (48 nm), then the lamellar domains in the block-copolymer films were oriented perpendicularly to the substrate and were perfectly registered with the chemical pattern of the SAM. The largest continuous area over which the chemical surface patterns were defect-free was larger than 8 μm by 5 μm. Slight mismatches in the periodicity of the patterned SAMs immediately influenced the alignment of the block copolymers on top, illustrating the level of control possible (and necessary) to arrange block copolymers in precisely defined patterns.

2.4. CONTROLLED DEWETTING

Self-organization during dewetting of thin films on deliberately tailored chemically[48,49] or topographically[50,51] heterogeneous substrates is of increasing promise for engineering of desired nanopatterns and micropatterns in thin films.[52] On a chemically heterogeneous substrate, dewetting is driven by the spatial gradient of microscale wettability, rather than by the nonwettability of the substrate itself. Theoretically, the process of patterned dewetting is well-understood,[53] with clear pattern formation predicted even with small differences in surface energy.[54,55] Inkjet printing on top of patterned surfaces was also recently modelled by Bucknall and co-workers. The final droplet shape was shown to depend on the droplet size relative to that of the stripes. In particular, when the droplet radius is of the same order as the stripe width, the final shape is determined by the dynamic evolution of the drop and shows a sensitive dependence on the initial droplet position and velocity.[56] Gau *et al.*[57] made elegant use of differential wetting to create free-standing strips of water.

Liquid 'virtual' microchannels on structured surfaces were built up using a wettability pattern consisting of hydrophilic stripes on a hydrophobic substrate. These channels undergo a shape instability at a certain amount of adsorbed volume, from a homogeneous state with a spatially constant cross section to a state with a single bulge. So far, most of the work on dewetting has focussed on producing micronscale features. However, by optimizing the 'contrast' (*i.e.*, wettability differences) and droplet shapes, smaller structures should be accessible. It would be very interesting to combine such nanostructuring strategy with inkjet printing, since the resolution of this technique is limited to typically 20–50 μm, due to the difficulties of controlling the flow and spreading of liquid inks on surfaces. An area of particular interest for the application of dewetting on patterned surfaces is the printing of (semi)conducting polymers for the fabrication of polymer transistors. This concept has been used successfully for patterning of source-drain electrodes of polymer FETs with channel lengths of 5 μm by inkjet printing.[58] Dewetting by dip coating has also been used to pattern the active semiconducting layer in transistor fabrication.[59] The performance of the resulting devices would greatly benefit from further reduction of channel length to submicrometer dimensions, because the mobility of the materials used is rather low, compared to inorganic counterparts. However, to achieve nanoscale dewetting a detailed understanding of the various factors that govern the interaction of droplets containing a solute of functional material with a patterned surface is required. As explained above, dewtting of pure water droplets has been investigated extensively. However, no experimental study towards the lower size limit of dewetting of solute-containing inks from narrow hydrophobic strips had been reported. We recently reported dewetting of the water-based conducting polymer poly(3,4-ethylenedioxythiophene)/poly(4-styrenesulfonate) (PEDOT/PSS) inks on patterned SiO_2 surfaces modified with a fluorinated SAM.[60] Submicrometer hydrophobic lines with widths varying from 250 nm to 20 μm were defined by electron beam lithography (EBL) (250 nm–1 μm) and optical lithography (2–20 μm), respectively. The smallest linewidth of 250 nm, which we were able to define by EBL, was limited by charging of the insulating substrate during electron beam exposure. In principle, nanocontact printing should provide even narrower lines, but this technique was not used in this study.[61]

Dewetting was studied by ink-jetting PEDOT/PSS water droplets of different concentrations with a droplet volume of $\sim$65 pl per drop on top of patterned surfaces (Figure 2.9). Drying of the ink droplets occurred in cleanroom air at 50% relative humidity. Experiments under higher humidity conditions showed significant increase in drying time, but little effect on the ability of the droplets to dewet.

Figures 2.9a and b compare dewetting of a 1:1 PEDOT/water ink droplet on top of substrates with different degree of hydrophilicity in the bare SiO_2 regions. On a substrate cleaned by oxygen plasma cleaning prior to deposition of the PMMA resist, dewetting from a 500 nm wide line is observed (Figure 2.9b). In contrast, if the substrate is only cleaned by washing in acetone and isopropanol, resulting in a higher contact angle and smaller droplet diameter (Figure 2.9a), even on a 700 nm wide line no complete dewetting is observed. Dewetting also depends on the relative position of the center of the droplet with respect to the hydrophobic barrier. If the hydrophobic line is close to the edge of the droplet, dewetting is possible even from very narrow hydrophobic lines (compare Figure 2.9a and c). Another key factor is the ink concentration and ink viscosity. Lower PEDOT/PSS concentrations enable dewetting on very narrow (250 nm) lines from which dewetting of more concentrated

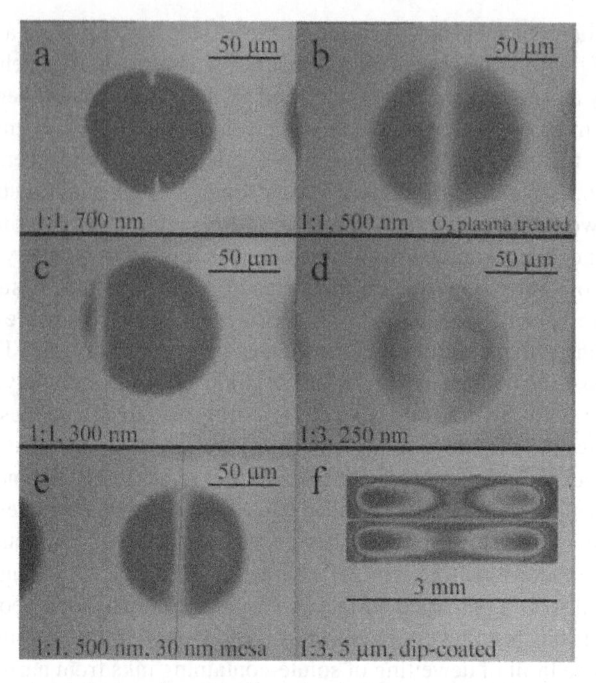

FIGURE 2.9. Inkjet printed droplets dewetting from chemically and topographically nanopatterned surfaces (reproduced with permission from reference [60].

solutions is not possible (compare Figures 2.9a and d). Finally, the use of a mesa of finite thickness also improves the ability to dewet significantly. On top of a 30 nm thick mesa a concentrated PEDOT/PSS solution is capable of dewetting from significantly narrower lines than on top of monolayer surface energy barriers (compare Figures 2.9e and a). Figure 2.9f shows a pattern formed from simple dipcoating a patterned surface and illustrates that both small and large patterns are accessible via controlled dewetting. The rationale behind this work was that droplets that land on top of the narrow fluorinated lines split into two during the drying of the ink, so defining the source and drain electrodes of the FET. To demonstrate the versatility of this approach, top-gate polymer FETs were fabricated employing dewetted PEDOT/PSS source and drain electrodes by spin coating a 50 nm polymer semiconductor layer of poly(9,9′-dioctyl-fluorene-co-bithiophene) (F8T2) and 1 μm insulating layer of PMMA, and finally inkjet printing a PEDOT/PSS top gate electrode.

2.5. POLYMER BRUSHES

Polymer brushes are polymers tethered to a surface via one end. The connection to the surface can be covalent or non-covalent, and the brushes can be made via *grafting to* or *grafting from* the surface. In the past few years, there has been considerable interest in the growth of polymer brushes *via* surface-initiated polymerisations from (patterned) initiator-functionalised SAMs.[62,63] For example, we have recently shown that surface confined Atom Transfer Radical Polymerisations (ATRP) in aqueous solvents leads to rapid and controlled

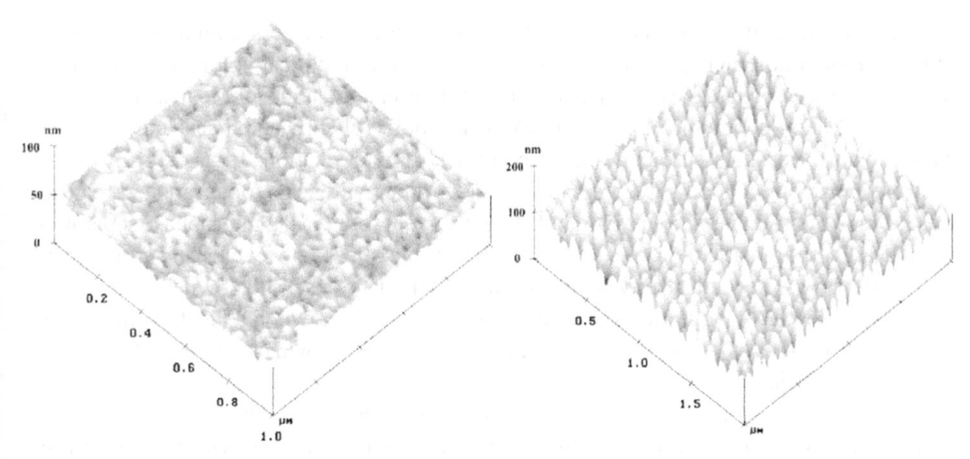

FIGURE 2.10. AFM image of the tethered PS-*b*-PMMA brushes with 23 nm thick PS layer and 14 nm thick PMMA layer after treatment with *left*: cyclohexane at 35 °C for 1 h and *right*: CH_2Cl_2 for 30 min and subsequently in CH_2Cl_2/cyclohexane 1:1 for 30 min, followed by CH_2Cl_2/cyclohexane 0.5:99.5 (reproduced with permission from reference.[68] Copyright 2000 American Chemical Society).

brush growth of a variety of polymers.[64] Polymer brushes are included in this chapter because the range of controlled polymerization strategies available to grow such brushes, enable the fabrication of nanostructured polymer surfaces in a completely different way. Phase separation has been discussed in detail in previous paragraphs, but when tethered to the surface, the phase separation has been predicted to show interesting characteristics.[65] Brittain and co-workers[66] synthesized PS-*b*-PMMA brushes via sequential carbocationic polymerisation and ATRP from initiator terminated SAMs.[67] The resulting polymer films hence consist of a bilayer of two polymers on top of each other, where the thickness of each layer can be controlled exactly, by varying the growth time of the different blocks. The behavior of these tethered diblock brushes, as explored by AFM, revealed the formation of various nanomorphologies which was dictated by the chemical and physical composition of the brush and the solvent it was exposed to (Figure 2.10).[68] A brush consisting of PS-b-PMMA showed a smooth featureless surface after treatment with CH_2Cl_2 and contact angles were in good agreement with previously reported values for PMMA thin films.

The same brush exhibited advancing contact angles close to that reported for PS after immersion in cyclohexane. Re-exposure of these films to CH_2Cl_2 resulted in featureless films with PMMA advancing and receding contact angles, showing that the rearrangement of the films is reversible. When the solvent was gradually changed from CH_2Cl_2 to cyclohexane, a very distinct morphology developed (Figure 2.10). The seemingly periodic nanomorphology is probably caused by the soluble block forming a layer around the insoluble core, resulting in an array of surface-immobilized micelles.[69] The length of each polymer block determines the size of the domains and the surface roughness, however the overall morphology of the surface is the same. ABA type triblock copolymer brushes showed similar reversible morphologies as the diblocks.[70,71] A not dissimilar effect was also observed when PMMA-b-PHEMA brushes grown using aqueous ATRP were compared to PMMA brushes grown under similar conditions. The PMMA surfaces were relatively featureless, while the thin PHEMA blocks on top formed nanoscale morphologies.[72]

Prokhorova and co-workers,[73] created polymer brush surfaces containing mixtures of two different polymers grown on the surface. Instead of a typical polymer blend phase-separation process, again the processes take place at the nanometer level because of tethering of chains. Mixed-copolymer brushes PMMA and poly (glycidyl methacrylate), PGMA, were grown form silica surfaces using free radical polymerization. It was shown that in chloroform vapour, a good solvent for both PGMA and PMMA, no topographical features were seen. After toluene vapour treatment, a poor solvent for PGMA, microphase separation occurred. Phase transitions were shown to be reversible after many cycles and could occur in as little as one second of expose to liquid solvent. They used this switching to move monodisperse silica nanoparticles around the polymer surface, which collected into islands after a number of cycles, and suggest that this system could provide surfaces with directed transport properties.

Minko and co-workers fabricated a mixed brush of poly(2-vinylpyridine), P2VP, and polyisoprene, PI.[74] The brush was illuminated through a photomask with UV light; illuminated areas became cross-linked and lost their responsive behaviour to stimuli, whilst non-irradiated areas retained their smart behaviour.[75] Upon a toluene wash, PI has the more favourable interactions with the solvent and comes to the surface making it hydrophobic – more so than the cross-linked background. An acidic aqueous wash (pH $= 2$) acts as a protonating agent for P2VP (PI being insoluble) causing the surface to become hydrophilic and ionised – more so than the cross-linked background. These washing processes are reversible, erasing and exposing the pattern at will.

We recently reported the synthesis of polylelectrolyte brushes from patterned surfaces.[76] These brushes carried cationic or anionic charges along the polymer backbone and have quite different properties from non-charged brushes.[77,78,79] In pure water, these brushes stretch out (regardless of grafting density) because of repulsion between the charged monomers. However, when salt is added to the solution, the charges become screened, and the monomers are allowed closer together. As a result, the brushes can relax into a more coiled conformation, thereby collapsing onto the surface. We have now studied these brushes by imaging brushes in solution using AFM. Figure 2.11 shows patterned

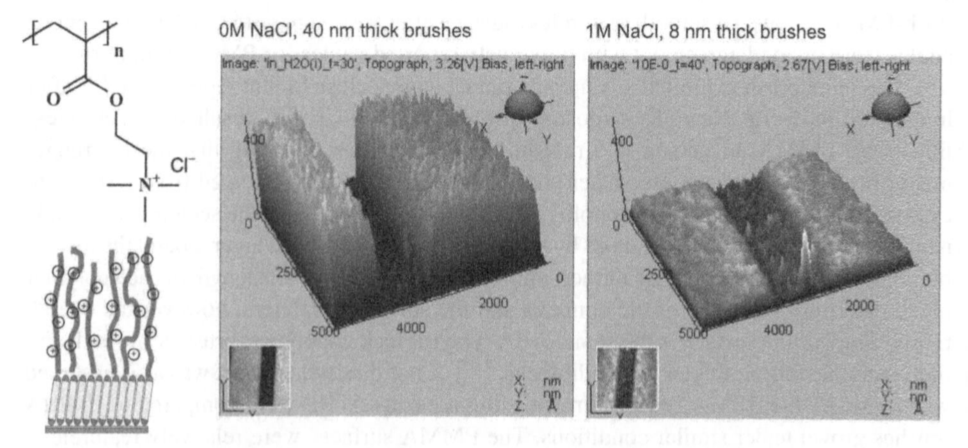

FIGURE 2.11. Collapse of cationic polyelectrolyte brushes upon addition of salt.

poly [2-(methacryloyloxy)ethyl] trimethylammonium chloride (PMETAC) brushes with an apparent thickness of around 40 nm in pure water. The patterning is introduced by micro-contact printing the initiator molecules from which the brushes grow. Upon addition of 1M NaCl, the thickness changed dramatically to around 10 nm, which is very close to their dry thickness. This collapse is completely reversible and after changing the solvent back to pure water, the brushes fully expand back to 40 nm. We do not have information on the timescales of switching, but within the time needed for the AFM to stabilize after adding a different solvent (approximately 10 minutes) all brushes seemed to have equilibrated into the new thickness.

These are only the first examples of switchable polymer brushes, which control surface topography and chemistry at the nanometer level *via* self-organisation and chain reorganization. The variations in combinations of different polymers, different molecular weights, external triggers, additional patterning and substrates are endless, and it is fully expected that these surfaces will play a major role in future microfluidic and micro (or nano) electromechanical systems (MEMS/NEMS).[80] Responsive surfaces will introduce a whole new level of complexity into these devices, allowing far better control over the internal chemistry. The idea of using chain collapse and expansion as a power source in such devices is a tempting idea which is currently being explored.

2.6. CONCLUSIONS

This review has highlighted a number of (very different) strategies to induce order at the 5 to 100 nm (and up to micron!) level in polymeric systems. Clearly, the design of intramolecular interactions is not nearly as elegant and precise as some of the work presented in other chapters in this book, where hydrogen bonding, Coulombic interactions, van der Waals forces, or ligand-metal binding are used to ensure precise 3D orientation of molecular building blocks. However, at the same time, the results in this chapter demonstrate that useful structures can be obtained and the resulting devices show improved characteristics as a result of the nanoscale organization.[81] This organization has been achieved partly by relying on 'cruder' intermolecular interactions such as phase separation and dewetting. As a result, the order is less precise and more imperfections can be seen. However, many polymeric devices do not need molecular scale perfection in order to enhance performance. At the same time, the results above have highlighted that by moving away from the molecular regime to the 5 to 100 nm regime, nanolithographic techniques can play a role to connect self-organizing systems with 'man-made' structures and to induce further ordering in the organization process as well. Polymers will therefore provide ideal building blocks for current and future nanotechnology.

ACKNOWLEDGMENTS

I would like to thank my collaborators in Cambridge for contributing to this work; in particular Guillaume Fichet (phase separation of polymer blends), Vicky Osborne (poly-elecrolyte brushes) and Dr. Hongwei Li (selective dewetting). All work on optoelectronic devices was done in collaboration with the groups of Prof. Friend and Prof. Sirringhaus in the Cavendish Laboratories, Department of Physics. The EPSRC, ICI and Cambridge Display Technology are gratefully acknowledged for financial support.

REFERENCES

1. P. Jordan, P. Fromme, H. T. Witt, O. Klukas, W. Saenger, and N. Krauss, *Nature* **411**, 909 (2001).
2. B. T. Wimberly, D. E. Brodersen, W. M. Clemons, Jr., R. J. Morgan-Warren, A. P. Carter, C. Vonrhein, T. Hartsch, and V. Ramakrishnan, *Nature* **407**, 327 (2000).
3. R. K. Soong, G. D. Bachand, H. P. Neves, A. G. Olkhovets, H. G. Craighead, and C. D. Montemagno, *Science* **290**, 1555 (2000).
4. R. A. L. Jones, L. J. Norton, E. J. Cramer, F. S. Bates, and P. Wiltzius, *Phys. Rev. Lett.* **66**, 1326 (1991).
5. M. Böltau, S. Walheim, S. Mlynek, G. Krausch, and U. Steiner, *Nature* **391**, 877 (1998).
6. G. Krausch, E. J. Kramer, M. H. Rafailovich, and J. Sokolov, *Appl. Phys. Lett.*, **64**, 2655 (1994).
7. J. J. M. Burroughes, D. D. C. Bradley, A. R. Brown, R. N. Marks, R. Mackay, R. H. Friend, P. L. Burn, and A. B. Holmes, *Nature* **347**, 539 (1990).
8. A. Arias, N. Corcoran, M. Banach, R. H. Friend, and W. T. S. Huck, *Appl. Phys. Lett.* **80**, 1695 (2002).
9. A. C. Arias, J. D. Mackenzie, R. Stevenson, J. J. M. Halls, M. Inbasekaran, E. P. Woo, D. Richards, and R. H. Friend, *Macromolecules* **34**, 6005 (2001).
10. J. J. M. Halls, A. C. Arias, J. D. Mackenzie, M. Inbasekaran, E. P. Woo, and R. H. Friend, *Adv. Mater.* **12**, 498 (2000).
11. A. W. Grice, D. D. C. Bradley, M. T. Bernius, M. Inbasekaran, W. W. Wu, and E. P. Woo, *Appl. Phys. Lett.* **73**, 629 (1998).
12. A. Budkowski, A. Bernasik, P. Cyganik, J. Raczkowska, B. Penc, B. Bergues, K. Kowalski, J. Rysz, and J. Janik, *Macromolecules* **36**, 4060 (2003).
13. A. K. Khandpur, S. Förster, F. S. Bates, I. Hamley, A. J. Ryan, K. Almdal, and K. Mortensen, *Macromolecules,* **28**, 8796 (1995).
14. F. S. Bates and G. H. Fredrickson, *Annu. Rev. Phys. Chem.* **41**, 525 (1990).
15. J. Y. Cheng, C. A. Ross, V. Z.-H. Chan, E. L. Thomas, R. G. H. Lammertink, and G. J.Vancso, *Adv. Mater.* **13**, 1174 (2001).
16. I. W. Hamley, *Nanotechnology* **14**, R39 (2003).
17. N. Hadjichristides, S. Pispas, and G. A. Floudas, *Block Copolymers, Synthetic Strategies, Physical Properties, and Applications* (Wiley Interscience, New Jersey, 2003).
18. G. Krausch and R. Magerle, *Adv. Mater.* **14**, 1579 (2002).
19. F. S. Bates and G. H. Fredrickson, *Phys. Today* **52**, 32 (1999).
20. A. Knoll, A. Horvat, K. S. Lyakhova, G. Krausch, G. J. A. Sevink, A. V. Zvelidnovsky, and R. Magerle, *Phys. Rev. Lett.* **89**, 035501 (2002).
21. H. Elbs, V. Drummer, G. Abetz, and G. Krausch, *Macromolecules* **35**, 5570 (2002).
22. A. Urbas, Y. Fink, and E. L. Thomas, *Macromolecules* **32**, 4748 (1999).
23. A. Urbas, R. Sharp, Y. Fink, E. L. Thomas, M. Xenidou, and L. Fetters *J. Adv. Mater.* **12**, 812 (2000).
24. T. Goldacker, V. Abetz, R. Stadler, I. Erukhimovich, and L. Leibler, *Nature* **398**, 137 (1999).
25. S. Horiuchi, T. Fujita, T. Hayakawa, and Y. Nakao, *Langmuir* **19**, 2963 (2003).
26. M. R. Bockstaller, Y. Lapetnikov, S. Margel, and E. J. J. Kramer, *Am. Chem. Soc.* **125**, 5276 (2003).
27. R. B. Thompson, Ginzburg, M. W. Matsen, and A. C. Balazs, *Science* **292**, 2469 (2001).
28. J. Huh, V. V. Ginzburg, and A. C. Balazs, *Macromolecules* **33**, 8085 (2000).
29. J. P. Spatz, P. Eibeck, S. Mößmer, M. Möller, T. Herzog, and P. Ziemann, *Adv. Mater.* **10**, 849 (1998).
30. M. Park, P. M. Chaikin, R. A. Register, and D. H. Adamson, *Appl. Phys. Lett.* **79**, 257 (2001).
31. M. Park, C. Harrison, P. M. Chaikin, R. A. Register, and D. H. Adamson, *Science* **276**, 1401 (1997).
32. T. L. Morkved, M. Lu, A. M. Urbas, E. E. Ehrichs, H M. Jaeger, P. Mansky, and T. P. Russell, *Science* **273**, 931 (1996).

33. E. Huang, L. Rockford, T. P. Russell, and C. J. Hawker, *Nature* **395**, 757 (1998).
34. T. Thurn-Albrecht, R. Steiner, J. DeRouchey, C. M. Stafford, E. Huang, Bal, M. M. Tuominen, C. J. Hawker, and T. P. Russell, *Adv. Mater.* **12**, 787–91 (2000).
35. T. Thurn-Albrecht, J. Schotter, G. A. Kästle, N. Emley, T. Shibauchi, L. Krusin-Elbaum, K. Guarini, B. C. T. lack, Tuominen, and M. T. Russell, *Science* **290**, 2126 (2000).
36. E. Jeoung, T. H. Galow, J. Schotter, M. Bal, A. Ursache, M. T. Tuominen, C. M. Stafford, T. P. Russell, and V. M. Rotello, *Langmuir* **17**, 6396 (2001).
37. H.-C. Kim, X. Jia, C. M. Stafford, D. Ha Kim, T. J. McCarthy, M. Tuominen, C. J. Hawker, and T. P. Russell, *Adv. Mater.* **13**, 795 (2001).
38. K. W. Guarini, C. T. Black, and S. H. I. Heung, *Adv. Mater.* **14**, 1290 (2002).
39. R. A. Segalman, H. Yokoyama, and E. J. Kramer, *Adv. Mater.* **13**, 1152 (2001).
40. R. A. Segalman, A. Hexemer, and E. J. Kramer, *Macromolecules* **36**, 6831 (2003).
41. S. H. Kim, M. J. Misner, T. Xu, M. Kimura, and T. P. Russell, *Adv. Mater.* **16**, 226 (2004).
42. D. Sundrani, S. B. Darling, and S. J. Sibener, *Nano Lett.* **2**, 273 (2004).
43. J. Y. Cheng, C. A. Ross, E. L. Thomas, H. I. Smith, and G. J. Vancso, *Adv. Mater.* **15**, 1599 (2003).
44. J. Y. Cheng, C. A. Ross, E. L. Thomas, H. I. Smith, and G. J. Vansco, *Appl. Phys. Lett.* **81**, 3657 (2002).
45. R. G. H. Lammertink, M. A. Hempenius, J. E. van den Enk, V. Z. H. Chan, E. L. Thomas, and G. J. Vancso, *Adv. Mater.* **12**, 98 (2000).
46. S. O. Kim, H. H. Solak, M. P. Stoykovich, N. J. Ferrier, J. J. de Pablo, and P. F. Nealey, *Nature* **424**, 411 (2003).
47. X. M. Yang, R. D. Peters, P. F. Nealey, H. H. Solak, and F. Cerrina, *Macromolecules* **33**, 9575 (2000).
48. A. Kumar and G. M. Whitesides, *Science* **263**, 60 (1994).
49. P. Lenz and R. Lipowsky, *Phys. Rev. Lett.* **80**, 1920 (1998).
50. Y. Cui, M. T. Björk, J. A. Liddle, Sönnichsen, B. C. Boussert, and A. P. Alivisatos, *Nano Lett.* **4**, 1093 (2004).
51. N. D. Denkov, O. D. Velev, P. A. Kralchevsky, I. B. Ivanov, H. Yoshimura, and K. Nagayama, *Nature* **361**, 26 (1993).
52. B. Zhao, J. S. Moore, and D. J. Beebe, *Science* **291**, 1023 (2001).
53. K. Kargupta and A. J. Sharma, *Chem. Phys.* **116**, 3042 (2002).
54. A. A. Darhuber, S. M. Troian, S. M. Miller, and S. J. Wagner, *Appl. Phys.* **87**, 7768 (2000).
55. K. Kargupta and A. Sharma, *Phys. Rev. Lett.*, **86**, 4536 (2001).
56. J. Léopoldes, A. Dupuis, D. G. Bucknall, and J. M. Yeomans, *Langmuir* **19**, 9818 (2003).
57. H. Gau, S. Herminghaus, P. Lenz, and R. Lipowsky, *Science* **283**, 46 (1999).
58. H. Sirringhaus, T. Kawase, R. H. Friend, T. Shimoda, M. Inbasekaran, W. Wu, and E. P. Woo, *Science* **290**, 2123 (2000).
59. R. Kagan, T. L. Breen, and L. L Kosbar, *Appl. Phys. Lett.*, **79**, 3536 (2001).
60. J. Z. Wang, Z. H. Zheng, H. W. Li, W. T. S. Huck, and H. Sirringhaus, *Nature Materials* **3**, 171 (2004).
61. H. W. Li, B. V. O. Muir, G. Fichet, and W. T. S. Huck, *Langmuir* **19**, 1963 (2003).
62. M. Husseman, E. E. Malmstrom, M. McNamara, M. Mate, O. Mecerreyes, D. G. Benoit, J. L. Hedrick, P. Mansky, E. Huang, T. P. Russell, and C. J. Hawker, *Macromolecules* **32**, 1424 (1999).
63. S. Edmondson, V. L. Osborne, and W. T. S. Huck, *Chem. Soc. Rev.* **33**, 14 (2004).
64. D. M. Jones, A. A. Brown, and W. T. S. Huck, *Langmuir* **18**, 1265 (2002).
65. E. B. Zhulina, C. Singh, and A. C. Balazs, *Macromolecules* **29**, 6338 (1996).
66. B. Zhao, W. J. Brittain, W. Zhou, and S. Z. D. Cheng, *Macromolecules* **33**, 8821 (2000).
67. B. Zhao and W. J. Brittain, *J. Am. Chem. Soc.* **121**, 3557 (1999).
68. B. Zhao, W. J. Brittain, W. Zhou, and S. Z. D. Cheng, *J. Am. Chem. Soc.* **122**, 2407 (2000).
69. X. Kong, T. Kawai, J. Abe, and T. Iyoda, *Macromolecules* **34**, 1837 (2001).
70. S. G. Boyes, W. J. Brittain, X. Weng, and S. Z. D. Cheng, *Macromolecules* **35**, 4960 (2002).
71. W. Huang, J.-B. Kim, G. L. Baker, and M. L. Bruening, *Nanotechnology* **14**, 1075 (2003).
72. D. M. Jones and W. T. S. Huck, *Adv. Mater.* **13**, 1256 (2001).
73. S. A. Prokhorova, A. Kopyshev, A. Ramakrishnan, H. Zhang, and J. Rühe, *Nanotechnology* **14**, 1098 (2003).

74. S. Minko, S. Patil, V. Datsyuk, F. Simon, K.-J. Eichhorn, M. Motornov, D. Usov, I. Tokarev, and M. Stamm, *Langmuir* **18**, 289 (2002).
75. L. Ionov, S. Minko, M. Stamm, J.-F. Gohy, R. Jerome, and A. Scholl, *J. Am. Chem. Soc.* **125**, 8302 (2003).
76. V. L. Osborne, D. M. Jones, and W. T. S. Huck, *Chem. Commun.* 1838 (2002).
77. E. B. Zhulina, J. Klein Wolterink, O. V. Borisov, *Macromolecules* **33**, 4945 (2000).
78. P. Pincus, *Macromolecules* **24**, 2912 (1991).
79. J. Klein and E. Kumacheva, *Science* **269**, 816 (1995).
80. D. J. Beebe, J. S. Moore, J. M. Bauer, Q. Yu, R. H. Liu, C. Devadoss, and B.-H. Jo, *Nature* **404**, 588 (2000).
81. H. Sirringhaus, P. J. Brown, R. H. Friend, M. M. Nielsen, K. Bechgaard, B. M. W. Langeveld-Vos, A. J. H. Spiering, R. A. J. Janssen, E. W. Meijer, P. Herwig, and D. M. de Leeuw, *Nature* **401**, 685 (1999).

3

Electronic Transport through Self-Assembled Monolayers

Wenyong Wang, Takhee Lee, and M. A. Reed

3.1. INTRODUCTION

The field of nanotechnology has made tremendous progress in the past decades, ranging from the experimental manipulations of single atoms and single molecules to the synthesis and possible applications of carbon nanotubes and semiconductor nanowires.[1–3] This remarkable research trend is driven partly by the human being's curiosities of exploring the ultimate small matter and partly by the microelectronics industry's need to go beyond the traditional photolithography-based top-down fabrication limitations. As the enormous literature has shown, nanometer scale device structures provide suitable testbeds for the investigations of novel physics in a new regime, especially at the quantum level, such as single electron tunneling or quantum confinement effect.[4,5] On the other hand, as the semiconductor device feature size keeps decreasing, the traditional top-down microfabrications will soon enter the nanometer range and further continuous downscaling will become scientifically and economically challenging.[6] This motivates researchers around the world to find alternate ways to meet the future increasing computing demands.

With a goal of utilizing individual molecules as the self-contained functioning nanoscale electronic components, molecular electronics was and remains as an active core part in the research field of nanotechnology.[2,3] Different from conventional solid state device fabrications, molecular electronics, like many other nanoscale electronic resolutions, adopt a bottom-up constructing strategy which starts with the ultimate building block of a single molecule or a single atom. This conceptually new approach considers the device behavior at the quantum level and allows engineering of electronic properties of the circuit components at the fundamental stage, thus addressing and overcoming the problems faced by the traditional fabrication method.[3]

The starting point of molecular electronics was a theoretical model of a unimolecular rectifier proposed in 1974, in which a single molecule consisting of an electron donor region

Departments of Electrical Engineering, Applied Physics, and Physics, Yale University, P.O. Box 208284, New Haven, CT 06520, U.S.A.

and an electron acceptor region separated by a σ bridge would behave as a molecular p-n junction.[7] However an experimental realization of such a molecular electronic device was hard to accomplish due to the difficulties of molecular chemical synthesis and solid state device microfabrication. Only with the recent developments in both fields could people realize such a molecular rectification using Langmuir-Blodgett (L-B) films.[8] In the mean time, Reed *et al.* proposed to use self-assembled conjugated oligomers as the active molecular electronic components instead of L-B films.[9,10]

Molecular self-assembly is an experimental method to spontaneously form highly ordered molecular monolayers on various substrate surfaces.[11,12] Earlier research in the field of self-assembly include, for example, Nuzzo and Allara's pioneering study of alkyl disulfide monolayers on gold surfaces[13] and the field has grown enormously in the past two decades. Being the major research object of this field, self-assembled monolayers (SAMs) have found their modern day applications in various areas such as nanoelectronics, surface engineering, biosensoring, etc, among many others.[11] In order to characterize the surface and bulk properties of SAMs, various analytical tools have been employed, such as scanning probe microscopy, X-ray photoelectron spectroscopy, etc.[11] Most importantly, due to the recent developments of advanced fabrication techniques, direct electronic conduction characterization of self-assembled monolayers or of individual molecules has become possible. Understanding the charge transport mechanism in such organic monolayers is critical before any possible device applications. One of the molecular systems that have been characterized is the alkanethiol. It is well known that alkanethiol can form robust SAM on gold surface and the conduction mechanism through alkanethiol monolayer has been reported to be tunneling. However, an unambiguous determination of the charge transport mechanism requires temperature- and molecular length-dependence studies along with the electrical characterizations. Without the temperature-dependent and length-dependent characterizations any claim on the tunneling transport mechanism is unsubstantiated. Together with theoretical modeling and calculations, experimental measurements help to answer fundamental questions regarding electronic properties of molecules at the quantum level and reveal novel device behaviors, thus guiding the modifications of molecular electrical properties and the chemical design of future nanoelectronics components.

3.2. MOLECULAR SELF-ASSEMBLY

Molecular self-assembly is a technique to form highly ordered, closely packed monolayers on various substrates via a spontaneous chemisorption process at the interface.[11,12] Earlier research done in this field includes the self-assembly of fatty acids monolayers on metal oxides,[14,15] SAMs of organosilicon derivatives on metal and semiconductor oxides,[16,17] and organosulfur SAMs on metal and semiconductor surfaces.[18,19] Among the organosulfur SAMs, the most thoroughly investigated and characterized one is alkanethiol SAM formed on Au(111) surfaces.[12]

Gold is known not having a stable oxide at its surface,[20] which makes it a good candidate substrate to study self-assembled monolayers. The deposition of an alkanethiol SAM on a gold surface is done by immersion of a clean gold substrate into the solution, and a spontaneous chemisorption at the interface produces gold thiolate species (See Figure 3.1):[11,12]

$$\text{RS-H} + \text{Au} = \text{RS-Au} + 0.5\,\text{H}_2$$

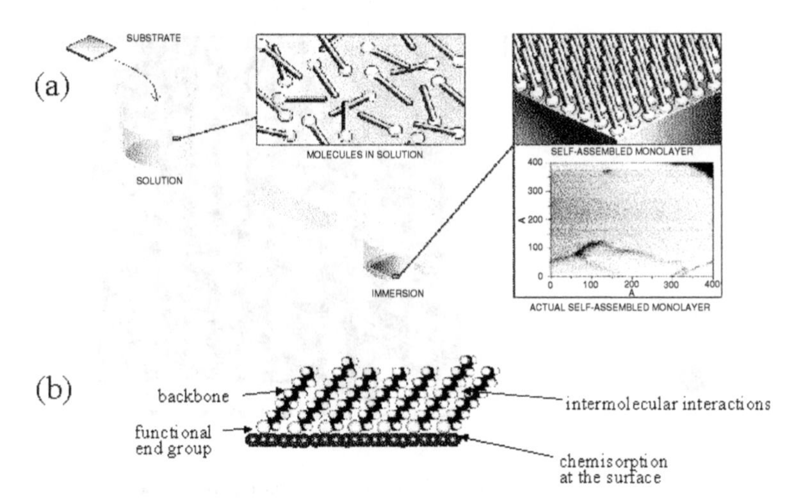

FIGURE 3.1. (a) Schematics illustrating self-assembly. Self-assembled monolayers are formed by immersing a substrate (*e.g.*, a piece of metal) into a solution of the surface-active material. The functional end groups of molecules chemically react with the substrate material spontaneously, forming a two-dimensional assembly. Its driving force includes chemical bond formation of functional end groups of molecules with the substrate surface and intermolecular interactions between the backbones. (b) Cross-sectional schematic of self-assembled monolayers formed on a substrate.

where R represents the backbone of the molecule and S is sulfur. This adsorption process is suggested to undergo two steps:[21] a very fast step, which takes minutes (depending on the thiol concentration) and gives ~90% of the film thickness, and a slow step, which lasts hours and reaches the final thickness and contact angles. Three forces are believed to determine this adsorption process and final monolayer structure: interactions between thiol head groups and gold lattice, dispersion forces between alkyl chains (van der Waals force, etc), and the interaction between the end groups.[21] The first adsorption step is governed by the head group-surface reaction while the second slow step depends on the chain-chain interactions, as well as surface mobility. The alkyl chain length also plays a role in the kinetics of the SAM layer formation, since the van der Waals force is a function of such chain length.[21]

Various surface analytical tools have been utilized to investigate the surface and bulk properties of the SAMs, such as X-ray photoelectron spectroscopy (XPS),[22] Fourier transform infrared spectroscopy (FTIR),[23] Raman spectroscopy,[24] scanning probe microscopy (SPM),[25] etc.

Studies have shown that the bonding of the thiolate group to the gold surface is very strong: the bond energy of RS-Au is ~40 kcal/mol (~1.7 eV).[12] Regarding the geometrical structure formed by alkanethiols on Au(111), STM topography studies revealed that alkanethiols adopt the commensurate underlying gold crystalline lattice characterized by a c(4 × 2) superlattice of a $(\sqrt{3} \times \sqrt{3})R30^\circ$, as shown in Figure 3.2.[25,26] The orientations of the alkanethiol SAMs on Au(111) surfaces have also been studied and FTIR results show that the alkyl chains usually are tilted ~30° from the surface normal.[27]

Studies have shown that defects, such as pinholes or grain boundaries, exist in the self-assembled monolayers.[11,25] In addition to the irregularities introduced during the self-assembly process, another cause of such defects is the roughness of the substrate surface. For

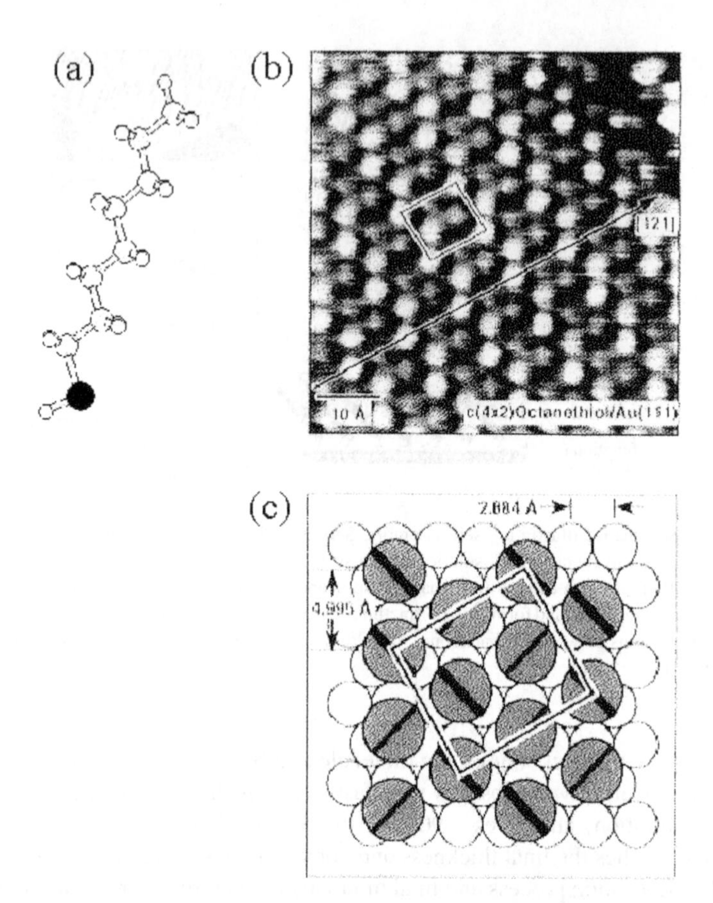

FIGURE 3.2. (a) Chemical structure of octanethiol. (b) A constant current STM image of octanethiol SAM on Au(111). Au reconstruction is lifted and alkanethiols adopt commensurate crystalline lattice characteriized by a c(4 × 2) superlattice of a $(\sqrt{3} \times \sqrt{3})R30^\circ$. (c) Model of commensuration condition between alkanethiol monolayer (large circles) and bulk-terminated Au surface (small circles). Diagonal slash in large circles represents azimuthal orientation of plane defined by all-trans hydrocarbon chain. (Reprint with permission from Ref.[25]: G. E. Poirier, *Chem. Rev.*, **97**, 1117–1127 (1997). Copyright 1997 American Chemical Society.)

example, although frequently called "flat" gold, there are grain boundaries existing on the Au surface layer, which introduces defects to the molecular monolayer.[25] The domain size of an alkanethiol SAM layer usually is in the order of several hundred angstroms.[25] However, the surface migration of thiolate-Au molecules, the so-called SAM annealing process, is found to be helpful for healing some of the defects.[11,25] Defects in the monolayers cause problems in the device applications of SAMs because such defects will cause short-circuit problem in a metal-SAM-metal junction.

The surface properties of alkanethiol SAMs have been studied thoroughly, and recently people have performed electrical characterizations and found the conduction mechanism through such self-assembled alkanethiol monolayers is tunneling due to its large HOMO-LUMO gap (HOMO: highest occupied molecular orbital, LUMO: lowest unoccupied molecular orbital) (>8 eV).[28] Another type of molecular system is the so-called

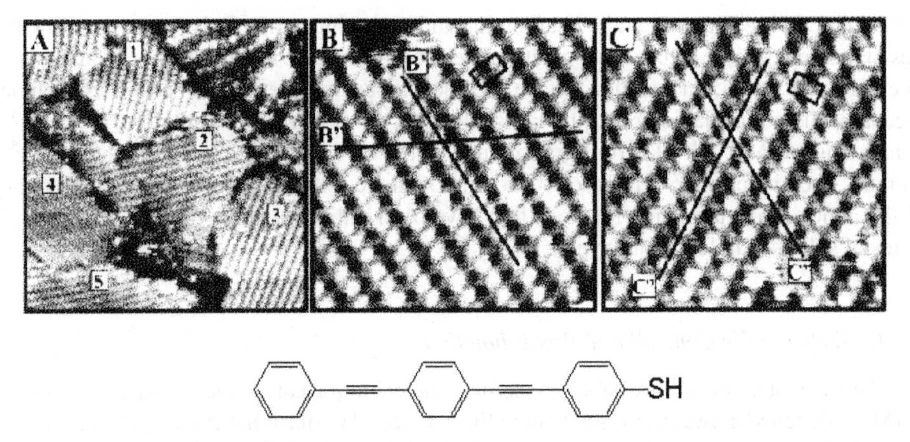

FIGURE 3.3. (A) 300 Å × 300 Å constant current STM image of 4-[4′-(phenylethynyl)-phenylethynyl]-benzenethiol (PPB-SH) (chemical structure shown below) on Au(111) showing five ordered domains. (B) A high-resolution STM image (80 Å × 80 Å) where bright rows of PPB-SH molecules (as oval shaped dots) are clearly visible. (C) A high-resolution STM image (80 Å × 80 Å) of another domain where the molecules appear as spherical dots. The rectangular unit cells are indicated in images (B) and (C). The cross-sectional profiles across lines (B′ B″ and c′, c″) are not shown here. (Reprinted with permission from Ref.[34]: Yang *et al., J. Phys. Chem. B*, **104**, 9059–9062 (2000). Copyright 2000 American Chemical Society.)

conjugated oligomers.[2] Molecular conjugation happens when the overlapping of the π orbitals is extended throughout the whole molecular structure and creates a delocalized electronic state[29] Compared with alkyl-chain systems, SAMs of conjugated molecules are more conductive, and are good candidates for nanoscale electronic applications.[2] One of the advantages of utilizing self-assembled conjugated molecules in electronic device applications is the flexibility in the engineering of the molecular structures.[2,3] SAMs can be designed to provide various functionalities via changing the molecular backbones and terminating groups. For example, molecular switches have been realized by adding different side-groups to a polyphenylene-based conjugated oligomer molecule.[30,31]

For conjugated molecules with a thiol head group, the driving force to form a SAM is still the thiol-Au bonding process.[32] However, the intermolecular van der Waals interaction is more complicated due to the rigid rod backbone structure and the various functional side-groups. Comparing to alkanethiol SAMs, it is hard to get STM images at good molecular resolution for conjugated molecule SAMs. However, studies have shown that, for example, 4-[4′-(phenylethynyl)-phenylethynyl]-benzenethiol can form highly ordered SAM on Au(111) surfaces[33,34] as shown in Figure 3.3. The SAM consists of domains of 50 to 200 Å, and due to the intermolecular interactions it does not have a simple commensurate structure. Unlike the alkanethiol SAMs, the titling angle of these arenethiol molecules is less than 5°.[34]

3.3. ELECTRICAL CHARACTERIZATION METHODS

A good understanding of the electron transport properties through self-assembled molecules is necessary before any device applications, however such transport measurements are experimentally challenging and intriguing particularly due to the difficulties of making reliable electrical contacts to the nanometer scale monolayers. Various experimental characterization techniques have been developed to achieve this goal, which include mechanically controllable break junction technique,[35–37] STM,[38–40] nanopore,[28,30,31,41,42] conducting AFM,[43–46] electromigration nanogap,[47–49] cross-wire tunnel junction,[50,51] mercury-drop junction,[52,53] nanorod[54] and others. In the following, we briefly review some of the major characterization methods that have been utilized to study the electrical properties of self-assembled monolayers.

3.3.1. Mechanically Controllable Break Junction

The mechanically controllable break junction technique can create a configuration of a SAM sandwiched between two stable metallic contacts, Two terminal current-voltage (I(V)) characterizations are possible to be performed on the scale of single molecules. For this technique, a metallic wire with a notch is mounted onto elastic bending beam and a piezo electric element is positioned at the bottom side underneath the notch. Pushing the piezo upward the strains imposed on the notch results in the wire breaking and thus two metallic electrodes are formed *in situ*. The wire breaking process is carried out in the solution of thiols, and after the breaking the solvent is allowed to evaporate and the two electrodes are brought back together to form the desired molecular junction. Using this technique Reed *et al.* measured the charge transport through a benzene-1,4-dithiol molecule at room temperature and observed a gap of ~0.7 V, which could be interpreted as a Coulomb gap or a mismatch between the Fermi levels of contacts and the lowest unoccupied molecular orbital (LUMO)[35] (Figure 3.4). Using a similar technique, Kergueris *et al.* and Reichert *et al.* have also performed conductance measurements through self-assembled molecules and concluded that measurements on a few or individual molecules were observed.[36,37]

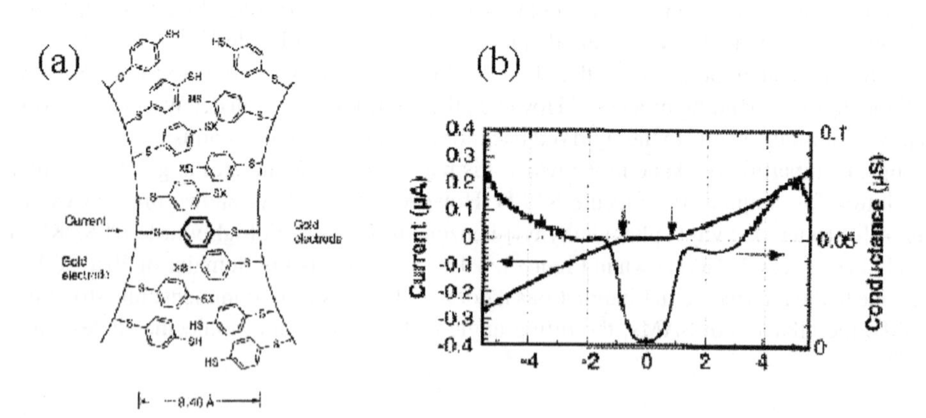

FIGURE 3.4. (a) Schematic of a benzene-1,4-dithiolate SAM between proximal gold electrodes formed in an mechanically controllable break junction. (b) Typical I(V) characteristics, which illustrate a gap of ~0.7 V, and the first derivative G(V), which shows a steplike structure. [Adapted from Ref.[35]: Reed *et al., science* **278**, 252–254 (1997).]

3.3.2. Scanning Probe Microscopy

Various scanning probe related techniques have been used for studies in molecular nanoelectronics, which include scanning tunneling microscopy (STM) and conducting atomic force microscopy (C-AFM). The STM has been used as an important tool in the studies on molecular systems due to its ability of not only imaging but also probing and manipulating of single atoms or molecules.[55–58] For example, the conduction measurement through a single molecule using STM was demonstrated by Bumm *et al.* who measured electron transport through individual benzenedithiol-derivative conjugated molecules inserted in a matrix of insulating alkanethiol SAMs on Au(111) substrate.[38] Using the insertion technique, Donhauser *et al.* demonstrated the conductance switching of single and bundled phenylene ethynylene oligomers isolated in a matrix of alkanethiol SAM.[39] The resistance of a single xylyldithiol ($HS\text{-}CH_2\text{-}C_6H_4\text{-}CH_2\text{-}SH$) molecule was estimated as $18 \pm 12\ M\Omega$ by Andres *et al.* from the electron transport measurement through gold nanoparticle/xylyldithiol SAM/Au by STM study.[59]

Similarly, tunneling characteristics through alkanethiol[45] and oligophenylene thiolate molecules[43] has been studied by Wold *et al.* using C-AFM technique. Cui *et al.* bound gold nanoparticles to octanedithiol in an octanethiol matrix and measured the conductance through gold nanoparticle/octanedithiol/Au using C-AFM. They observed that the measured I(V) curves are integer multiples of a single fundamental curve (Figure 3.5), and estimated the resistance of a single octanedithiol molecule (unconjugated type) as $900 \pm 60\ M\Omega$, which is less than that of a single xylyldithiol molecule (conjugated type).[44]

3.3.3. Nanopore

In the so-called nanopore device, as illustrated in Figure 3.6(a), a number of molecules ($\sim$several thousands) are sandwiched between two metallic contacts. This testbed provides a stable device structure and makes cryogenic measurements possible. As reported previously,[41,42] this single nanometer scale pore is created via e-beam lithography and reactive ion etching on a suspended Si_3N_4 membrane. Metal (usually gold) is then thermally evaporated onto the topside form the top metallic contacts. The device is then transferred into a molecular solution to deposit the SAM layer. Next the second metallization on the bottom side is performed to form the metal-SAM-metal junction. During the bottom side metallization process, liquid nitrogen is usually kept flowing through a cooling stage in order to avoid thermal damage to the molecular layer. This technique reduces the kinetic energy of evaporated metal atoms at the surface of the monolayer, thus preventing them from punching through the monolayer. For the same reason the evaporation rate is kept very low. Using this nanopore technique, Zhou *et al.* measured 4-thioacetylbiphenyl SAMs and observed a rectifying I(V) behavior that is attributed to the asymmetry of the molecule-metal contacts.[42] Chen *et al.* measured molecules containing a redox center (2′-amino-4-ethynylphenyl-4′-ethynylphenyl-5′-nitro-1-benzenethiol) and observed large on-off ratios and negative differential resistance (NDR) at cryogenic temperature (Figure 3.6b).[30] At temperature of 60 K, the current density of the device is $50\ A/cm^2$, the NDR is $2400\ \mu\Omega\cdot cm^2$, and the peak-to-valley ratio is 1030:1. Measurements on a similar molecule that has only a nitro redox center (4,4′-di(ethynylphenyl)-2′-nitro-1-benzenethiol) using the same technique

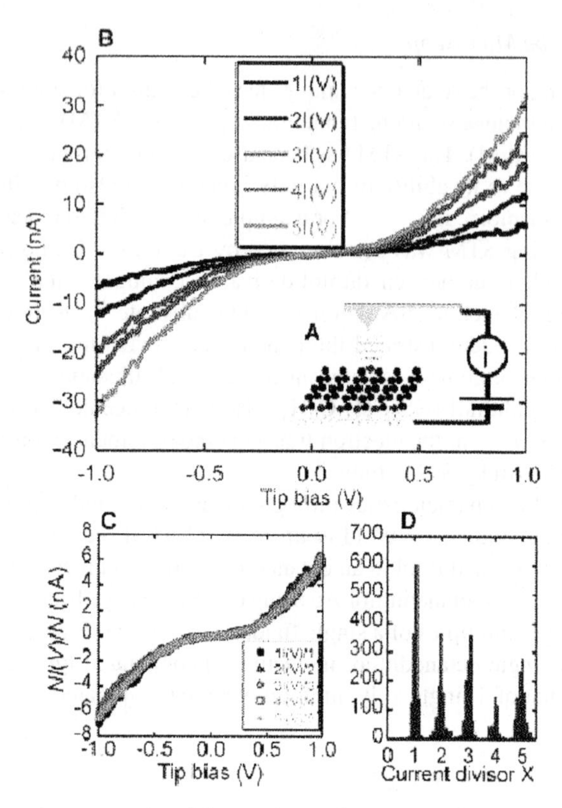

FIGURE 3.5. (A) Schematic representation of the experiment. (B) Five representative I(V) curves measured with the apparatus diagrammed in (A). (C) I(V) curves from (B) divided by 1, 2, 3, 4, and 5, implying the I(V) curve depends only on the number of molecules attached to the nanoparticle. (D) Histogram of values of a divisor. (Reprinted with permission from Ref.[44]: Cui *et al., Science* **294**, 571–574 (2001). Copyright 2001, American Association for the Advancement of Science.)

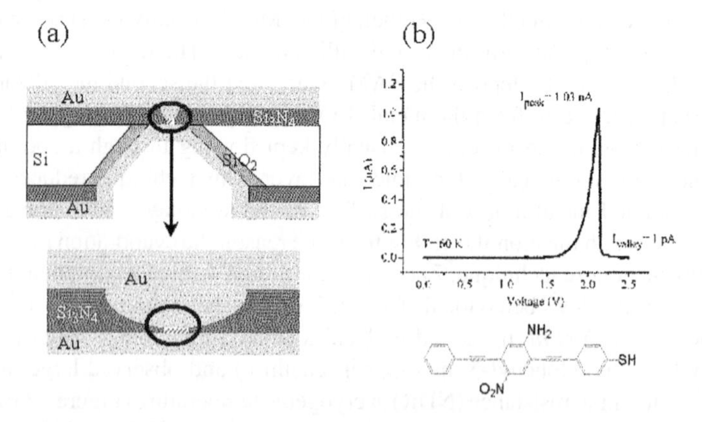

FIGURE 3.6. (a) Cross-sectional schematics of a silicon wafer with a nanopore etched through a suspended silicon nitride membrance. SAM is formed between sandwiched Au eletrodes in the pore area (circled). (b) I(V) characteristics of a Au-2′-amino-4-ethynylphenyl-4′-ethynylphenyl-5′-nitro-1-benzenethiolate-Au (chemical structure shown below) molecular junction device at 60 K. The peak current density is ~50 A/cm^2, the NDR is ~2400 μΩ· cm^2, the peak-to-valley ratio is 1030:1. [Adapted from Ref.[30]: Chen *et al., Science* **286**, 1550–1552 (1999).]

showed NDR results at both room and cryogenic temperatures, with a peak-to-valley ratio of 1.5:1 at 300 K.[31] Reed *et al.* also reported electronically programmable memory devices utilizing self-assembled monolayers of the above mentioned molecules.[60]

3.3.4. Electromigration Nanogap

A nanogap structure has been fabricated via standard e-beam lithography on a bilayer resist and subsequent lift-off process. The SAMs are deposited after the fabrication of the gap.[61] However it is difficult to achieve 1 ∼ 2 nm scale gaps with current e-beam processing and another method developed to create nanoscale gaps is to use the metal atoms electromigration properties.[47] In this method, a thin gold wire with a width of several hundred nanometers is created via e-beam lithography and shadow evaporation.[47] Bias is then applied and the high current passing through this nanowire causes the gold atoms to migrate and thus creates a small gap, usually about 1–2 nm wide. Molecules are deposited at room temperature before the breaking of the gold wire at cryogenic temperatures.[48] The advantage of this technique is that a third gating electrode can be introduced to the two terminal I(V) measurements. Using this electromigration nanogap technique Park *et al.* measured $[Co(tpy\text{-}(CH_2)_5\text{-}SH)_2]^{2+}$ and $[Co(tpy\text{-}SH)_2]^{2+}$ molecules at cryogenic temperatures. For the longer molecule $[Co(tpy\text{-}(CH_2)_5\text{-}SH)_2]^{2+}$ at temperatures less than 100 mK the Coulomb blockade behavior was observed (Figure 3.7). For the shorter molecule $[Co(tpy\text{-}SH)_2]^{2+}$ logarithmic temperature dependence and magnetic field splitting of the zero bias differential conductance peak were observed, indicating that this differential conductance peak is due to the existence of the Kondo effect.[48] Similar Kondo resonance in a single molecular transistor was also observed by Liang *et al.* using a similar test structure on a molecular system of $[(N,N',N''\text{-trimethyl-1,4,7-triazacyclononane})_2\text{-}V_2(CN)_4(\mu\text{-}C_4N_4)]$.[49]

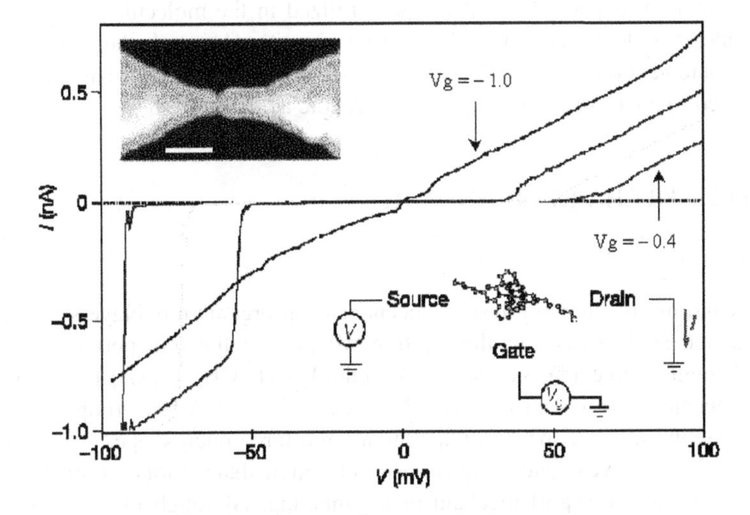

FIGURE 3.7. I(V) curve of a $[Co(tpy\text{-}(CH2)5\text{-}SH)2]^{2+}$ single-eletron transistor at different gate voltages (Vg) from −0.4 V to −1.0 V with $\Delta Vg \approx 0.15$ V. Upper inset, a topographic AFM image of the electrodes with a gap (scale bar: 100 nm). Lower inset, a schematic diagram of the device. (Reprinted with permission from Ref.[48], Park *et al., Nature* **417**, 722–725 (2002). Copyright 2002 Nature.)

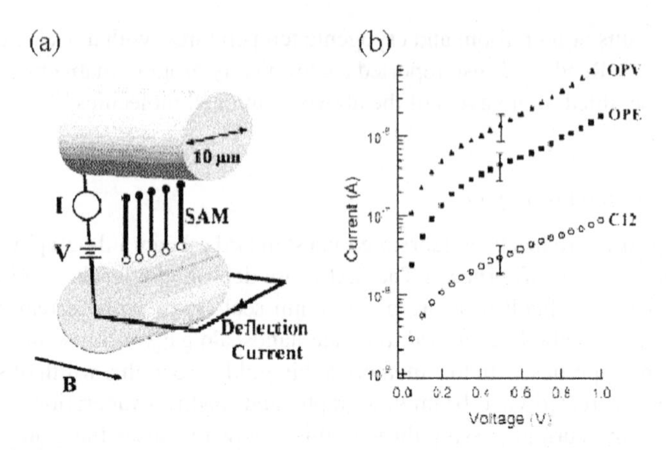

FIGURE 3.8. (a) Schematic representation of the crossed-wire tunnel junction (not to scale). (b) Plots of I(V) (current in logarithmic scale) for junctions formed from three molecular compounds. (Reprinted with permission from Ref.[50]: Kushmetrick *et al.*, *J. Am. Chem. Soc.* **124**, 10654–10655 (2002). Copyright 2002 American Chemical Society.)

3.3.5. Cross-Wire Tunnel Junction

The cross-wire tunnel junction can be formed by mounting two wires in such a manner that the wires are in a crossed geometry with one wire perpendicular to the applied magnetic field, as schematically shown in Figure 3.8(a). The junction separation is then controlled by deflecting this wire with the Lorentz force generated from a direct current.[50,51,62] Using this testbed Kushmerick *et al.* studied dodecanedithiol (C12 dithiol), oligo(phenylene ethynylene)dithiol (OPE), and oligo(phenylene vinylene)dithiol (OPV) molecules and observed conductance differences due to molecular conjugation and molecular length differences (Figure 3.8b).[50]

Other important experimental techniques utilized in the molecular transport studies include the mercury-drop junctions[52,53] which have been used to study the transport through alkanethiols, and nanorods (a metallic nanowire-SAM-metallic nanowire junction)[54] which have been used to measure 16-mercaptohexadecanoic acid.

3.4. MOLECULAR CHARGE TRANSPORT

3.4.1. Background

Understanding the charge transport mechanism in organic monolayers is crucial before any device applications. In this section we present the transport studies on one of the fundamental molecular systems. Alkanethiol ($CH_3(CH_2)_{n-1}SH$) has been studied extensively because it forms a robust SAM on Au surfaces.[11] A few groups have utilized scanning tunneling microscope,[63] conducting atomic force microscope,[43–46] or mercury-drop junctions[52,53] to investigate charge transport through alkanethiols at room temperature and claimed that the transport mechanism is tunneling. Although the electron conduction is expected to be tunneling when the Fermi levels of contacts lie within the HOMO-LUMO gap of a short length molecule as is for the case of these alkanethiols,[64] in the absence of temperature-dependent current-voltage (I(V,T)) characterizations such a claim is unsubstantiated since other conduction mechanisms (such as thermionic or hopping conduction) can contribute and complicate the analysis.

In the following section we present the detailed measurement results of tunneling characteristics through alkanethiol SAMs.

3.4.2. Temperature Dependent Current-Voltage Measurement

Electronic transport measurements on alkanethiol SAMs were performed using the nanopore device as illustrated in Figure 3.6(a), similar to one reported previously.[41,42] The molecule deposition is done in a ~5 mM alkanethiol solution for typically 24 hours inside a nitrogen filled glove box with an oxygen level of less than 100 ppm. Three molecules of different molecular lengths: octanethiol ($CH_3(CH_2)_7SH$; denoted as C8, for the number of CH_2 units), dodecanethiol ($CH_3(CH_2)_{11}SH$; denoted as C12), and hexadecanethiol ($CH_3(CH_2)_{15}SH$; denoted as C16) were used to form the active molecular components. As a representative example, the chemical structure of octanethiol is shown in Figure 3.2(a).

In Table 3.1, possible conduction mechanisms are listed with their characteristic current, temperature- and voltage-dependencies.[65,66] Based on whether thermal activation is involved, the conduction mechanisms fall into two distinct categories: (i) thermionic or hopping conduction which has temperature-dependent I(V) behavior and (ii) direct tunneling or Fowler-Nordheim tunneling which does not have temperature-dependent I(V) behavior. For example, thermionic and hopping conductions have been observed for 4-thioacetylbiphenyl SAMs[42] and 1,4-phenylene diisocyanide SAMs.[67] On the other hand, the conduction mechanism is expected to be tunneling for alkanethiols.[64] Charge transport studies on Langmuir-Blodgett alkane films have previously been reported.[68,69] For example, Figure 3.9 shows I(V) curves measured on Langmuir-Blodgett monolayer of fatty acid ($CH_3(CH_2)_{17}COOH$) on Al at 295 K and 77 K.[69] The temperature-dependent I(V) behavior in this figure is due to impurity-dominated transport, which indicates that impurity density also plays an important role in conduction mechanisms even for the large HOMO-LUMO gap molecules where tunneling conduction is expected. A recent study on eicosanoic acid ($CH_3(CH_2)_{18}COOH$) between Pt electrodes at 50–300 K also exhibited temperature-dependent I(V) characteristics,[70] where the measured I(V) data does not fit temperature-independent tunneling model, thus again implying that the transport is complicated by

TABLE 3.1. Possible conduction mechanisms.

Conduction Mechanism	Characteristic Behavior	Temperature Dependence	Voltage Dependence
Direct Tunneling*	$J \sim V \exp\left(-\frac{2d}{\eta}\sqrt{2m\Phi}\right)$	none	$J \sim V$
Fowler-Nordheim Tunneling	$J \sim V^2 \exp\left(-\frac{4d\sqrt{2m}\Phi^{3/2}}{3q\eta V}\right)$	none	$\ln\left(\frac{J}{V^2}\right) \sim \frac{1}{V}$
Thermionic Emission	$J \sim T^2 \exp\left(-\frac{\Phi - q\sqrt{qV/4\pi\varepsilon d}}{kT}\right)$	$\ln\left(\frac{J}{T^2}\right) \sim \frac{1}{T}$	$\ln(J) \sim V^{1/2}$
Hopping Conduction	$J \sim V \exp\left(-\frac{\Phi}{kT}\right)$	$\ln\left(\frac{J}{V}\right) \sim \frac{1}{T}$	$J \sim V$

*Note that this characteristic of direct tunneling is valid for the low bias regime (see Eq. 3.3a).

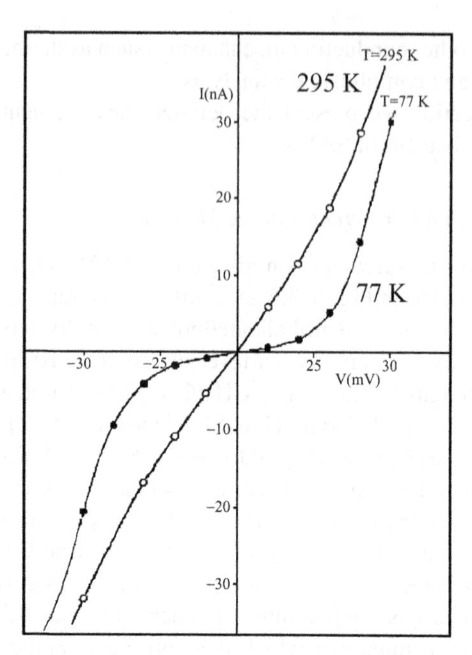

FIGURE 3.9. I(V) characteristics at 295 K and 77 K for Langmuir-Blodgett monolayer of fatty acid ($CH_3(CH_2)_{17}COOH$) on Al. Junction area is 0.2 cm^2. Reprinted with permission from Ref.[69]: Polymeropoulos *et al., J. Chem. Phys.* **69**, 1836–1847 (1978). Copyright 1978 American Institute of Physics.

defects formed in the Langmuir-Blodgett film. I(V) measurements on alkanethiol SAMs have been reported,[43,46,52,63,71–74] however all of these measurements were performed at fixed temperature (300 K) which is insufficient to prove tunneling as the dominant mechanism. Without temperature-dependent current-voltage characterization, other conduction mechanisms (such as thermionic or hopping conduction) cannot be excluded. Reported here are I(V) measurements in a sufficiently wide temperature range of 300 to 80 K and with a resolution of 10 K to determine the mechanism in self-assembled alkanethiol molecular systems.

Figure 3.10 shows a representative I(V,T) characteristic of dodecanethiol (C12) measured with the device structure as shown in Figure 3.6(a). Positive bias corresponds to electrons injected from the physisorbed Au contact (bottom contact in Figure 3.6a) into the molecules. By using the contact area of 45 ± 2 nm in diameter (estimated from SEM statistical study,[28] a current density of $\sim 1{,}500 \pm 200$ A/cm^2 at 1.0 Volt is determined. No significant temperature dependence of the characteristics (from V = 0 to 1.0 Volt) is observed over the range from 300 to 80 K. An Arrhenius plot (ln(I) versus $1/T$) of this is shown in Figure 3.11(a), exhibiting little temperature dependence in the slopes of ln(I) versus $1/T$ at different bias and thus indicating the absence of thermal activation. Therefore, we conclude that the conduction mechanism through alkanethiol is tunneling. Based on the applied bias as compared with the barrier height (Φ_B), the tunneling through a SAM layer can be categorized into either direct ($V < \Phi_B/e$) or Fowler-Nordheim ($V > \Phi_B/e$) tunneling. These two tunneling mechanisms can be distinguished by their distinct voltage dependencies (see Table 3.1). Analysis of $\ln(I^2/V)$ versus $1/V$ (in Figure 3.11b) shows no

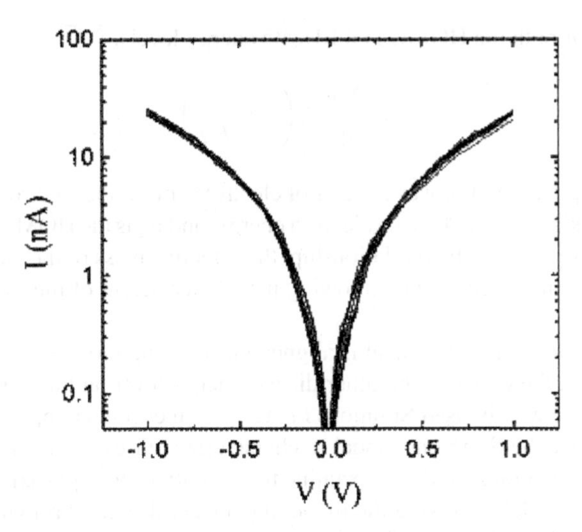

FIGURE 3.10. Temperature-dependent I(V) characteristics of dodecanethiol (C12). I(V) data at temperatures from 300 to 80 K with 20 K steps are plotted on a log scale.

significant voltage dependence, indicating no obvious Fowler-Nordheim transport behavior in this bias range (0 to 1.0 Volt) and thus determining that the barrier height is larger than the applied bias, *i.e.*, $\Phi_B > 1.0\,\text{eV}$.

3.4.3. Tunneling Characteristics Through Alkanethiols

To describe the transport through a molecular system having HOMO and LUMO energy levels, one of the applicable models is the Franz two-band model.[75–81] This model provides a non-parabolic energy-momentum E(k) dispersion relationship by considering

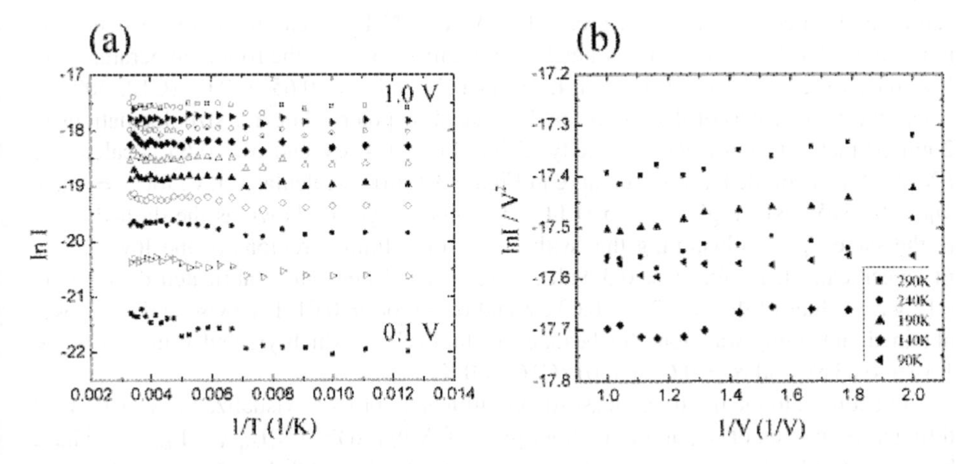

FIGURE 3.11. (a) Arrhenius plot generated from the I(V) data in Figure 3.10, at voltages from 0.1 to 1.0 Volt with 0.1 Volt steps. (b) Plot of $\ln(I^2/V)$ versus $1/V$ at selected temperatures.

the contributions of both the HOMO and LUMO energy levels:[75–77]

$$k^2 = \frac{2m^*}{\eta^2} E\left(1 + \frac{E}{Eg}\right) \tag{3.1}$$

where k is the imaginary part of wave vector of electrons, m* is the electron effective mass, h (= $2\pi\hbar$) is Planck's constant, E is the electron energy, and E_g is the HOMO-LUMO energy gap. From this non-parabolic E(k) relationship, the effective mass of the electron tunneling through the SAM can be deduced by knowing the barrier height of the metal-SAM-metal junction.[76,77]

When the Fermi level of the metal is aligned close enough to one energy level (either HOMO or LUMO), the effect of the other distant energy level on the tunneling transport is negligible, and the widely used Simmons model[82] is an excellent approximation.[83,84] In the following we use the Simmons model to characterize our experimental I(V) data, and later compare it to the Franz model to examine the validity of the approximation.

The Simmons model expressed the tunneling current density through a barrier in the tunneling regime of $V < \Phi_B/e$ as[52,82]

$$J = \left(\frac{e}{4\pi^2 h\, d^2}\right)\left\{\left(\Phi_B - \frac{eV}{2}\right)\exp\left[-\frac{2(2m)^{1/2}}{h}\alpha\left(\Phi_B - \frac{eV}{2}\right)^{1/2} d\right]\right.$$
$$\left. - \left(\Phi_B + \frac{eV}{2}\right)\exp\left[-\frac{2(2m)^{1/2}}{h}\alpha\left(\Phi_B + \frac{eV}{2}\right)^{1/2} d\right]\right\} \tag{3.2}$$

where m is electron mass, d is barrier width, Φ_B is barrier height, V is applied bias, and α is a unitless adjustable parameter that is introduced to modify the simple rectangular barrier model or to account for an effective mass.[46,52,82] $\alpha = 1$ corresponds to the case for a rectangular barrier and bare electron mass, and has been previously shown not to fit I(V) data well for some alkanethiol measurements at fixed temperature (300 K).[52]

From Eq. 3.2 by adjusting two parameters Φ_B and α, a nonlinear least square fitting can be performed to fit the measured C12 I(V) data.[85] By assuming a device size of 45 nm in diameter, the best fitting parameters (minimized χ^2) for the room temperature C12 I(V) data were found to be $\Phi_B = 1.42 \pm 0.04$ eV and $\alpha = 0.65 \pm 0.01$ (C12, 300 K), where the error ranges of Φ_B and α are dominated by potential device size fluctuations of 2 nm estimated from device size study. Using $\Phi_B = 1.42$ eV and $\alpha = 0.65$, a calculated I(V) for C12 is plotted as a solid curve in Figure 3.12(a). A calculated I(V) for $\alpha = 1$ and $\Phi_B = 0.65$ eV (which gives the best fit at low bias range) is shown as the dashed curve in the same figure, illustrating that with $\alpha = 1$ only limited regions of the I(V) can be fit (specifically here, for $V < 0.3$ Volt). A second independently fabricated device with C12 gave values of $\Phi_B = 1.37 \pm 0.03$ eV and $\alpha = 0.66 \pm 0.01$. Likewise, a data set was obtained and fitting was done for hexadecanethiol (C16), which yielded values of $\Phi_B = 1.40 \pm 0.03$ eV and $\alpha = 0.68 \pm 0.01$ (C16, 300 K).

In addition to the nonlinear least square fittings, in order to visualize the variations of fitting as a function of Φ_B and α, contour plots of $\Delta(\Phi_B, \alpha) = (\Sigma|I_{exp,V} - I_{cal,V}|^2)^{1/2}$ have been generated, where $I_{exp,V}$ are experimental current values at different bias and $I_{cal,V}$ are calculated ones using Eq. 3.2. Figure 3.12(b) is such a contour plot generated for the C12

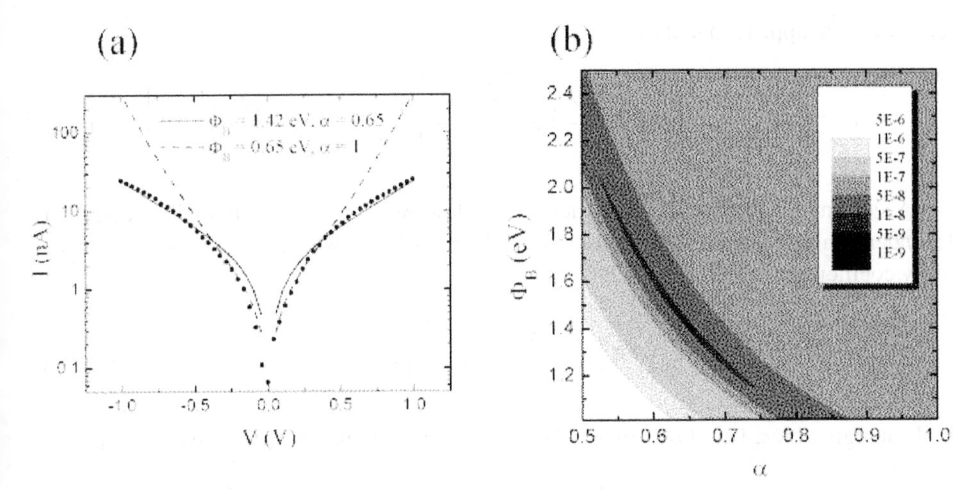

FIGURE 3.12. (a) Measured C12 I(V) data (circular symbols) is compared with calculated one (solid curve) using the optimum fitting parameters of $\Phi_B = 1.42$ eV and $\alpha = 0.65$. Calculated I(V) from the simple rectangular model ($\alpha = 1$) with $\Phi_B = 0.65$ eV is also shown as dashed curve. Current is on log scale. (b) Contour plot of Δ (Φ_B, α) value for C12 I(V) data as a function of Φ_B and α, where the darker region corresponds to a better fitting.

I(V) data (300 K). In this plot, darker regions correspond to a more accurate fitting. The best fitting condition from this contour plot was determined to be $\Phi_B = 1.42$ eV and $\alpha = 0.65$ by taking the median of Φ_B and α values for minimized $\Delta(\Phi_B, \alpha)$ (specifically, $<4.5 \times 10^{-9}$, the most accurate 0.04% of the sample).

Although the physical meaning of α is not unambiguously defined, it provides a way of applying the tunneling model of a rectangular barrier to tunneling either through a nonrectangular barrier,[52] a proposed effective mass (m*) of the tunneling electrons through the molecules,[46,76,77] (*i.e.*, for $\alpha = 0.65$, m* would be 0.42 m here), or a combination of both. Note that the I(V) data can be fit to arbitrary accuracy over the entire bias range by allowing a slight bias dependence of α(or Φ_B).

Nonlinear least square fittings on C12 I(V) data at all temperatures allow us to determine $\{\Phi_B, \alpha\}$ over the entire temperature range and show that Φ_B and α values are temperature-independent in our temperature range (300 to 80 K). For the first C12 sample reported, a value of $\Phi_B = 1.45 \pm 0.01$ eV and $\alpha = 0.64 \pm 0.01$ was obtained ($1\sigma_M$ (standard error)).

3.4.4. Length Dependence of Tunneling Through Alkanethiols

Equation 3.2 can be approximated in two limits: low bias and high bias as compared with the barrier height Φ_B. For the low bias range, Eq. 3.2 can be approximated as[82]

$$J \approx \left(\frac{(2m\Phi_B)^{1/2} e^2 \alpha}{h^2 d}\right) V \exp\left[-\frac{2(2m)^{1/2}}{\mathrm{h}}\alpha(\Phi_B)^{1/2} d\right] \tag{3.3a}$$

To determine the high bias limit, we compare the relative magnitudes of the first and second exponential terms in Eq. 3.2. At high bias, the first term is dominant and thus the current density can be approximated as

$$J \approx \left(\frac{e}{4\pi^2 \mathrm{h}\, d^2}\right)\left(\Phi_B - \frac{eV}{2}\right)\exp\left[-\frac{2(2m)^{1/2}}{\mathrm{h}}\alpha\left(\Phi_B - \frac{eV}{2}\right)^{1/2} d\right] \qquad (3.3b)$$

According to the Simmons model, in the low bias regime the tunneling current is dependent on the barrier width d as $J \propto (1/d)\exp(-\beta_o d)$, where β_o is bias-independent decay coefficient:

$$\beta_0 = \frac{2(2m)^{1/2}}{\mathrm{h}}\alpha(\Phi_B)^{1/2} \qquad (3.4a)$$

while at higher bias, $J \propto (1/d^2)\exp(-\beta_v d)$, where β_v is bias-dependent decay coefficient:

$$\beta_V = \frac{2(2m)^{1/2}}{\mathrm{h}}\alpha\left(\Phi_B - \frac{eV}{2}\right)^{1/2} = \beta_0\left(1 - \frac{eV}{2\Phi_B}\right)^{1/2} \qquad (3.4b)$$

At high bias β_V decreases as bias increases (Eq. 3.4b), which results from barrier lowering effect due to the applied bias.

We define the high bias range somewhat arbitrarily by comparing the relative magnitudes of the first and second exponential terms in Eq. 3.2. Using $\Phi_B = 1.42$ eV and $\alpha = 0.65$ obtained from nonlinear least square fitting of the C12 I(V) data, the second term becomes less than ~10% of the first term at ~0.5 Volt that is chosen as the boundary of low and high bias ranges.

To determine the β values for alkanethiols used in this study, three alkanethiols of different molecular length, octanethiol (C8), dodecanethiol (C12), and hexadecanethiol (C16) were investigated to generate length-dependent I(V) data. Figure 3.13 is a log plot of tunneling current densities multiplied by molecular length (Jd at low bias and Jd^2 at high bias) as a function of the molecular length for these alkanethiols.[86] The molecular lengths used in this plot are 13.3, 18.2, and 23.2 Å for C8, C12, and C16, respectively (each molecular length was determined by adding an Au-thiol bonding length to the length of molecule).[43] Note that these lengths implicitly assume "through-bond" tunneling, that is, along the tilted molecular chains between the metal contacts.[43]

As seen in Figure 3.13, the tunneling current shows exponential dependence on molecular length. The β values can be determined from the slope at each bias and are plotted in Figure 3.14. The error bar of an individual β value in this plot was obtained by considering both the device size uncertainties and the linear fitting errors.

According to Eq. 3.4b, β_{V^2} depends on bias V linearly in the high bias range. The inset in Figure 3.14 is a plot of β_{V^2} versus V in this range (0.5 to 1.0 Volt) along with linear fitting of the data. From this fitting, $\Phi_B = 1.32 \pm 0.18$ eV and $\alpha = 0.63 \pm 0.03$ were obtained from the intercept and the slope, respectively, consistent with the more precise values obtained from the nonlinear least square fitting in the previous section. The Φ_B (square symbols) and α (circular symbols) values obtained by the C12 and C16 I(V) data fittings and $\beta_{V^2}-V$ linear fitting are summarized in Figure 3.15. The combined values are

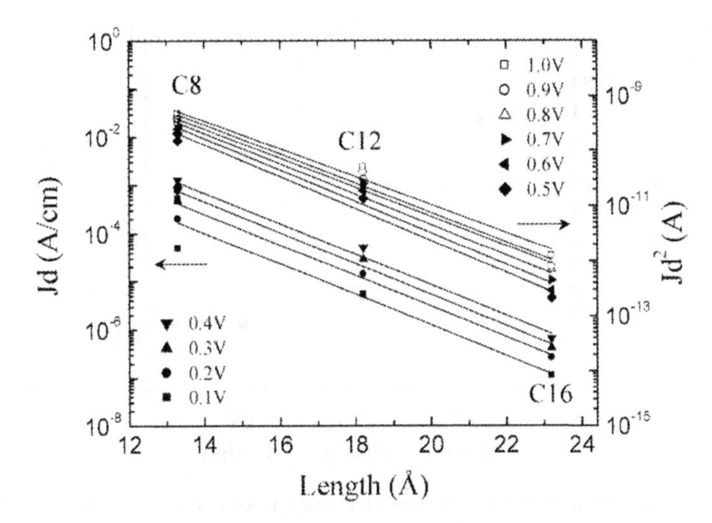

FIGURE 3.13. Log plot of tunneling current densities multiplied by molecular length d at low bias and by d^2 at high bias (symbols) versus molecular lengths. The lines through the data points are linear fittings.

$\Phi_B = 1.39 \pm 0.01$ eV ($1\sigma_M$) and $\alpha = 0.65 \pm 0.01$ ($1\sigma_M$). Using Eq. 3.4a, we can derive a zero field decay coefficient β_0 of 0.79 ± 0.01 Å^{-1}.

β values for alkanethiols obtained by various experimental techniques have previously been reported.[43,46,52,63,71–74] In order to compare with these reported β values, we also performed length-dependent analysis on our experimental data according to the generally

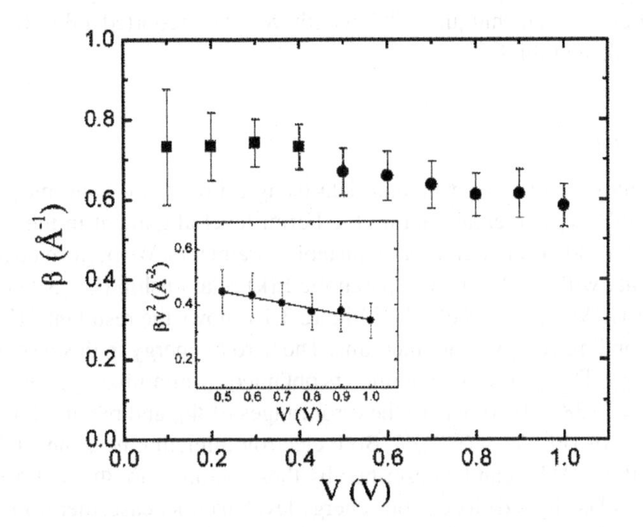

FIGURE 3.14. Plot of β versus bias in the low bias range (square symbols) and high bias ranges (circular symbols). The inset shown a plot of β_{V^2} versus bias with a linear fitting.

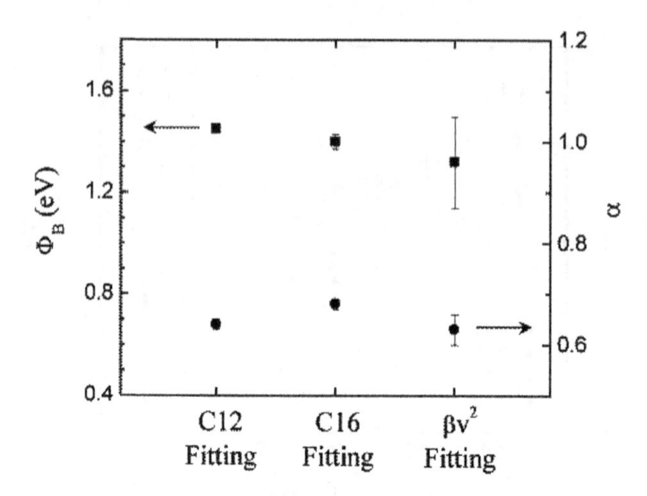

FIGURE 3.15. Summary of Φ_B (square symbols) and α (circular symbols) values obtained from alkanethiol I(V) fittings and the β_{V^2}-V fitting.

used equation:[43,46,52,63,87]

$$G = G_o \exp(-\beta\, d) \tag{3.5}$$

This gives a β value from 0.83 to 0.72 $Å^{-1}$ in the bias range from 0.1 to 1.0 volt, which is comparable to results reported previously. For example, Holmlin *et al.* reported a β value of 0.87 $Å^{-1}$ by mercury drop experiments,[52] and Wold *et al.* have reported β of 0.94 $Å^{-1}$ and Cui *et al.* have reported β of 0.64 $Å^{-1}$ for various alkanethiols by using the conducting atomic force microscope technique.[43,46] Recently Xu *et al.* reported a β of 0.8 $Å^{-1}$ (1.0 per methylene) for alkanedithiols.[88]

3.4.5. Franz Model

We have analyzed our experimental data using a Franz two-band model.[75–81] Since there is no reliable experimental data on the Fermi level alignment in these metal-SAM-metal systems, Φ_B and m* are treated as adjustable parameters. We performed a least squares fitting on our data with the Franz non-parabolic E(k) relationship (Eq. 3.1) using an alkanethiol HOMO-LUMO gap of 8 eV.[89–92] Figure 3.16 shows the resultant E(k) relationship and the corresponding energy band diagrams. The zero of energy in this plot was chosen as the LUMO energy. The best fitting parameters obtained by minimizing χ^2 were $\Phi_B = 1.55 \pm 0.59$ and $m^* = 0.38 \pm 0.20$, where the error ranges of Φ_B and m* are dominated by the error fluctuations of β $[-k^2 = (\beta/2)^2]$. Both electron tunneling near the LUMO and hole tunneling near the HOMO can be described by these parameters. $\Phi_B = 1.55$ eV indicates that the Fermi level is aligned close to one energy level in either case, therefore the Simmons model is a valid approximation. The previous best fits obtained from Simmons model of $\Phi_B = 1.39$ eV and $\alpha = 0.65$ (corresponding to $m^* = 0.42$ m for the rectangular barrier case) are in reasonable agreement.

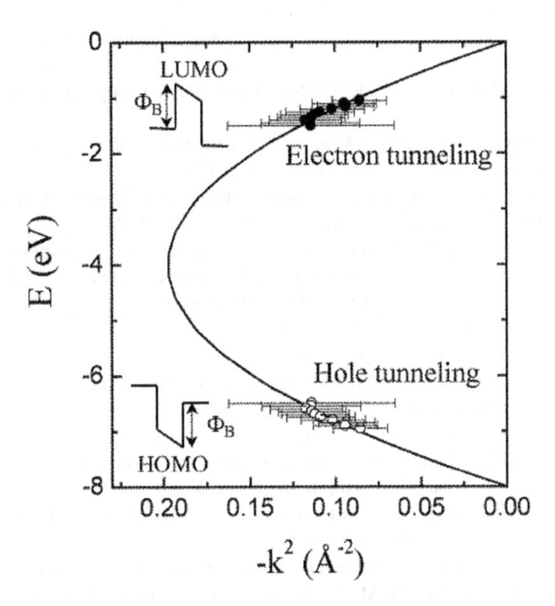

FIGURE 3.16. E(k) relationship (symbols) generated from the length-dependent measurement data for alkanethiols. Solid and open symbols correspond to electron and hole, respectively. The insets show the corresponding energy band diagrams. The solid curve is the Franz two-band expression for $m^* = 0.38$ m.

3.5. SUMMARY AND OUTLOOK

In this chapter we have presented an introductive review on the subject of molecular self-assembly and the various electrical characterization techniques that have been utilized to study self-assembled monolayers. A thorough understanding of the fundamental charge transport mechanisms in different molecular systems is crucial for any device application or circuit design purposes and this task is in no small part challenging. As an example, the studies on large HOMO-LUMO gap alkanethiol SAMs have been performed and tunneling was determined as the main conduction mechanism along with such tunneling characteristics as barrier height and tunneling decay coefficient. In the field of molecular electronics other critical issues such as the molecule—metal contact properties still exist (*i.e.*, the electronic properties of isolated molecules are strongly affected by the metal contacts and their configurations), and more effort is being made to address them.[93] Another area needs to be explored is on the interface of nanoscale elements with existing microscale electronics. However molecular electronics holds a bright future and together with other aspects of nanoscale technologies it has a potential to make fundamental changes in future electronics applications.

REFERENCES

1. G. Timp (Ed.), *Nanotechnology* (Springer-Verlag, Berlin, Germany, 1999).
2. A. Aviram and M. A. Ratner (Eds.), *Molecular Electronics, Science and Technology* (The Annals of the New York Academy of Sciences, The New York Academy of Sciences, New York, USA, 1998), Vol. 852.

3. M. A. Reed and T. Lee (Eds.), *Molecular Nanoelectronics* (American Scientific Publishers, Stevenson Ranch, USA, 2003).
4. H. Grabert, J. M. Martinis, and M. H. Devoret (Eds.), *Single Charge Tunneling* (Plenum, New York, USA, 1991).
5. M. A. Reed and A. C. Seabaugh, in *Molecular and Biomolecular Electronics*, edited by R. R. Birge (American Chemical Society, Washington DC, USA, 1994).
6. International Technology Roadmap for Semiconductors (2004), http//public.itrs.net/
7. A. Aviram and M. A. Ratner, *Chem. Phys. Lett.* **29**, 277–283 (1974).
8. R. M. Metzger, B. Chen, U. Ho1pfner, M. V. Lakshmikantham, D. Vuillaume, T. Kawai, X. Wu, H. Tachibana, T. V. Hughes, H. Sakurai, J. W. Baldwin, C. Hosch, M. P. Cava, L. Brehmer, and G. J. Ashwell, *J. Am. Chem. Soc.* **119**, 10455–10466 (1997).
9. M. A. Reed, U. S. Patent No. 5,475,341 (1995).
10. M. A. Reed, U. S. Patent No. 5,589,629 (1996).
11. A. Ulman, *An Introduction to Ultrathin Organic Films from Langmuir-Blodgett to Self-Assembly* (Academic Press, Boston, USA, 1991).
12. A. Ulman, *Chem. Rev.* **96**, 1533–1554 (1996).
13. R. G. Nuzzo and D. L. Allara, *J. Am. Chem. Soc.* **105**, 4481–4483 (1983).
14. R. G. Nuzzo and D. L. Allara, *Langmuir* **1**, 45–52 (1985).
15. H. Ogawa, T. Chihera, and K. Taya, *J. Am. Chem. Soc.* **107**, 1365–1369 (1985).
16. J. J. Sagiv, *J. Am. Chem. Soc.* **102**, 92 (1980).
17. N. Tillman, A. Ulman, J. S. Schildkraut, and T. L. Penner, *J. Am. Chem. Soc.* **110**, 6136–6144 (1988).
18. E. B. C. D. Troughton, Bain, G. M. Whitesides, D. L. Allara, M. D. Porter, *Langmuir* **4**, 365–385 (1988).
19. C. W. Sheen, J. X. Shi, J. Martensson, A. N. Parikh, D. L.Allara, *J. Am. Chem. Soc.* **114**, 1514–1515 (1992).
20. G. A. Somorjai, *Chemistry in Two Dimensions – Surfaces* (Cornell University Press, Ithaca, NY, USA, 1982).
21. C. D. Bain, E. B. Troughton, Y-T. Tao, J. Evall, G. M. Whitesides, and R. G. Nuzzo, *J. Am. Chem. Soc.* **111**, 321–335 (1989).
22. M. W. Walczak, C. Chung, S. M. Stole, C. A. Widrig, and M. D.Porter, *J. Am. Chem. Soc.* **113**, 2370 (1991).
23. R. G. Nuzzo, B. R. Zegarski, and L. H. Dubois, *J. Am. Chem. Soc.* **109**, 733–740 (1987).
24. C. A. Widrig, C. Chung, and M. D. Porter, *J. Electroanal. Chem.* **310**, 335 (1991).
25. G. E. Poirier, *Chem. Rev.* **97**, 1117–1127 (1997).
26. G. E. Poirier and M. J. Tarlov, *Langmuir* **10**, 2853–2856 (1994).
27. M. D. Porter, T. B. Bright, D. L. Allara, and C. E. D. Chidsey, *J. Am. Chem. Soc.* **109**, 3559–3568 (1987).
28. W. Wang, T. Lee, and M. A. Reed, *Phys. Rev. B* **68**, 035416 (2003).
29. G. Solomons and C. Fryhle, *Organic Chemistry* (John Wiley & Sons Inc, New York, 2000).
30. J. Chen, M. A. Reed, A. M. Rawlett, and J. M. Tour, *Science* **286**, 1550–1552 (1999).
31. J. Chen, W. Wang, M. A. Reed, A. M. Rawlett, D. W. Price, and J. M. Tour, *Appl. Phys. Lett.* **77**, 1224–1226 (2000).
32. J. M. Tour, L. Jones, D. L. Perason, J. J. S. Lamda, T. P. Burgin, G. M. Whitesides, D. L. Allara, A.N. Parikh, and S. V. Atre, *J. Am. Chem. Soc.* **117**, 9529–9534 (1995).
33. A. Dhirani, P. H. Lin, P. Guyot-Sionnest, R. W. Zehner, and L. R. Sita, *J. Chem. Phys.* **106**, 5249–5253 (1997).
34. G. Yang, Y. Qian, C. Engtrakul, L. R. Sita, and G. Liu, *J. Phys. Chem. B* **104**, 9059–9062 (2000).
35. M. A. Reed, C. Zhou, C. J. Muller, T. P. Burgin, and J. M. Tour, *Science* **278**, 252–254 (1997).
36. C. Kergueris, J.-P. Bourgoin, S. Palacin, D. Esteve, C. Urbina, M. Magoga, and C. Joachim, *Phys. Rev. B.* **59**, 12505–12513 (1999).
37. J. Reichert, R. Ochs, D. Beckmann, H. B. Weber, M. Mayor, and H. v. Löhneysen, *Phys. Rev. Lett.* **88**, 176804 (2002).
38. L. A. Bumm, J. J. Arnold, M. T. Cygan, T. D. Dunbar, T. P. Burgin, L. Jones II, D. L. Allara, J. M. Tour, P. S. Weiss, *Science* **271**, 1705–1707 (1996).

39. Z. J. Donhauser, B. A. Mantooth, K. F. Kelly, L. A. Bumm, J. D. Monnell, J. J. Stapleton, D. W. Price Jr. A. M. Rawlett, D. L. Allara, J. M. Tour, and P. S. Weiss, *Science* **292**, 2303–2307 (2001).
40. M. Dorogi, J. Gomez, R. Osifchin, R. P. Andres, and R. Reifenberger, *Phys. Rev. B* **52**, 9071–9077 (1995).
41. K. S. Ralls, R. A. Buhrman, and T. C. Tiberio, *Appl. Phys. Lett.* **55**, 2459–2461 (1989).
42. C. Zhou, M. R. Deshpande, M. A. Reed, L. Jones II, and J. M. Tour, *Appl. Phys. Lett.* **71**, 611–613 (1997).
43. D. J. Wold, R. Haag, M. A. Rampi, and C. D. Frisbie, *J. Phys. Chem. B*, **106**, 2813–2816 (2002).
44. X. D. Cui, A. Primak, X. Zarate, J. Tomfohr, O. F. Sankey, A. L. Moore, T. A. Moore, D. Gust, G. Harris, and S. M. Lindsay, *Science* **294**, 571–574 (2001).
45. D. J. Wold and C. D. Frisbie, *J. Am. Chem. Soc.* **122**, 2970–2971 (2000).
46. X. D. Cui, X. Zarate, J. Tomfohr, O. F. Sankey, A. Primak, A. L. Moore, T. Moore, A. D. Gust, G. Harris, and S. M. Lindsay, *Nanotechnology* **13**, 5–14 (2002).
47. H. Park, A. K. L. Lim, J. Park, A. P. Alivisatos, and P. L. McEuen, *Appl. Phys. Lett.* **75**, 301–303 (1999).
48. J. Park, A. N. Pasupathy, J. I. Goldsmith, C. Chang, Y. Yaish, J. R. Petta, M. Rinkoski, J. P. Sethna, H. D. Abruna, P. L. McEuen, and D. C. Ralph, *Nature* **417**, 722–725 (2002).
49. W. Liang, M. P. Shores, M. Bockrath, J. R. Long, and H. Park, *Nature* **417**, 725–729 (2002).
50. J. G. Kushmerick, D. B. Holt, S. K. Pollack, M. A. Ratner, J. C. Yang, T. L. Schull, J. Naciri, M. H. Moore, and R. Shashidhar, *J. Am. Chem. Soc.* **124**, 10654–10655 (2002).
51. J. G. Kushmerick, D. B. Holt, J. C. Yang, J. Naciri, M. H. Moore, and R. Shashidhar, *Phys. Rev. Lett.* **89**, 086802 (2002).
52. R. Holmlin, R. Haag, M. L. Chabinyc, R. F. Ismagilov, A. E. Cohen, A. Terfort, M. A, Rampi, and G. M. Whitesides, *J. Am. Chem. Soc.* **123**, 5075–5085 (2001).
53. M. A. Rampi, and G. M. Whitesides, *Chem. Phys.* **281**, 373–391 (2002).
54. J. K. N. Mbindyo, T. E. Mallouk, J. B. Mattzela, I. Kratochvilova, B. Razavi, T. N. Jackson, and T. S. Mayer, *J. Am. Chem. Soc.* **124**, 4020–4026 (2002).
55. G. Binnig, H. Rohrer, C. Gerber, and H. Weibel, *Phys. Rev. Lett.* **49**, 57 (1982).
56. D. M. Eigler and E. K. Schweizer, *Nature* **344**, 524 (1990).
57. H. J. Lee and W. Ho, *Science* **286**, 1719–1722 (1999).
58. A. J. Heinrich, C. P. Lutz, J. A. Gupta, and D. M. Eigler, *Science* **298**, 1381–1387 (2002).
59. R. P. Andres, T. Bein, M. Dorogi, S. Feng, J. I. Henderson, C. P. Kubiak, W. Mahoney, R. G. Osifchin, and R. Reifenberger, *Science* **272**, 1323–1325 (1996).
60. M. A. Reed, J. Chen, A. M. Rawlett, D. W. Price, and J. M. Tour, *Appl. Phys. Lett.* **78**, 3735–3737 (2001).
61. D. R. Lombardi, *Ph.D. thesis*, Yale University, USA, 1997.
62. S. Gregory, *Phys. Rev. Lett.* **64**, 689–692 (1990).
63. L. A. Bumm, J. J. Arnold, T. D. Dunbar, D. L. Allara, and P. S. Weiss, *J. Phys. Chem. B* **103**, 8122–8127 (1999).
64. M. A. Ratner, B. Davis, M. Kemp, V. Mujica, A. Roitberg, and S. Yaliraki, in *Molecular Electronics, Science and Technology*, edited by A. Aviram, M. A. Ratner (The Annals of the New York Academy of Sciences, The New York Academy of Sciences, New York, USA, 1998), Vol. 852.
65. S. M. Sze, *Physics of Semiconductor Devices* (Wiley, New York, USA, 1981).
66. The models listed in Table 3-1 apply to solid state insulators with one band.
67. J. Chen, L. C. Calvet, M. A. Reed, D. W. Carr, D. S. Grubisha, and D. W. Bennett, *Chem. Phys. Lett.* **313**, 741–748 (1999).
68. B. Mann and H. Kuhn, *J. Appl. Phys.* **42**, 4398–4405 (1971).
69. E. E. Polymeropoulos and J. Sagiv, *J. Chem. Phys.* **69**, 1836–1847 (1978).
70. D. R. Stewart, D. A. A. Ohlberg, and R. S. Williams (unpublished results).
71. F. F. Fan, J. Yang, L. Cai, D. W. Price, S. M. Dirk, D. V. Kosynkin, Y. Yao, A. M. Rawlett, J. M. Tour, and A. J. Bard, *J. Am. Chem. Soc.* **124**, 5550–5560 (2002).
72. K. Slowinski, H. K.Y. Fong, M. Majda, *J. Am. Chem. Soc.* **121**, 7257–7261 (1999).
73. R. L. York, P. T. Nguyen, and K. Slowinski, *J. Am. Chem. Soc.* **125**, 5948–5953 (2003).
74. J. F. Smalley, S. W. Feldberg, C. E. D. Chidsey, M. R. Linford, M. D. Newton, and Y. Liu, *J. Phys. Chem.* **99**, 13141–13149 (1995).

75. W. Franz, *Handbuch der Physik*, edited by S. Flugge (Springer-Verlag, Berlin, Germany, 1956), Vol. 17, p. 155.
76. C. Joachim and M. Magoga, *Chem. Phys.* **281**, 347–352 (2002).
77. J. K. Tomfohr and O. F. Sankey, *Phys. Rev. B* **65**, 245105 (2002).
78. G. Lewicki and C. A. Mead, *Phys. Rev. Lett.* **16**, 939–941 (1966).
79. R. Stratton, G. Lewicki, and C. A. Mead, *J. Phys. Chem. Solids.* **27**, 1599–1604 (1966).
80. G. H. Parker and C. A. Mead, *Phys. Rev. Lett.* **21**, 605–607 (1968).
81. B. Brar, G. D. Wilk, and A. C. Seabaugh, *Appl. Phys. Lett.* **69**, 2728–2730 (1996).
82. J. G. Simmons, *J. Appl. Phys.* **34**, 1793–1803 (1963).
83. J. G. Simmons, *J. Phys. D* **4**, 613 (1971).
84. J. Maserjian and G. P. Petersson, *Appl. Phys. Lett.* **25**, 50–52 (1974).
85. Nonlinear least square fittings were performed using Microcal Origin 6.0.
86. Both bias values were used for the C8 data to compensate for an observed asymmetry, and are plotted in Figure 3.13.
87. M. P. Samanta, W. Tian, S. Datta, J. I. Henderson, and C. P. Kubiak, *Phys. Rev. B* **53**, R7626-R7629 (1996).
88. B. Xui and N. J. Tao, *Science* **301**, 1221–1223 (2003).
89. Although the HOMO-LUMO gap of alkyl chain type molecules has been reported (see Ref. [90–92]), there is no experimental data on the HOMO-LUMO gap for Au/alkanethiol SAM/Au system. 8 eV is commonly used as HOMO-LUMO gap of alkanethiol.
90. C. Boulas, J. V. Davidovits, F. Rondelez, and D. Vuillaume, *Phys. Rev. Lett.* **76**, 4797–4800 (1996).
91. M. Fujihira and H. Inokuchi, *Chem. Phys. Lett.* **17**, 554–556 (1972).
92. S. G. Lias, J. E. Bartmess, J. F. Liebman, J. L. Holmes, R. D. Levin, and W. G. Mallard, Gas-phase ion and neutral thermochemistry, *J. Phys. Chem. Ref. Data* **17**(1), 24 (1988).
93. V. Nazin, X. H. Qiu, and W. Ho, *Science* **302**, 77–81 (2003).

4

Nanostructured Hydrogen-Bonded Rosette Assemblies

Self-Assembly and Self-Organization

Mercedes Crego-Calama, David N. Reinhoudt, Juán J. García-López, and Jessica M.C.A. Kerckhoffs

4.1. SELF-ASSEMBLY IN SOLUTION

Self-assembly[1] has become a promising option for the construction of molecular nanoscale devices.[2,3] Well-defined nanostructures, also termed "supramolecular aggregates", are formed by self-assembly of a limited number of well-defined building blocks with strong affinity for each other. They are formed via reversible noncovalent interactions such as hydrophobic and electrostatic effect, π–π stacking, hydrogen bonds and/or metal coordination.[4] These noncovalent systems, generally highly dynamic on the human time scale, are distinctly different from the non-reversible covalent molecules, and they offer some advantages. The advantage of noncovalent synthesis is that noncovalent bonds are formed spontaneously and reversibly under conditions of thermodynamic equilibrium, with the possibility of error correction and without undesired side products. Furthermore, it does not require harsh chemical reagents or conditions. For instance, we have developed the self-assembly of nanosized molecular structures as large as $\sim 5.5 \times 3.1 \times 2.7$ nm, via molecular recognition between complementary hydrogen-bonding building blocks, that are otherwise inaccessible via *traditional* covalent synthesis. These hydrogen-bonded aggregates form spontaneously under thermodynamically controlled conditions, which give these nanostructures their ability to "proofread" and correct mistakes.

The choice of molecular shape and size and especially the arrangement of hydrogen bonds are essential for the build-up of two- and three-dimensional architectures. The

Laboratory of Supramolecular Chemistry and Technology, MESA$^+$ Research Institute, University of Twente, P. O. Box 217, 7500 AE Enschede, The Netherlands.

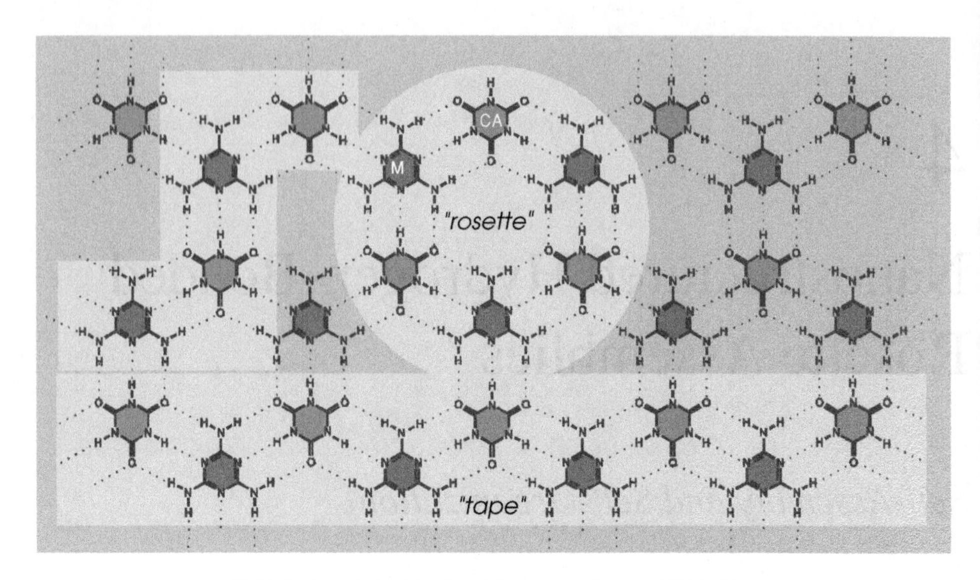

FIGURE 4.1. 2-D Melamine (M)-isocyanuric acid (CA) lattice.

stability and selectivity of the association depend on the nature and position of the hydrogen-bonding donor and acceptor sites. A well-known hydrogen-bonded system is based on isocyanuric acid and melamine. The three orthogonal ADA (Acceptor-Donor-Acceptor) hydrogen-bonding arrays of isocyanuric acid (CA) are complementary to the three DAD hydrogen-bonding arrays of melamine (M). Mixing both components in a 1:1 ratio leads to an infinite 2D lattice held together by hydrogen bonds (Figure 4.1).

Lehn[5] and Whitesides[6,7] identified this complementary hydrogen bonding in the melamine-cyanuric lattice and showed that depending on the preorganization of the melamines and the steric interactions between the building blocks, selectively the cyclic mono rosette instead of tapes could be formed. Recently, our group[8] showed by model calculations that it is indeed possible to promote the rosette formation over tape formation by steric repulsion in the linear tapes, but the effect is not as large as believed before.

4.1.1. Noncovalent Synthesis of Hydrogen-Bonded Assemblies

We wanted to explore the limits of the self-assembly approach for the synthesis of well-defined structures with nanoscale dimensions by further increasing the size of the assemblies. Our group has shown that calix[4]arenes diametrically substituted with two melamine units at the upper rim are very good platforms for the formation of double rosette assemblies.

In the presence of two equivalents of 5,5-diethylbarbiturate (DEB) or isocyanuric acid derivatives (CYA), dimelamines **1a–d** form stable double rosette assemblies **1a–d**$_3$•(DEB)$_6$ in apolar solvents such as chloroform, benzene or toluene even at 10^{-4} M concentration (Figure 4.2).[9,10] They consist of 9 different components held together by 36 hydrogen bonds. X-ray diffraction studies of **1b**$_3$•(DEB)$_6$ showed that the two rosette layers are tightly stacked on top of each other with an interatomic separation of 3.5 Å at the edges and

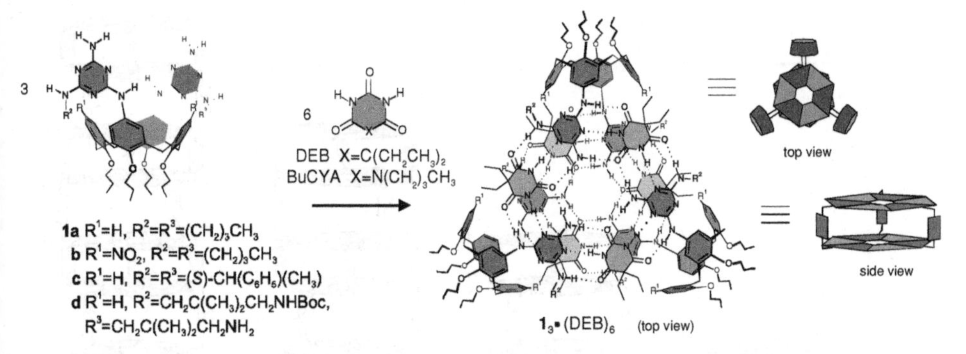

FIGURE 4.2. Formation of the double rosette assemblies 1_3•$(DEB)_6$, from its building blocks calix[4]arene dimelamine **1** and DEB.

3.2 Å at the centre of the rosette. The assemblies have a height of 1.2 nm and a width of ~3.1 nm.

These double rosettes can be easily extended to tetrarosettes by connecting two calix[4]arene dimelamines by a flexible linker unit X (Figure 4.3).[11,12] A tetrarosette assembly consists of fifteen components, three tetramelamines **2** and twelve barbiturate/cyanurate derivatives held together by 72 hydrogen bonds and it has a size of ~3.0 × 3.1 × 2.7 nm. The thermodynamic equilibrium for the assembly $\mathbf{2}_3$•$(DEB)_{12}$ is reached within seconds after mixing the two different components. Similarly, hexarosettes $\mathbf{3}_3$•$(DEB)_{18}$ are prepared

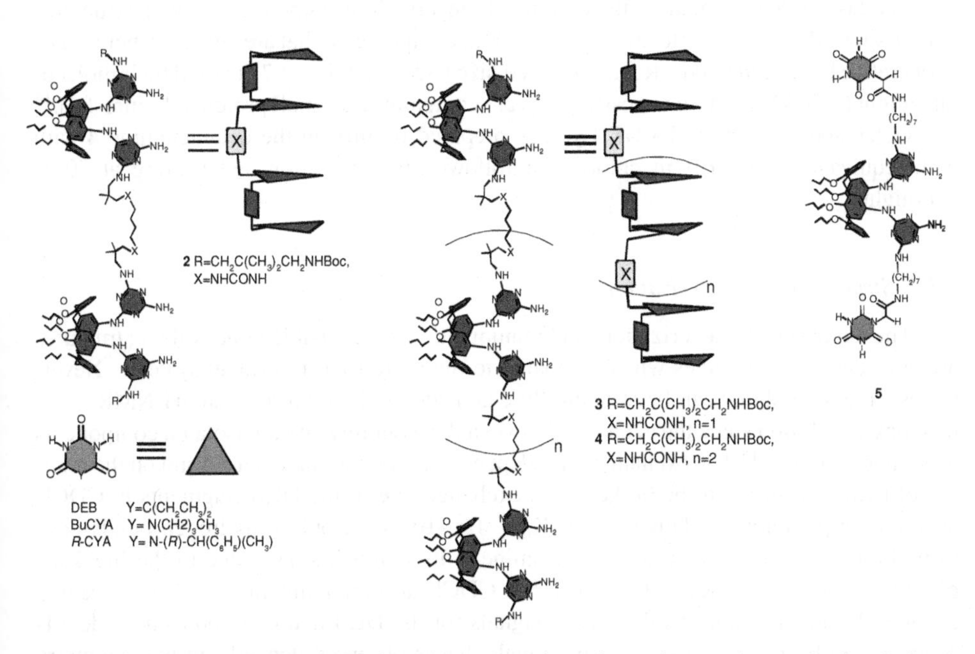

FIGURE 4.3. Molecular structures and schematic representations of tetra- hexa- and octamelamines **2**, **3** and **4**, respectively, and di(melamine-cyanurate) **5**. The molecular structure of DEB and different isocyanuric acid derivatives (CYA) are also shown.

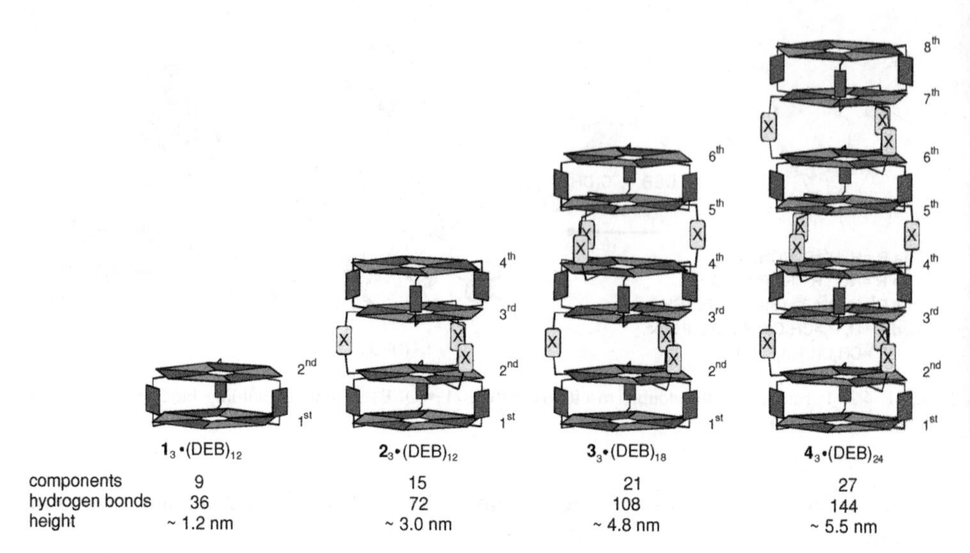

FIGURE 4.4. Schematic representation of the double, tetra-, hexa-, and octarosette. The corresponding heights, number of components, and hydrogen bonds of the assemblies are also shown.

from three calix[4]arene dimelamines connected by two flexible linker units X and six equivalents DEB.[13] The 21-component assembly is held together by 108 hydrogen bonds and has a size of $\sim 3.1 \times 4.8 \times 2.7$ nm.

The lastest development is the completely reversible formation of the nanostructures comprising 8 different rosette layers (27 different components) that are held together by 144 cooperative hydrogen bonds. Regarding their size ($\sim 5.5 \times 3.1 \times 2.7$ nm) and their molecular weight ($\sim$20 kDa), these nanostructures are comparable to small proteins (Figure 4.4).[14]

Octarosette assembly $\mathbf{4}_3$•(DEB)$_{24}$ was prepared by mixing the octamelamine **4** with eight equivalents of DEB in chloroform followed by sonication and heating at 50 °C overnight.

4.1.2. Structural Characterization

The isolation, characterization, and manipulation of reversible noncovalent structures are considerable challenges when compared to the more robust covalent systems. Nevertheless, the assemblies can be structurally characterized by conventional ^{1}H NMR spectroscopy in solution since the exchange between the assembly and the isolated components is relatively slow.[15,16] The exchange rates depends *i.e.*, on the solvent used and on the number of hydrogen bonds to be broken. The exchange rate of the DEB fragments in $CDCl_3$ at room temperature is relatively fast ($K_d \sim 10^2$–10^3 s^{-1}), only 6 hydrogen bonds have to be broken. The exchange of the melamine units is much slower due to the breakage of 12 hydrogen bonds ($K_d \sim 10^1$–10^0 s^{-1} in $CDCl_3$ at room temperature).[16] In the region between 13 and 16 ppm, the diagnostic signals for the DEB hydrogen-bonded imide NH protons are observed. The number of signals that is observed depends on the symmetry of the assembly and the number of rosette layers. For the double rosette assemblies, three different isomers are possible; the D_3 (staggered) and C_{3h} (symmetrical eclipsed) isomers

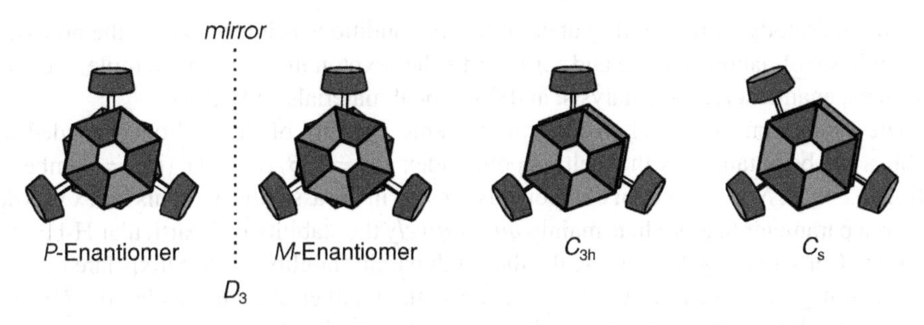

FIGURE 4.5. Conformational isomers of the double rosette assembly.

both giving two signals for the imide NH protons, while the C_s (unsymmetrical eclipsed) isomer gives six different signals (Figure 4.5). In principle, all possible isomers can potentially be formed when cyanurate derivatives are used.[17] However, double rosette assemblies with DEB preferentially form the D_3 isomer.

For the tetrarosette assemblies with the melamines in the staggered D_3 conformation four signals are observed for the imide NH protons; two signals for the protons in the first rosette floor, which are identical to the corresponding two signals for the fourth floor, and two different signals for the second and third floors (Figure 4.4). In analogy, the hexarosette give rise to six different proton signals (identical signals for the 1st and 6th, the 2nd and 5th, and the 3rd and 4th floors). For the octarosette eight partially overlapping signals are observed. The chemical shift of imide NH protons for the first and the eight floors (outer floors) are significantly different for these for the second to the seventh floors (inner floors).

MALDI-TOF mass spectroscopy using the Ag^+-labeling technique[18,19] gives additional evidence for the formation of the hydrogen-bonded nanostructures. Positively charged assemblies are generated in a nondestructive way by coordination of Ag^+ to a cyano group or by complexation of Ag^+ between two phenyl rings.

Furthermore, when one of the building blocks is chiral, the assembly formation can also be studied using CD spectroscopy. In the absence of any other element of chirality, these hydrogen-bonded assemblies in the D_3 conformation are formed as racemic mixture of the P- and M-enantiomers (Figure 4.5). After the introduction of a chiral substituent to either the melamine or barbiturate/cyanurate, two diastereomeric assemblies with either the P- or M-helicity can be formed.[20] The degree of the chiral induction depends on the difference in free energy at the thermodynamic equilibrium and is mainly effected by the proximity of the chiral groups to the core of the assembly and the solvent. The chiral hydrogen-bonded assemblies display highly characteristic CD spectra while the melamine or barbiturate/cyanurate are CD silent.

4.1.3. Thermodynamic and Kinetic Stability

The formation of synthetic hydrogen-bonded structures is typically achieved under thermodynamic control. Therefore, product formation is usually quantitative because erroneous structures can dissociate and recombine to give the correct assembly. However, it is expected that a large increase in the number of hydrogen bonds will lead to a kinetic control in the self-assembly process limiting the use of noncovalent synthesis of nanostructures.

Thus, the thermodynamic stability at equilibrium conditions is important for the noncovalent synthesis of nanostructures and for their further exploitation as, for example, receptor molecules, enantioselective catalysts, and functional materials and devices.

The estimation of the relative thermodynamic stability of the hydrogen-bonded assemblies can be obtained by the melting point index $I_{Tm} = HB/(N - 1)$, (HB = number of hydrogen bonds, N = number of components present in the assembly).[21] This index was defined as a parameter that predicts mainly *qualitatively* the stability of a particular H-bonded assembly. Consequently, to increase the thermodynamic stability of the calix[4]are rosettes the first strategy that we have used is to increase the number of hydrogen bonds (HB) per components (N) resulting in a larger I_{Tm}. Synthetically the strategy is to covalently link multiple melamines, which results in the formation of tetra-, hexa-, and octamelamines. Theoretically based on the melting point index, the thermodynamic stability of the octarosette ($I_{Tm} = 5.5$), hexarosette ($I_{Tm} = 5.4$), and tetrarosette ($I_{Tm} = 5.1$) assemblies should be higher than that of the double rosette assemblies ($I_{Tm} = 4.5$) because the number of HB per building block increases more rapidly that the number of components N. Experimentally, ^{1}H NMR and CD spectroscopy studies of the different rosette assemblies with polar solvents are used to measure the stability in solution, and the χ value *i.e.*, the % of polar solvent at which 50% of the assembly is still present, has been defined to asses the thermodynamic stability of the assemblies. Because the thermodynamic stability reflects the difference in energy ΔG between the assembly and the free components, polar solvents such as DMSO and methanol can be added to the assembly to decrease the enthalpy and reach $\Delta G = 0$.

The χ value is much more accurate than the I_{Tm} because the comparison of the latter for different assemblies is only meaningful when these are structurally related in terms of rigidity. For example, replacement of the barbiturate for cyanurates units in the assemblies does not change the I_{Tm} value, while it is known to significantly increase the thermodynamic stability as a result of the increased H-bond strength of cyanurates.[22]

Whitesides *et al.* showed that a single rosette ($HB = 18$; $N = 6$) is not stable in polar solvents and dissociates to the separate building blocks after addition of DMSO ($\chi_{DMSO} < 5\%$).[21] Initially, our experiments using ^{1}H NMR showed that the calix[4]arene based double rosettes are far more stable that single ones, and in analogy, tetra- and hexarosettes are more stable that double rosettes. However, more precise results can obtained using CD spectroscopy. As pointed out before, the presence of chiral centers induce one single handedness in the double, tetra-, and hexarosette assemblies. Therefore, the thermodynamic stabilities of the different assemblies were also investigated by CD spectroscopy in $CDCl_3$ in the presence of increasing amounts of polar solvents, such as THF, DMSO, and MeOH.[23,24] The results showed that the thermodynamic stability of the tetrarosette $\mathbf{2}_3$•(R-CYA)$_{12}$ (80% assembly formation at 60% MeOH) and hexarosette $\mathbf{3}_3$•(R-CYA)$_{18}$ (100% assembly formation at 40% MeOH) is significantly higher compared to double rosette assembly $\mathbf{1d}_3$•(R-CYA)$_6$ (~30% assembly formation at 50% MeOH).[25]

Furthermore, CD was used to assess the stability of assemblies having the same I_{Tm} but with slightly different chemical structure. CD titrations with THF of chiral double rosette assemblies **1c**•$_3$(DEB)$_6$ and **1c**•$_3$(BuCYA)$_6$ in $CHCl_3$ showed values of $\chi_{THF} = \sim 10\%$ and $\chi_{THF} = \sim 95\%$, respectively, as predicted taking in account the higher strength of the cyanurate H-bonds when compared with barbiturates.

The second strategy to increase the thermodynamic stability of these assemblies is based in the reduction of the number of particles (N) maintaining the number of hydrogen bonds constant. The covalent linkage between the barbiturate or cyanurate and the dimelamine moieties reduces the number of particles involved in the double rosette formation from 9 to 3. Several systems were studied where the structure of the linker connecting the melamine and the cyanurate or barbiturate units was sequentially varied. The highest stability was observed for the di(melamine-cyanurate) assembly $\mathbf{5}_3$ in which the components are connected through an *n*-heptylamidomethyl linker. At 70% DMSO present, 50% of assembly $\mathbf{5}_3$ ($\chi_{DMSO} = 70\%$) is still formed. The reference assembly $\mathbf{1a}_3$•(BuCYA)$_6$, wherein both building blocks, namely melamine and cyanuric acid derivatives, were not covalently linked exhibited an $\chi_{DMSO} = 40\%$, already after addition of 40% DMSO only 50% of the assembly was present.

Beside the dependence of the thermodynamic stability of the assemblies on the number and the strength of the hydrogen bonds formed and the number of particles present in the assembly, we have observed that in general the thermodynamic stability also depends on the steric strain in the assembly and the inference of hydrogen bond donating and accepting groups with the rosette platform.[23] The position at which the hydrogen bond donating and accepting functionalities are introduced also plays a major role.

Alongside the thermodynamic stability, we have also studied the kinetic stability of hydrogen-bonded rosette nanostructures. Chiral amplification and racemerization studies for enantiomerically enriched double, tetra-, and hexarosettes[26] by CD spectroscopy were used for the determination of the relative kinetic stabilities of these rosettes.[13,27] The chiral amplification studies show that solvent polarity and temperature strongly affect the kinetic stabilities of these assemblies. For example, the activation energy for the dissociation of a tetramelamine from a tetrarosette assembly, a process that involves the breakage of 24 hydrogen bonds, was determined at 98.7 ± 16.6 kJ.mol^{-1} in chloroform and 172.8 ± 11.3 kJ.mol^{-1} in benzene. Moreover, racemization studies with enantiomerically enriched assemblies reveal a strong dependence of the kinetic stability on the number and strength of the hydrogen bonds involved in assembly formation. The half-life time of double rosette $\mathbf{1b}_3$•(BuCYA)$_6$ (in the presence of barbiturate) is 8.4 min in chloroform at 50 °C, while the half-life time of the tetrarosette $\mathbf{2}_3$•(BuCYA)$_{12}$ is 5.5 h, 40 times higher compared to double rosette assemblies. The higher racemization half-life of the tetrarosettes over the double rosettes is the direct result of the increased number of hydrogen bonds that need to be broken in order for racemerization to occur (24 versus 12 hydrogen bonds). As expected, the racemization half-life of the hexarosette $\mathbf{3}_3$•(BuCYA)$_{18}$ (36 hydrogen bonds to be broken) is even higher, 150 h in chloroform at 50 °C, which approach the racemization half-life times of covalent molecules. Extrapolation of the data obtained for the double, tetra-, and hexarosette results in a predicted racemization half-life time of more than 200 h for the octarosette.

4.2. SELF-ORGANIZATION OF ROSETTE ASSEMBLIES ON SURFACES

Currently, the self-organization, *i.e.*, self-assembly of entities with high degree of confinement, of building blocks into noncovalent polymeric nanostructures[11] has become a promising alternative in nanotechnology for the construction of molecular scale devices via the so-called *bottom up* approach.[2,3,28–30] Recent progress has enabled several groups to arrange supramolecular aggregates in two dimensions on surfaces by clever design of selective and directional intermolecular connections between building blocks.[31,33] These findings suggest that careful placement of functional groups that are able to participate in directed noncovalent interactions will allow the rational design and construction of more complex supramolecular architectures on solid surfaces. For this purpose, several prerequisites have to be fulfilled. In first place, the components involved in the final molecular construction sets should serve as platforms for the incorporation of further functional groups, in such a way that the final molecular construction sets will inherit the recognition and electronic and/or magnetic properties of the new functionalities introduced.[34] Secondly, the molecules should self-organize on the substrates by self-assembly. Thus, the systems require reversibility of the connecting events, *i.e.*, kinetic liability and weak bonding.

The multicomponent nature of hydrogen-bonded nanostructures based on rosette motifs as well as the multiple sites available for further chemical modifications at building blocks make these structures a very interesting candidates for the precise spatial arrangement of functional diversity onto solid supports.

4.2.1. 2-D Arrangements of the Hydrogen-Bonded Rosettes

We have carried out the study of the adsorption on surfaces of finite double rosette $\mathbf{1}_3 \bullet (\text{DEB})_6$ assemblies and the polymeric nanostructures $[\mathbf{1}_3 \bullet \mathbf{6}_3]_n$ (Figure 4.6)[35] by means of atomic force microscopy.[36–41]

In all the cases the TM-AFM (Tapping Mode Atomic Force Microscopy) images show that the hydrogen-bonded assemblies self-organize on regular patterns of perfectly aligned nanorods when deposited on HOPG (Highly Oriented Pyrolitic Graphite) from different solvents (Figure 4.7), and that the mutual orientation of the rods in different domains is correlated with the symmetry directions of the graphite lattice. However, the driving force is

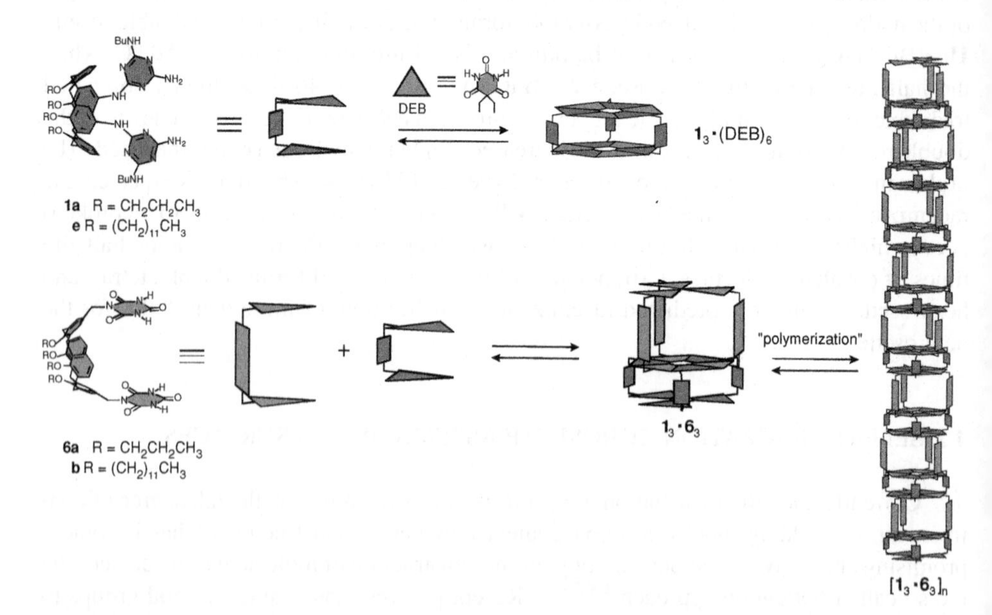

FIGURE 4.6. Schematic representation of the finite $\mathbf{1}_3 \bullet (\text{DEB})_6$ and polymeric $[\mathbf{1}_3 \bullet \mathbf{6}_3]_n$ nanostructures based on rosette motifs.

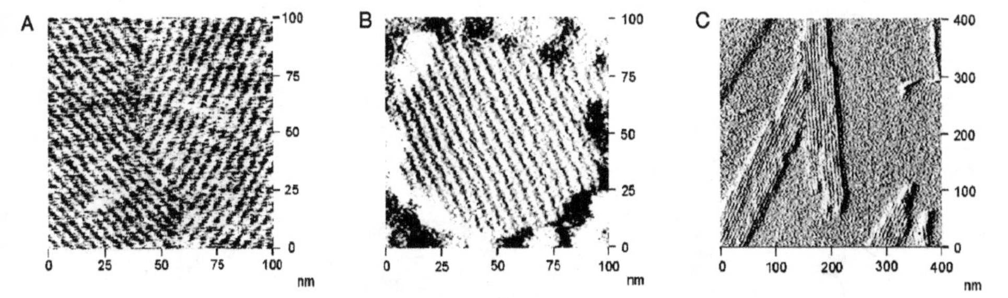

FIGURE 4.7. TM-AFM images of nanorod domains: (A) $[\mathbf{1a}_3\bullet(\mathrm{DEB})_6]_n$ in chloroform; (B) $[\mathbf{1a}_3\bullet\mathbf{6a}_3]_n$ in DMSO; (C) $[\mathbf{1e}_3\bullet\mathbf{6a}_3]_n$ in THF. (Reproduced with permission from: *J. Am. Chem. Soc.*, **1999**, *121*, 7154–7155. Copyright 1999 Am. Chem. Soc.).

different for the finite assemblies than for the polymeric structures. In case of the assemblies formed by depositing on graphite an equimolecular solution of the calix[4]arene derivatives **1** and **6**, the nanorods are due to the infinite polymerization of building blocks through the $\mathbf{1}_3\bullet\mathbf{6}_3$. While, in the case of the finite assemblies $\mathbf{1}_3\bullet(\mathrm{DEB})_6$, the polymerization process is not possible and the nanorods are most likely formed by the face to face arrangement of multiple disk-like assemblies $\mathbf{1}_3\bullet(\mathrm{DEB})_6$, a process that is driven by solvophovic interactions.

Analysis of the TM-AFM data demonstrates that there is a good correlation ($<20\%$ difference) between the experimentally determined and calculated parameters for four of the five nanostructures (Table 4.1). The best correlation was found for assemblies $[\mathbf{1}_3\bullet(\mathrm{DEB})_6]_n$ ($<10\%$), which is the result of a most densely packed structure. It is remarkable that in the case of assembly $[\mathbf{1a}_3\bullet\mathbf{6a}_3]_n$, the only one that is formed from the polar solvent DMSO, exhibits a significant distance between nanorods (n = 1.5). The value suggests that the nature of the solvent plays an important role.

Similarly to double rosettes $\mathbf{1}_3\bullet(\mathrm{DEB})_6$, nanorod domains are observed for tetrarosette assembly $\mathbf{2}_3\bullet(\mathrm{DEB})_{12}$, in addition to a bulk crystalline phase, a granular phase and a featureless gas-like or liquid-like phase.[42] In this case a reproducible heart-to-heart distance (h) of 4.6 ± 0.1 nm was found for their adsorption on HOPG surfaces (Figure 4.8). Furthermore, higher resolution TM-AFM images allow observing the presence of a smaller periodicity. The raw data shown in Figure 4.8A suggest the presence of oblique-elongated features along the rows. The quantitative analysis of the two dimensional fast Fourier transform in

TABLE 4.1. TM-AFM Data for the Hydrogen-Bonded Rodlike Nanostructures.

Assembly	$[\mathbf{1a}_3\bullet\mathbf{6a}_3]_n$	$[\mathbf{1e}_3\bullet\mathbf{6a}_3]_n$	$[\mathbf{1e}_3\bullet\mathbf{6b}_3]_n$	$[\mathbf{1a}_3\bullet(\mathrm{DEB})_6]_n$	$[\mathbf{1e}_3\bullet(\mathrm{DEB})_6]_n$
Solvent	DMSO	THF	$CHCl_3$	$CHCl_3$	$CHCl_3$
L [nm]	2.9 ± 0.1	4.8 ± 0.1	4.8 ± 0.1	2.9 ± 0.1	4.8 ± 0.1
d (calc) [nm][b]	3.4 ± 0.1	5.5 ± 0.1	5.5 ± 0.1	3.4 ± 0.1	5.5 ± 0.1
h (exptl) [nm][a]	4.9 ± 0.2	5.0 ± 0.2	4.6 ± 0.2	3.8 ± 0.2	5.3 ± 0.2
n (exptl) [nm]	1.5 ± 0.2	$-0\ (-0.5)$	$-0\ (-0.9)$	0.4 ± 0.2	$-0\ (-0.2)$

[a] The heart-to-heart distance (h) is experimentally determined from the images by averaging the distance between 10–15 adjacent nanorods. This parameter is related to the others via $h = d + n$.
[b] The diameter of the nanorods (d) is calculated via the relation $d = 2L/(3)^{0.5}$ from the crystal structure of $\mathbf{1b}_3\bullet(\mathrm{DEB})^6$.

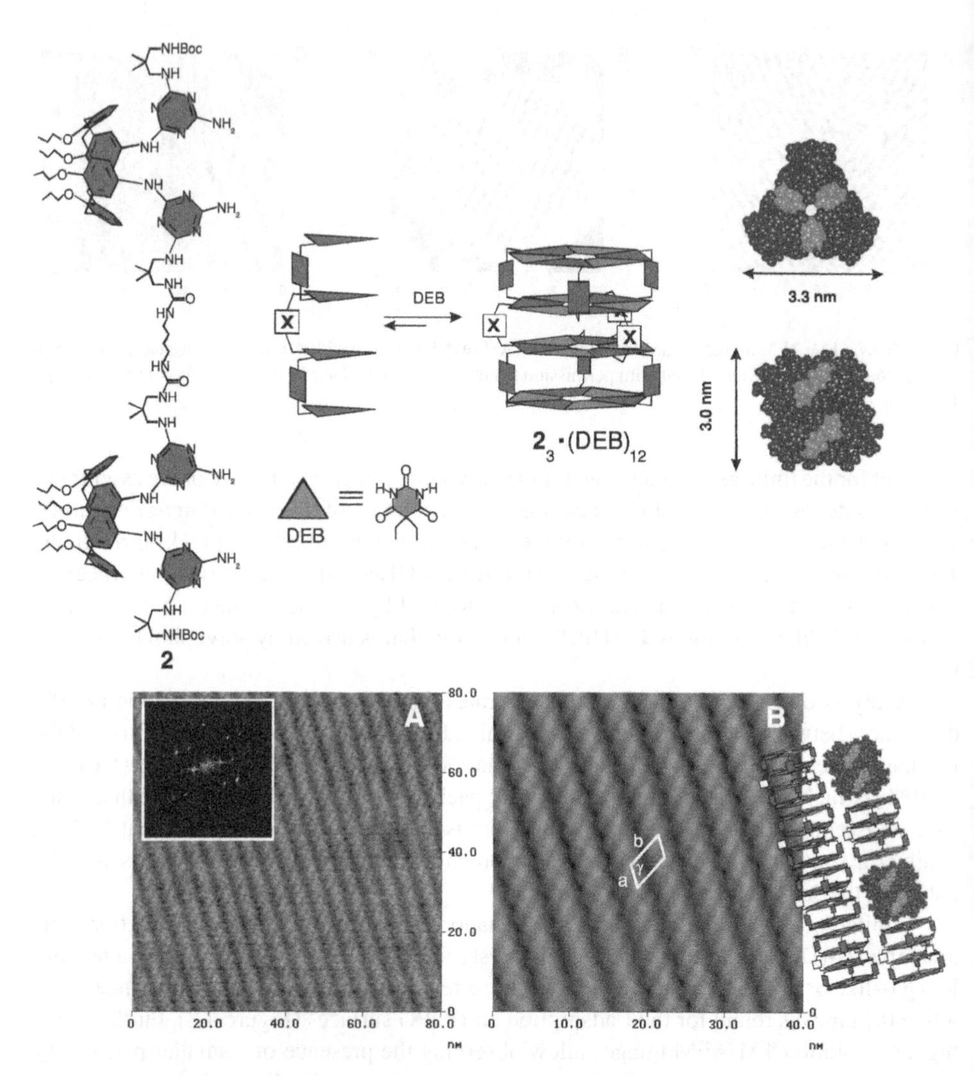

FIGURE 4.8. Schematic representation of the tetrarosette assembly. (A) Unfiltered high resolution tapping mode AFM phase image of the nanorod domain structure; (inset) fast Fourier transform. (B) Fourier filtered section of raw data shown in A and unit cell of the lattice structure. (Reproduced with permission from: *Proc. Natl. Acad. Sci. USA* **2002**, *99*, 5024–5027. Copyright 2002 National Academy of Sciences, U.S.A.).

Figure 4.8A (inset) reveals an oblique lattice structure with $a = 2.5 \pm 0.3$ nm, $b = 5.0 \pm 0.1$ nm and $\gamma = 122 \pm 3°$. This value compares reasonably well with the value obtained for the gas phase minimized nanostructure, which ranges between 8.6–9.9 nm^2, depending on the orientation. Moreover, if the possible spreading of the alkyl side chains of the rosettes due to the strong interaction of the methylene units with the graphite and the concomitant flattening of the nanostructures is taken into account, the area requirements would agree even better.

4.2.2. Growth of Individual Rosette Nanostructures

Hitherto, we have shown the self-organization of noncovalent hydrogen bonded assemblies on surfaces; however, the ultimate challenge in nanotechnology is the controlled growth and positioning of single nanostructures onto solid supports. With this aim, the growth of single double rosette nanostructures has been studied on gold monolayers.[43] For this purpose, first a calix[4]arene dimelamine **7** is embedded into hexanethiol self-assembled monolayer[44] on gold, followed by the growth of single assemblies on the monolayer via an exchange reaction between calix[4]arene dimelamines **7** and calix[4]arene dimelamines constituting a preformed double rosette assembly $\mathbf{1a}_3\bullet(\mathrm{DEB})_6$ in solution (Figure 4.9), similarly as it occurs in solution studies.[45] The high thermodynamic stability of the H-bonded rosettes assemblies combined with a relatively low kinetic stability allows the exchange, and consequently the growth process.

The monolayers corresponding to the insertion of calix[4]arene dimelamine **7** and the grown assembly $\mathbf{1a}_2\bullet\mathbf{7}(\mathrm{DEB})_6$ were studied by TM-AFM. In the first case, TM-AFM images show single isolated features with an average height of 1.1 ± 0.2 nm, in good agreement with the height obtained from the CPK model (0.9 nm) of the adsorbate **7**. Contrary, the TM-AFM images corresponding to the growth of the assemblies show two different size features (Figure 4.10). The height of the largest size (3.5 nm, Figure 4.10c) agrees well with the expected size for the assembly $\mathbf{1a}_2\bullet\mathbf{7}(\mathrm{DEB})_6$ considering the crystal structure of a similar double rosette assembly.[12] Therefore they are attributed to mixed assemblies $\mathbf{1a}_2\bullet\mathbf{7}(\mathrm{DEB})_6$ obtained through reversible exchange of components **1a** and **7**. The height of the smallest features (0.9 nm, Figure 4.10c) corresponds to single isolated molecules of **7** which were

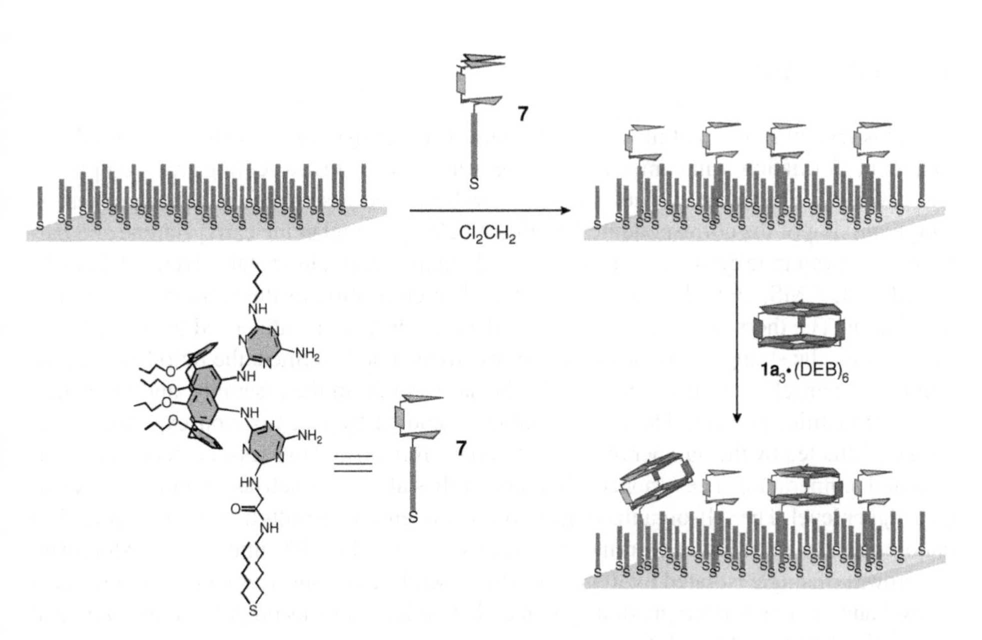

FIGURE 4.9. Schematic representation of the methodology followed for the growth of assemblies $\mathbf{1a}_2\bullet\mathbf{7}\bullet(\mathrm{DEB})_6$ on gold monolayers.

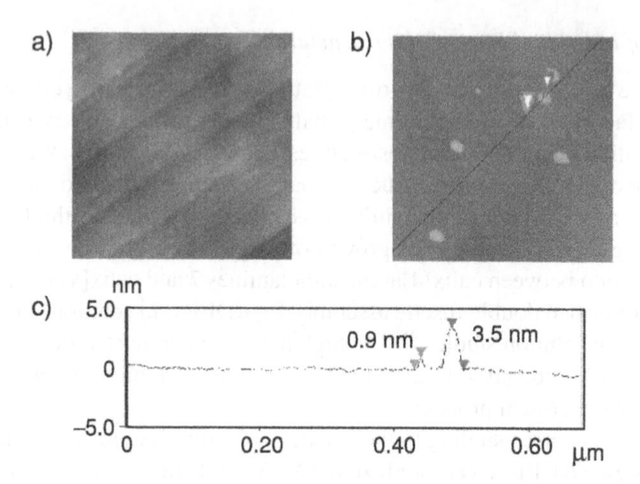

FIGURE 4.10. a) TM-AFM image in air (500 × 500 nm) of hexanethiol monolayers on Au (111) containing calix[4]arene dimelamine **7**. b) TM-AFM image (500 × 500 nm) after treatment with a solution of assembly $\mathbf{1a}_3\bullet(\mathrm{DEB})_6$. c) Comparison of heights. The height profile corresponds to the line drawn in the AFM image 9b. Color scale from dark to yellow: Z = 10 nm. (Reproduced with permission from: *J. Chem. Soc., Chem. Comm.* **2003**, 352–353. Copyright 2003 The Royal Society of Chemistry).

not involved in exchange reactions with assemblies $\mathbf{1a}_3\bullet(\mathrm{DEB})_6$. Statistical analysis of the size of the features, reveals an average height of 3.4 ± 1.4 nm that correlates perfectly with previous findings for the size of this hydrogen bonded nanostructure.

4.3. CONCLUSIONS

This review shows that noncovalent synthesis using hydrogen bonds can be used as a tool for the formation of nanometer-sized rosette nanostructures presenting great complexity and diversity. The facile synthesis of these well-defined assemblies is accomplished by simply mixing of the corresponding building blocks, *i.e.*, melamine calix[4]arene and barbiturate or cyanurate derivatives under thermodynamic control in apolar solvents. ^{1}H NMR, MALDI-TOF MS, and CD studies allow the characterization of the nanosize assemblies in solution. The thermodynamic stability of these assemblies is influenced by a number of factors, *i.e.*, the strength and the number of hydrogen bonds formed, the ratio between the number of particles and hydrogen bonds, the steric strain in the assembly, and the nature of functionalities present. The kinetic stability, studied by racemization experiments, is strongly affected by the number of hydrogen bonds formed and to a lesser extent by the solvent and temperature. These noncovalent assemblies also show self-assembling behaviour on a higher level. The self-organization on rosettes assemblies results in a highly regular 2-D nanorod arrays at the HOPG-air interface as revealed by TM-AFM microscopy. Moreover, the growth of single isolated hydrogen bonded assemblies on surfaces can be achieved *via* an exchange reaction from previously embedded building blocks on gold monolayers and assemblies preformed in solution.

These results demonstrate the potential application of hydrogen-bonded assemblies in the *bottom up* approach to nanotechnology, and together with recent synthetic improvements, form the basis for a general strategy to develop nanostructures on solid surface with molecular precision and specific functions.

ACKNOWLEDGMENTS

We thank the authors of the numerous rosette publications, which names are on the reference list, the EU for the MCR Grant to Dr. J. J. García-López and the Dutch Technology Foundation for financial support (project number TST 4624) to Dr. J.M.C.A. Kerckhoffs. Dr. M. Crego-Calama's research has been made possible by a fellowship of the Royal Netherlands Academy of Arts and Science.

REFERENCES

1. D. N. Reinhoudt and M. Crego-Calama, in Supramolecular Chemistry and Self-Assembly (special issue), *Science* **295**, 2403–2407 (2002).
2. E. A. Chandross and R. D. Miller, in Nanostructures (special issue) *Chem. Rev.* **99**, issue 7 (1999).
3. T. Ito and S. Okazaki, Pushing the limits of lithography, *Nature* **406**, 1027–1031 (2000).
4. J.-M. Lehn, *Angew. Chem. Int. Ed. Engl.* **27**, 89–112 (1988).
5. J.-M. Lehn, M. Mascal, A. DeCian, and J. Fischer, *J. Chem. Soc. Chem. Commun.* 479–480 (1990).
6. C. T. Seto and G. M. Whitesides, *J. Am. Chem. Soc.* **112**, 6409–6044 (1990).
7. J. A. Zerkowski, C. T. Seto, D. A. Wierda, and G. M. Whitesides, *J. Am. Chem. Soc.* **112**, 479–480 (1990).
8. A. G. Bielejewska, C. E. Marjo, L. J. Prins, P. Timmerman, F. de Jong, D. N. Reinhoudt, *J. Am. Chem. Soc.* **123**, 7518–7533 (2001).
9. R. H. Vreekamp, J. P. M. van Duynhoven, M. Hubert, and D. N. Reinhoudt, *Angew. Chem. Int. Ed. Engl.* **35**, 1215–1218 (1996).
10. P. Timmerman, R. Vreekamp, R. Hulst, W. Verboom, D. N. Reinhoudt, K. Rissanen, K. A. Udachin, and J. Ripmeester, *Chem. Eur. J.* **3**, 1823–1832 (1997).
11. K. A. Jolliffe, P. Timmerman, and D. N. Reinhoudt, *Angew. Chem. Int. Ed.* **38**, 933– 937 (1999).
12. V. Paraschiv, M. Crego-Calama, T. Ishi-I, C. J. Padberg, P. Timmerman, and D. N. Reinhoudt, *J. Am. Chem. Soc.* **124**, 7638–7639 (2002).
13. L. J. Prins, E. E. Neuteboom, V. Paraschiv, M. Crego-Calama, P. Timmerman, and D. N. Reinhoudt, *J. Org. Chem.* **67**, 4808–4820 (2002).
14. V. Paraschiv, M. Crego-Calama, R. Fokkens, C. J. Padberg, P. Timmerman, and D. N. Reinhoudt, *J. Org. Chem.* **66**, 8297–8301 (2001).
15. More detailed information about the characterization of double rosette assemblies can be found in D. N. Reinhoudt, P. Timmerman, F. Cardullo, and M. Crego-Calama, in *Supramolecular Science: Where it is and Where it is Going*, edited by R. Ungaro and E. Dalcanale (Kluwer Academic Publishers, Dordrecht, 1999), pp. 181–195.
16. P. Timmerman and L. J. Prins, *Eur. J. Org. Chem.* **17**, 3191–3205 (2001).
17. L. J. Prins, K. A. Jolliffe, R. Hulst, P. Timmerman, and D. N. Reinhoudt, *J. Am. Chem. Soc.* **122**, 3617–3627 (2000).
18. K. A. Jolliffe, M. Crego-Calama, R. Fokkens, N. M. M. Nibbering, P. Timmerman, and D. N. Reinhoudt, *Angew. Chem. Int. Ed.* **1998**, 37, 1247–1251.
19. P. Timmerman, K. A. Jollife, M. Crego-Calama, J. -L. Weidmann, L. J. Prins, F. Cardullo, B. H. M. Snellink-Ruël, R. Fokkens, N. M. M. Nibbering, S. Shinkai, and D. N. Reinhoudt, *Chem. Eur. J.* **2000**, 6, 4104–4115.
20. L. J. Prins, J. Huskens, F. de Jong, P. Timmerman, and D. N. Reinhoudt, *Nature* **398**, 498–502 (1999).
21. M. Mammen, E. E. Simanek, and G. M. Whitesides, *J. Am. Chem. Soc.* **118**, 12614–12623 (1996).

22. M. Mascal, P. S.. Fallon, A. S. Batsanov, B. R. Heywood, S. Champ, and M. J. Colclough, *Chem. Soc. Chem. Commun.* 805–806 (1995).
23. J. M. C. A. Kerckhoffs, M. Crego-Calama, I. Luyten, P. Timmerman, and D. N. Reinhoudt, *Org. Lett.*, **2**, 4121–4124 (2000).
24. M. Crego-Calama et al., unpublised results.
25. The tetra- and hexarosette assemblies start to precipitate at higher volume fractions of MeOH.
26. Enantiomerically enriched assemblies were prepared via substitution of chiral barbiturates by achiral cyanurates, see also Ref. 27
27. L. J. Prins, F. de Jong, P. Timmerman, and D. N. Reinhoudt, *Nature* **408**, 181–184 (2000).
28. A. ten Wolde (Ed.), *Nanotechnology, Towards a Molecular Construction Kit* (Netherlands Study Center for Technology Trends, The Hague, 1998), pp. 225–229.
29. Special issue: Nanotechnology, A C&EN Special Report, *Chem. & Eng. News*, **78**(42), 24–42 (2000).
30. Nanotechnology, *Sci. Am.* **285**, 32–91 (2001).
31. M. Böhringer, K. Morgenstern, and W. D. Schneider, *Phys. Rev. Lett.* **83**, 324–327 (1999).
32. T. Yokoyama, S. Yokoyama, T. Kamikado, Y. Okuno, and S. Mashiko, *Nature* **413**, 619–621 (2001).
33. K.W. Hipps, L. Scudiero, D.E. Barlow, and M.P. Cooke, *J. Am. Chem. Soc.* **124**, 2126–2127 (2002).
34. J. Michl, (Ed.), *Modular Chemistry* (Kluwer, Dordrecht, 1997).
35. H.-A. Klok, K.A. Jolliffe, C.L. Schauer, L.J. Prins, J.P. Spatz, M. Möller, P. Timmerman, and D.N. Reinhoudt, *J. Am. Chem. Soc.* **121**, 7154–7155 (1999).
36. For an introduction on surface scanning techniques and applications see: G. Binnig, C.F. Quate, and Ch. Gerber, *Phys. Rev. Ltt.* **56**, 930–933 (1986).
37. D. A. Bonnell, *Scanning Tunneling Microscopy and Spectroscopy: Theory, Techniques and Applications* (VCH, New York, 1993).
38. G. J. Leggett, in *Surface Analysis: the Principal Techniques*, edited by J. C. Vickerman (Wiley, Chichester, 1997) pp. 393–449.
39. S. N. Magonov and M.-H. Whangbo, *Surface Analysis with STM and AFM* (VCH, New York, 1996).
40. E. Meyer, *Prog. Surf. Sci.* **41**, 3–49 (1992).
41. R. Wiesendanger, *Scanning Probe Microscopy and Spectroscopy: Methods and Applications* (Cambridge University Press, Cambridge, 1994).
42. H. Schönherr, V. Paraschiv, S. Zapotoczny, M. Crego-Calama, P. Timmerman, C. W. Frank, G. J. Vancso, and D. N. Reinhoudt, *Proc. Natl. Acad. Sci. USA* **8**, 5024–5027 (2002).
43. J. J. García-López, S. Zapotoczny, P. Timmerman, F. C. J. M. van Veggel, G. J. Vancso, M. Crego-Calama, and D. N. Reinhoudt, *Chem. Commun.* 352–353 (2003).
44. O. Chailapakul, L. Sun, C. Xu, and R. M. Crooks, *J. Am. Chem. Soc.* **115**, 12459–12467 (1993).
45. M. Crego-Calama, R. Hulst, R. Fokkens, N. M. M. Nibbering, P. Timmerman, and D. N. Reinhoudt, *Chem. Commun.* 1021–1022 (1998).

5

Self-Assembled Molecular Electronics

Dustin K. James and James M. Tour

5.1. INTRODUCTION

When Moore's Law hits the solid-state fabrication "brick wall," researchers in the field of molecular electronics want to be there to pick up the pieces, using self-assembly as one of the tools. Of course, Gordon Moore, one of the founders of Intel, did not actually posit a Law, he made a prediction that the number of components "crammed" onto integrated circuits would double every year.[1] This prediction was later modified to a doubling every 18 months, and has held true long past the 1975 end date Moore originally used, an accuracy that convinced industry pundits to refer to his prediction as a Law.

The "brick wall" that the semiconductor industry is facing in its ever-increasingly faster march to smaller circuitry is based on physical and monetary limitations.[2–4] For instance, charge leakage becomes a problem when silicon oxide, used as insulation between conducting layers, is about 3 atoms deep. The industry's development of high-k dielectric layers to act as barriers to charge leakage may solve the problem and illustrates the fact that the brick wall appears to be moving further and further out; but the solution adds additional levels of complexity to the semiconductor fabrication process.[5] Another related physical limitation is that the thinning of silicon, for further miniaturization of circuits, destroys the band structures necessary for movement of electrons. Again, the semiconductor industry may be innovating their way around the problem, since IBM recently announced that a functioning silicon transistor with a 6 nm thick gate had been produced using ultra-thin silicon-on-insulator (SOI) technology.[6] There are additional physical limitations related to the lithographical methods used in laying down the circuits. New lithographic tools being developed cost millions of US dollars and their use would necessitate remodeling fabs or the construction of new ones because the tools are too large and complex to be retrofitted into most existing fabs.

Intel's Fab 22, in Chandler, Arizona, cost $2 billion to build in 2001. The company recently announced it would spend another $2 billion to upgrade the fab for 300 mm wafers

Rice University Chemistry Dept. & Center for Nanoscale Science & Technology. 6100 Main Street, Houston, TX 77005 tel: 713-348-6246, tour@rice.edu with 65 nm feature sizes.[7] Further changes in processing equipment may necessitate additional investments. One could easily project that the cost of fabs built by the end of this decade could be in the $10 to $25 billion range.

The use of molecular switches as components in the circuits of the future is attractive since one kg of molecules could contain 10^{23} or more components, depending on the molecular weight of the molecule. This is an incredibly large number of components, more than the number of solid-state transistors ever produced in the history of the semiconductor industry. The chemistry used to produce these molecules can be carried out in standard chemical plant reactors using well-characterized reactants and reagents.

The problem in using these molecular components arises when we start to try to place them in known positions with each end of the molecule connected in a known manner to the circuit. As of the time of this writing, no efficient method besides self-assembly exists for the individual placement of billions of molecules reproducibly in known positions. It is thus easy to understand why so much research has been conducted on self-assembly as it relates to molecular electronics. According to Whitesides,

> "A self-assembling process is one in which humans are not actively involved, in which atoms, molecules, aggregates of molecules and components arrange themselves into ordered, functioning entities without human intervention."[8]

Whitesides reviewed the principles of molecular self-assembly over a decade ago,[9] including the possibility of using self-assembly to make semiconductor devices. We recently reviewed our extensive work in the field of molecular electronics[10] as well as the sub-field of molecular wires,[11] both of which include self-assembly as a necessary step in the construction of devices. A review of the physical characteristics of molecular electronics devices has recently appeared,[12] along with a review of the genesis of molecular electronics.[13] The semiconductor industry has realized that research achievements in the molecule electronics and self-assembly fields could enable the development of methods to bypass existing roadblocks.[14]

The amount of published research in the field of self-assembly has exploded in recent years. A computerized search of the literature using "self-assembled" as the key phrase returned over 14,000 references (another 29,000 references were returned when the "concept" of self-assembled was searched). There were 2,619 hits in 2002, 2,336 hits in 2001, and 1,971 hits in 2000. The first time the phrase "self-assembled" appeared was in a review of primitive protein models in 1969.[15] Refining the search by including "molecular electronics" with "self-assembled" produced 356 hits, with the first reviews appearing in 1981[16] and 1991.[17] Since the thoroughness of any literature search is limited by the search terms used, we have used a variety of methods to find relevant literature through March 2003 for the present work.

In this chapter we will review some of the landmark publications in self-assembled molecular electronics. We will also review newer work from our own lab as well as others in the field.

5.2. THE FIELD EMERGES

In early work to lay the foundation for the use of self-assembly in construction of electronic devices from molecules, self-assembled monolayers (SAMs) of various

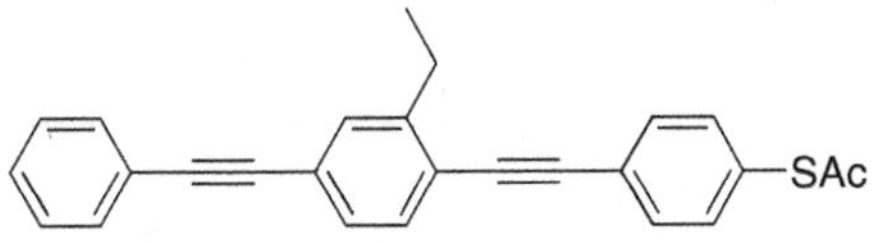

FIGURE 5.1. After deprotection and deposition on the surface of Au, a SAM of this OPE molecule had a tilt angle of the long molecular axis <20° from the normal to the substrate surface.

thiol-containing molecules were formed on the surface of Au and analyzed using ellipsometry, X-ray photoelectron spectroscopy, and infrared external reflectance spectroscopy.[18] It was found that the thiol moieties dominated the adsorption on the Au sites, with the direct interaction of the conjugated π-systems with the Au surface being weaker. The tilt angle of the long molecular axis of a thiol-terminated oligo(phenylene ethynylene) (OPE) system shown in Figure 5.1 was found to be <20° from the normal to the substrate surface. In situ deprotection of the thiol moiety via deacylation of the thioacetyl group with NH_4OH allowed for formation of the SAM without isolation of the oxidatively unstable free thiol.

A series of OPEs[19] and a series of oligo(2,5-thiophene ethynylene)s (OTEs)[20] of increasing lengths were synthesized via solution and solid phase chemistry to explore the physical and electronic characteristics of the molecules. The working theory was that conductance occured through the overlapping molecular orbitals of OPEs[21,22] and OTEs. Latter work has concentrated on OPEs in order to maximize molecular orbital overlap.

The thiol moieties that have been used to attach the molecules to metal surfaces have been referred to as "alligator clips" since they form very strong bonds, 1.8–2 eV.[23] Theoretical work using density functional theory (DFT) has indicated that the best alligator clip would be S followed by Se and Te, however, a direct aryl-metal bond might be best.[24] Recent work done in air and ultra-high vacuum (UHV) scanning tunneling microscopy (STM) on SAMs formed from S- or Se-terminated terthiophene molecules indicated that Se provides a better coupling link than S whatever the tunneling conditions.[25]

The conductance of a molecular junction was measured in 1997.[26] As shown in Figure 5.2, two Au wires were covered with SAMs of benzene-1,4-dithiol in THF. The wires were bent until they broke, and the broken ends were brought together in picometer increments

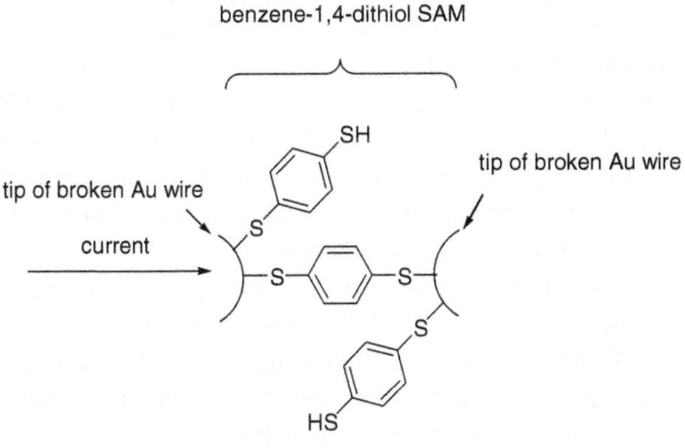

FIGURE 5.2. The measurement of conductance through a single molecule of benzene-1,4-dithiol using a break junction experiment.

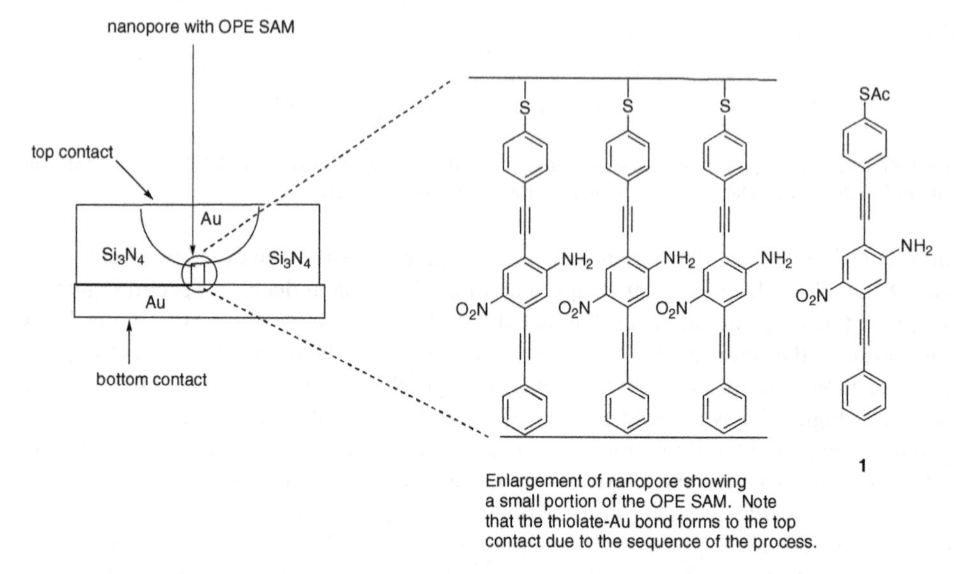

FIGURE 5.3. The nanopore test bed structure containing a SAM of functionalized OPE **1**.

via a lateral piezoelectric crystal, until the onset of conductance was measured. The spacing between the tips of the wires was set to about 8.0 Å using calibrated piezo voltage measurements, in agreement with the calculated molecule length of 8.46 Å. That the conductance of a single molecule was measured, as illustrated in Figure 5.2, was supported by the experimental data. A large body of theoretical data exists on the subject that has recently been reviewed.[27]

In 1999 large ON-OFF ratios and negative differential resistance (NDR) were measured in molecular electronic devices constructed using functionalized OPEs and a nanopore test bed.[28]

The nanopore testbed, shown in Figure 5.3, was constructed by etching, via e-beam, a small hole 30 to 50 nm in diameter in a silicon nitride membrane. The conditions of the etch were such that a bowl-shaped geometry was produced, with the hole at the bottom of the bowl. The bowl was then filled with evaporated Au and the device was placed in a solution of the functionalized OPE **1**. After allowing the SAM to form under basic conditions for 48 h, the device was removed from the solution, quickly rinsed, and placed on a liquid nitrogen cooling stage for the deposition of the bottom Au electrode via evaporation. The device was then diced into individual chips that were bonded onto packaging sockets. The electrical characteristics of the packaged testbeds were measured in a variable temperature cryostat using a semiconductor parameter analyzer.

Figure 5.4 shows the NDR peak measured in a nanopore test bed device containing a SAM of **1** at 60 K. Note that at about 1.75 V the SAM becomes conductive to a peak of 1.08 nA at about 2.1 V. The conductance then sharply drops to about 1 pA at 2.2 V. The SAM therefore acted as an electrical switch, turning ON then OFF depending on the applied voltage. The peak-to-valley ratio (PVR) was about 1000:1.

A SAM of **1** in a two-terminal cell provided electronically programmable and erasable memory with long bit retention times.[29]

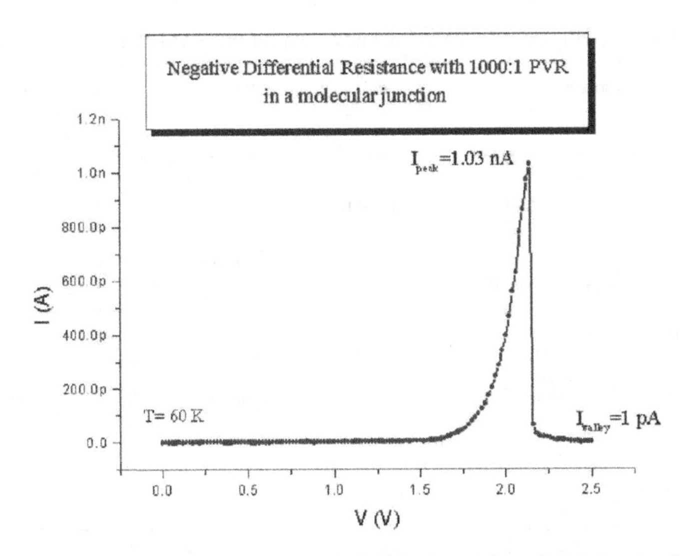

FIGURE 5.4. $I(V)$ Characteristics of a nanopore test bed device containing a SAM of molecule **1** at 60 K. The peak current density is $\sim$50 A/cm^2 and the peak-to-valley ratio of the NDR response is 1030:1.

5.3. RECENT WORK IN SELF-ASSEMBLED MOLECULAR ELECTRONICS

The years 2000–2003 have seen a great deal of progress made in the field of molecular electronics. *Science Magazine* named molecular electronics the "breakthrough of the year" in 2001.[30] In the following sections we will review the more recent work.

5.3.1. Organic Chemistry of Molecular Switches is Refined

The beauty of organic chemistry is that practitioners can easily make hundreds of different molecules by changing the structure of one partner in a multi-step synthesis. The problem is that the pace of synthesis can easily overrun the ability of the device engineers to test the molecules in the appropriate testbeds. Nanopores are not easy to fabricate and the yield of working devices is not very high. The development of testbeds that can be fabricated reproducibly in high yields is work that in ongoing in many labs.

We have synthesized molecules with regioisomeric variations in the placement of the nitro group and with quinone functionality,[31] heterocyclic and porphyrin-based molecules,[32] pyridine-based molecules,[33] molecules with diazonium alligator clips,[34] biphenyl- and fluorenyl-based molecules,[35] nitrile terminated molecules,[36] previously-published molecules with improved syntheses,[37] and combinatorial methods of synthesizing tens to hundreds of molecule types at a time.[38]

5.3.2. Physical and Electrical Characterization of SAMs

We recently synthesized a series of oligo(phenylene vinylene) (OPV) molecules, shown in Figure 5.5.[39] Using known assembly procedures[40] SAMs were made of compounds **2** and **3** to assess their efficacy in surface adhesion. The thioacetate is easily deprotected to

TABLE 5.1. The ellipsometry SAM thickness data for the SAMs formed from **2** and **3**.

compound	solvent[a]	acid	base	thickness (nm)	
				found[b]	calc[c]
2	THF		NH_4OH	2.2	2.05
	CH_2Cl_2/MeOH	H_2SO_4		2.03	2.05
3	THF		NH_4OH	1.3	2.05
	CH_2Cl_2/MeOH	H_2SO_4		3.2	2.05

All assembly times are 24 h.
[a]The ratio of mixed solvents is 2:1.
[b]The value measured by ellipsometry.
[c]The theoretical thickness calculated without the consideration of the tilt angle in the SAM, i.e. the Au-S-Ar bond angle is assumed to be 180°.

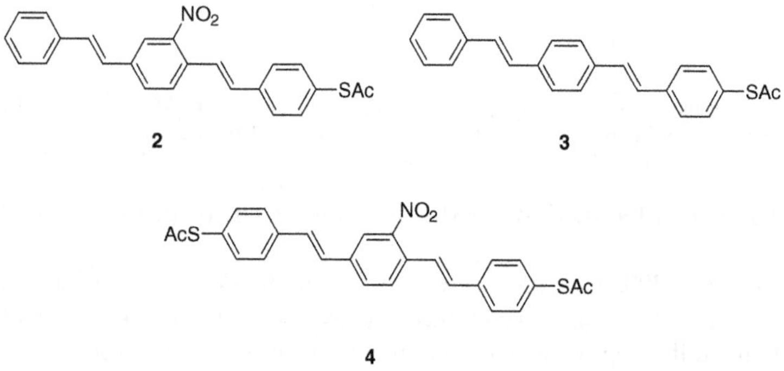

FIGURE 5.5. Three oligo(phenylene vinylene) molecules 2–4 synthesized as potential molecular electronic components.

the free thiol or thiolate by deacylation with H_2SO_4 or NH_4OH, respectively, and then the molecules were allowed to assemble on a gold surface via formation of Au-S-Ar bonds.

Table 5.1 shows the results of the chemical assembly of compounds **2** and **3** using the two different deprotection techniques. As shown in the Table, under both basic and acid catalyzed conditions, mononitro compound **2** formed SAMs that are consistent with its theoretical calculated thickness. On the other hand, SAMs of compound **3** formed thinner layers under basis conditions and thicker layers under acid conditions when compared to the theoretical values. The thinner layer might be due to the formation of bond angles smaller than 180° or to incomplete SAM formation; the thicker layer is likely a multi-layer.

Dithiol-containing non-functionalized and functionalized OPEs **5** and **6** (Figure 5.6) have been inserted into dodecanethiol SAMs for conducting atomic force microscopy (cAFM) measurements.[41] An epitaxial Au{111} surface was covered with a SAM of dodecanethiol. The SAM was exposed to a solution of either **5** or **6** in THF in the presence of NH_4OH (for base-catalyzed removal of the acetyl protection groups) for 30 min to allow the molecules to insert into the naturally occurring defect sites of the SAM. Scanning tunneling microscopy (STM) was used to image the SAM before and after the insertion. In the images obtained after insertion, the inserted dithiol OPEs appeared brighter than the surrounding, less conductive dodecanethiol-containing regions. The dithiol was used so that one end of

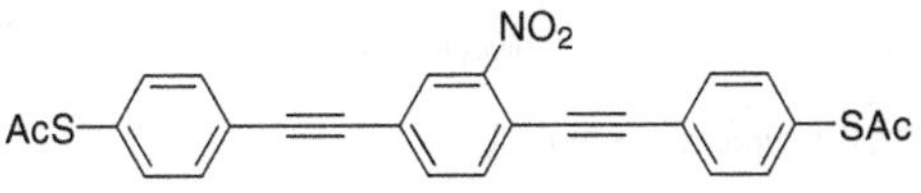

FIGURE 5.6. The dithiol-containing OPEs **5** and **6** that were inserted into dodecanethiol SAMs. Au nanoparticles were attached to the top unreacted thiol terminus of the molecule and a Au-coated AFM probe was used to make contact with the molecule and measure its conductance.

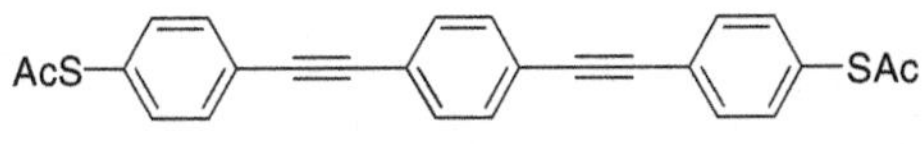

the molecule would bond with the Au surface while the other end would project up above the SAM, available for reaction with Au nanoparticles.

The SAM containing the inserted OPEs was rinsed and placed in a THF solution of Au nanoparticles for 12 h. After a final rinse, the assemblies were evaluated by STM and cAFM. The STM analysis indicated that the density of the Au nanoparticles was about the same as the density of the inserted molecules. Using cAFM, NDR peaks were observed for the molecule **5** with a PVR of 2:1, but repeated scanning caused a degradation of the effect, possibly due to the large current flowing through the molecules in the presence of atmospheric oxygen.

Conducting probe atomic force microscopy (CP-AFM), in which a SAM is formed on a Au metal surface, and a Au coated AFM tip is used to contact the top of the SAM (a different procedure than that above) has been used to measure conductance of alkanethiolate monolayers.[42] No NDR effects were seen.

A series of SAMs formed on Au from mono- and dithiol conjugated aromatic molecules was characterized by cyclic voltammetry, grazing incidence Fourier transform infrared spectroscopy, contact angle measurement, and ellipsometry.[43] The analyses indicated that the molecular orientation of conjugated phenylene- and thophene-based dithiols became less tilted with respect to the surface normal as the chain length of the organic molecules increased.

By studying single and bundles of OPE molecules isolated in alkanethiolate SAMs by STM over long periods of time (Figure 5.7), the height and conductance data indicated that switching of molecules ON and OFF was related to tilting in the OPE molecules.[44] The tighter packed the SAM, the less the molecules switched ON and OFF, indicating that close crystal packing in the SAM prevented the tilting necessary to produce the switching activities. The tilting was assessed through height differences between the ON and OFF molecules; both protruded through the top of the SAM but the ON molecule tended to be higher than the OFF molecule.

The conductance of SAMs produced from an alkanethiol and various aromatic and OPE thiols was measured using a tuning fork-based scanning probe microscope (SPM).[45] The experimental set-up is illustrated in Figure 5.8. NDR effects were seen in some of the SAMs from the OPEs while the alkanethiolate SAM did not exhibit NDR behavior.

The conductance of SAMs formed from an alkanedithiol, a OPE dithiol, and an OPV dithiol was measured using a crossed-wire experimental apparatus, in which a SAM of the

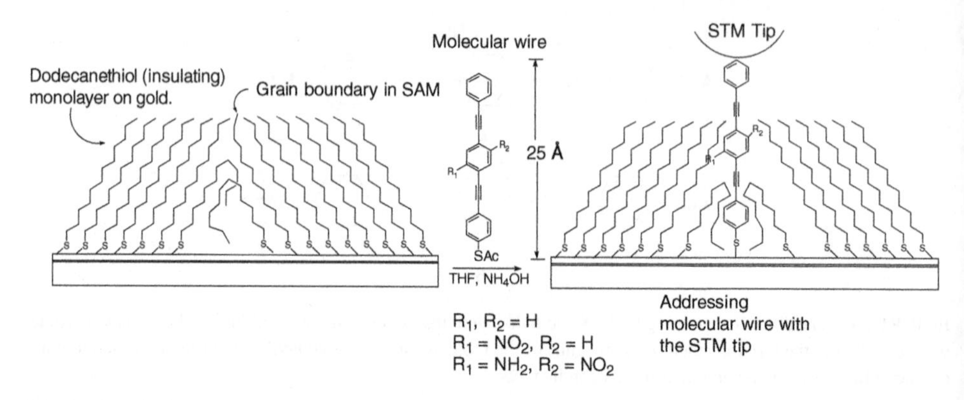

FIGURE 5.7. OPE molecules were inserted into the defect sites of a dodecanethiol SAM on Au, and the molecules were imaged by STM to determine height and conductance.

molecule of interest were formed on one 10 μm Au wire, and another 10 μm Au wire was brought in contact with the first wire by passing a dc deflection current through it.[46] The alkanedithiol conducted little current; the OPV was more conductive than the OPE. The difference between the OPV and the OPE was attributed to the former's higher planarity and to smaller bond-length alternation.

5.3.3. Langmuir-Blodgett Monolayers in Molecular Electronic Devices

Besides self-assembly on the surface of metals, another example of self-assembly is the formation of Langmuir-Blodgett monolayers on the surface or at the interface of liquids.[47,48] One approach to molecular-based electronic devices that has made use of self-assembled Langmuir-Blodgett films is the crossbar-based defect-tolerant approach to molecular computing.[49] Rotaxane molecules that depend on physical dislocations within the molecule to produce switching functionality have been designed such that they will

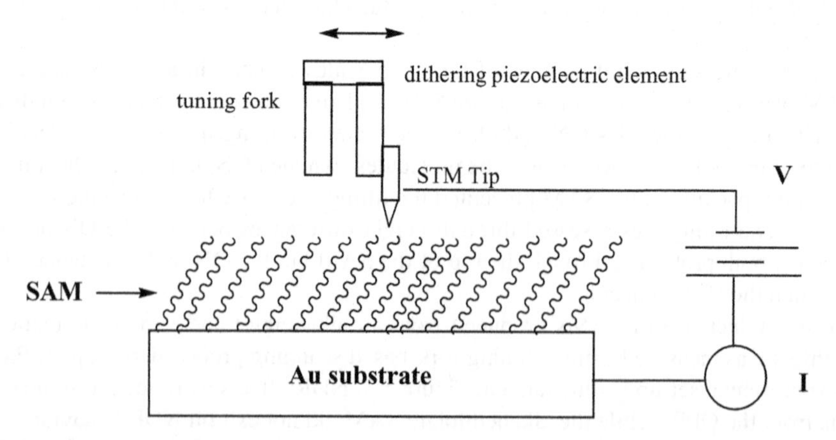

FIGURE 5.8. The experimental setup for measuring the conductance of SAMs using a tuning fork-based scanning probe microscope (SPM).

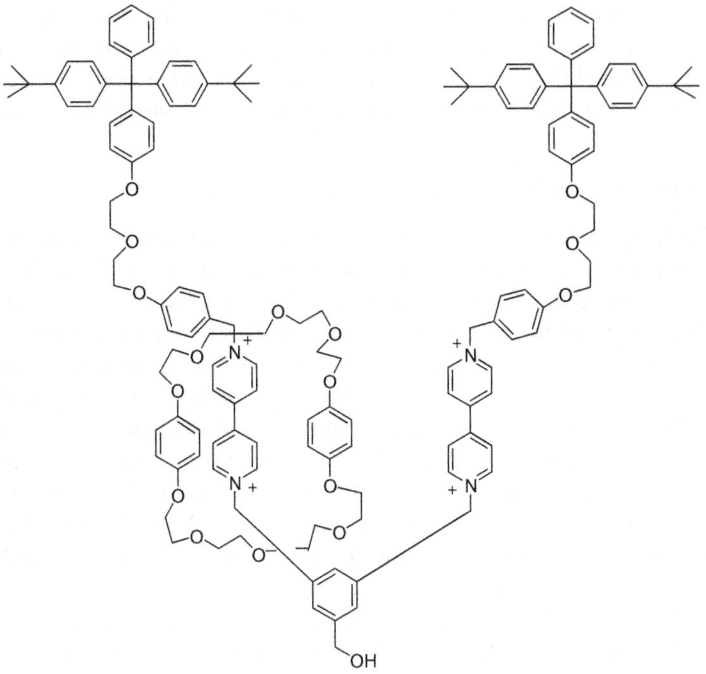

FIGURE 5.9. A rotaxane molecule used as a component of a Langmuir-Blodgett film in the crossbar-based defect-tolerant approach to molecular computing.

assemble into Langmuir-Blodgett films.[50] The films are placed between junctions of nano-sized crossbars where the assembled molecules are thought to act as switches.

One rotaxane molecule is shown in Figure 5.9. This molecule is not expected to demonstrate bistable switching, but is a candidate for a molecular switch that can be configured one time. Note that both ends of the molecule are non-polar hydrocarbon moieties while the center of the molecule is polar. Molecules that have both a non-polar and a polar end, for easier formation of Langmuir-Blodgett films, have been synthesized and tested in devices.[51]

However, the successful use of processes dependent on Langmuir-Blodgett films for semiconductor applications is in doubt due to the presence of water and other detrimental contaminants in the films produced. The internal dislocations required for switching by the molecules may not be facile since the molecules are expected to be packed into semi-crystalline phases in the Langmuir-Blodgett films. In addition, new developments in the field may obviate the need for such films as part of the device.

Single-walled carbon nanotubes (SWNTs) had been considered for the crossbar components of the defect-tolerant molecular computers but they have been found to be too difficult to handle due to their insolubility and their tendency to form bundles or ropes. Instead, metallic nanowires have become the materials of choice used in the construction of the crossbar devices, with ultrahigh-density lattices and circuits being built, having groups of nanowires 8 nm in diameter and 16 nm apart in layers perpendicular to each other to create nanowire junction densities of 10^{11} per cm^2.[52] The process does not depend on self-assembly but rather on molecular beam epitaxy.

It is probable that rather than using self-assembly to add any active molecules to the device architecture, the nanowires used to construct the devices will be precoated with active molecules via spin coating as shown by Lieber in his recent work.[53] Specially doped silicon nanowires have also shown interesting properties in crossbar-based devices.[54]

5.3.4. The Chemistry of Self-Assembled Monolayers

Once the SAM has been formed on the surface of a metal, it is possible to modify it by adding different atoms or molecules, or by adding additional layers. All of these activities are expected to change the electrical and physical characteristics of the SAM, with the intention of leading to useful molecular electronic devices or other materials.

Ozonation of single-walled carbon nanotubes (O_3-SWNTs) produces oxygenated functional groups, e.g., carboxylic acid, ester and quinone moieties.[55] The electrical resistance of the O_3-SWNTs depends on the degree of oxidation and is about 20–2000 times higher than that of pristine SWNTs owing to the deformation of the π-conjugation structure along the tube. The O_3-SWNTs are easily dispersed in dimethylformamide (DMF) and show enhanced solubility in other polar solvents like water and ethanol. The O_3-SWNT suspensions consist of individual and shortened tubes due to the separation of nanotube bundles and the shortening that results from oxidation.

Towards fabrication of SWNT-based molecular electronic devices, two methods have been used to assemble the O_3-SWNTs on functionalized SAMs of OPEs, as shown in Figure 5.10. The first, termed "chemical assembly", is based on a condensation reaction between the carboxylic acid functionalities of O_3-SWNTs and the amine functionalities of SAMs to form amides. The results show that O_3-SWNTs coat the amino-terminated SAM with a high degree of surface coverage. The second method is based on physical adsorption via layer-by-layer (LBL) deposition with bridging of metal cations, i.e., Fe^{3+} on carboxylate terminated SAMs or Cu^{2+} on thiol-terminated SAMs. The oxidatively shortened O_3-SWNTs are shown to be perpendicular to the surface with random adsorption of longer tubes. The patterned nanotube assemblies may be useful in hybridized electronic devices, where device functions can be modified by the orientation and stacking of SWNTs, and the properties of the SAM.

Complexes of Ru^{II} can be reversibly oxidized to Ru^{III} analogues both chemically and electrochemically. This oxidation is accompanied by a striking change in their optical properties. Ru^{II} complexes have absorption maxima in the 580–640 nm region associated with their strong, low energy metal-to-ligand charge transfer (MLCT) electronic transitions. Ru^{III} analogues are transparent in the whole visible region with little second-order optical nonlinearity. A three-step process led to novel SAMs of redox-switch dipolar ruthenium (III/II) pentaamine(4,4′-bipyridinium) complexes.[56]

The SAMs were formed using 11-mercapto undecanoic acid on ultrathin, optically transparent Pt films that made analysis by transmission UV/Vis spectroscopy easier. The carboxylic acid was converted to the acid chloride using $SOCl_2$, and the resulting SAM of acyl chloride molecules was exposed to a solution of $[Ru^{II}(NH_3)_5\text{-}4,4'\text{-bipyridine}](PF_6)_2$ in DMF at 45° C for 20 h. The quaternized product is shown in Figure 5.11. Analysis indicated about 20% of the acid chloride moieties formed a complex. The remaining acid chloride moieties were converted back to the carboxylic acid on work-up.

Comparison of absorption spectra before and after quaternization provided unambiguous proof that the quaternization had occurred. Immersion of the SAM shown in

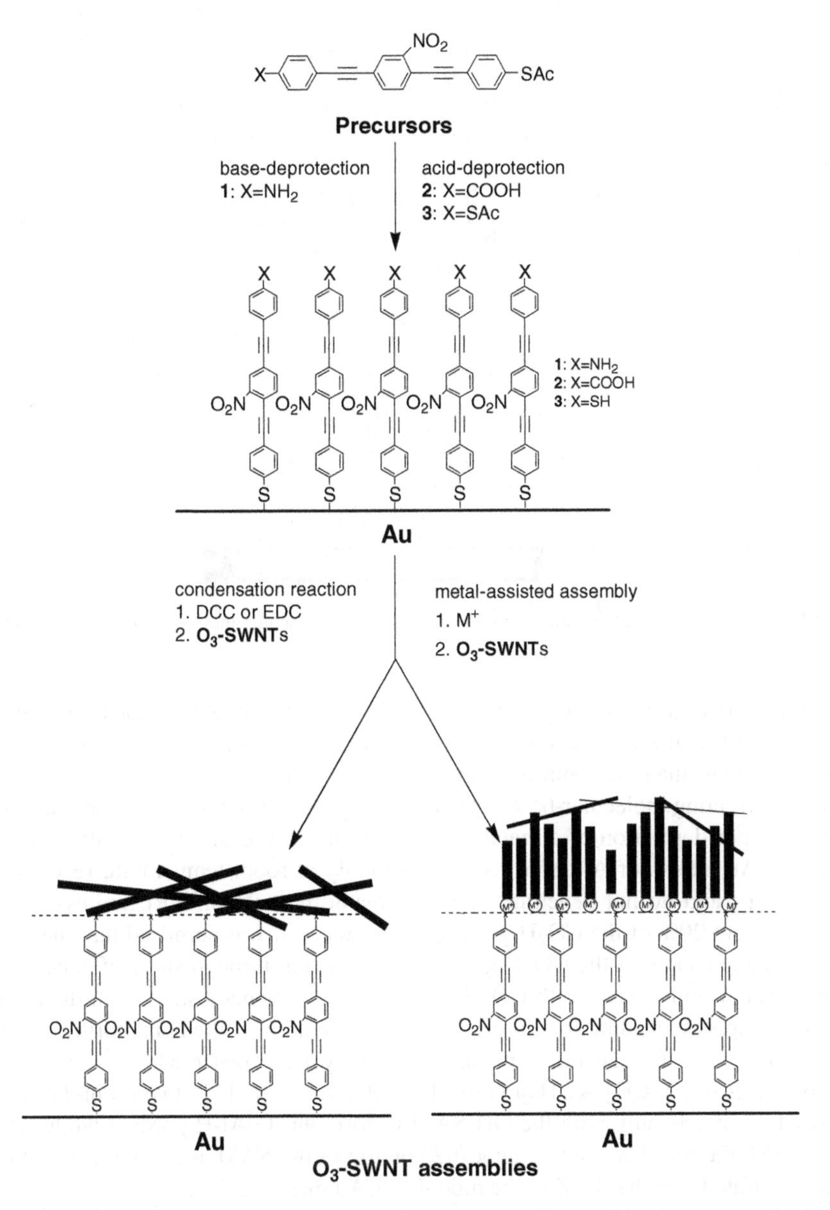

FIGURE 5.10. OPE SAMs and subsequent O_3-SWNT Assemblies. The depiction is not to scale since the O_3-SWNTs are in reality much longer than the OPEs. Note that when X = SH, the nitro group could be pointing toward the Au surface or away from the Au surface.

Figure 5.11 in oxidative solutions of $Ce_4(SO_4)_3$ or reducing solutions of H_2NNH_2 produced a cycle of complete disappearance/restoration of the MLCT optical absorption band, thereby establishing that the SAM could be used as a reversible molecular switch.

A different approach to placing a Ru redox site in a SAM was to first synthesize a Ru complex with an alkanethiol tail. The product was diluted with underivatized

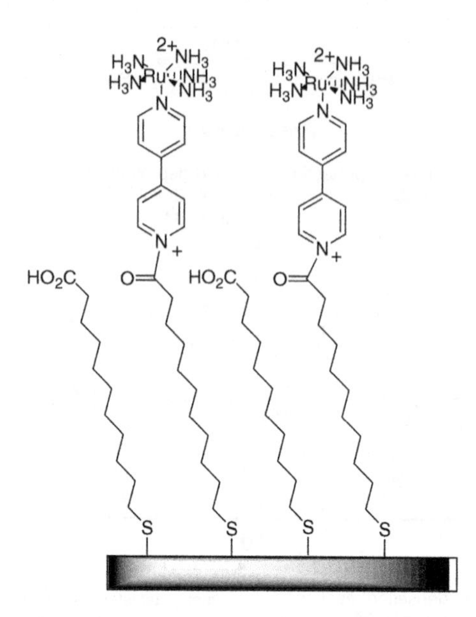

FIGURE 5.11. The Ru^{II} complex formed in three synthetic steps on an ultrathin, optically transparent Pt layer.

alkanethiol and the mixture was used to form a SAM in which the Ru complex was embedded among unfunctionalized molecules.[57] This sequence gave a redox center that was thought to be more robust than the commonly used ferrocene redox probe.[58]

In constructing molecular-based electronic devices, it is common to apply a top electrode via vapor deposition of a metal such as Au or Al. For SAMs of $HS(CH_2)_{15}CO_2H$ formed on Au, atoms of Al deposited incrementally at room temperature react with the CO_2H groups and do not penetrate into the organic monolayer.[59] Regardless of the Al coverage, about 20% of the CO_2H groups did not react. It was surmised that the growing Al film shielded some of the CO_2H groups from reaction through steric effects. It is also possible that Al only reacts with CO_2H groups in the H-bonded state, with disruption of the H-bonding state leading to a larger energy of activation at the neighboring group. No reaction of Al with the CH_2 backbone of the monolayer was observed.

For SAMs of $HS(CH_2)_{16}OH$ or $HS(CH_2)_{16}OCH_3$ on Au{111}, it was found that vapor-deposited Al inserts only with the OH SAM to form an -O-Al-H product while with the OCH_3 SAM, the metal was either deposited on top of the SAM as an overlayer, or the Al atoms penetrated into the SAM to the monolayer/Au interface.[60]

Bombardment of a hexadecanethiol SAM on Au or Ag by 800 eV He^+ ions first produces local disordering followed by removal and/or scission of molecules of the SAM that remain upright.[61] Loss of H_2 leads to formation of olefins, and the hydrocarbon chains continue to fragment until the sub-layer of strongly chemisorbed S atoms is uncovered. The rate of destruction of the SAM on Au was double that of the rate of destruction of the SAM on Ag.

The formation of an alkanethiolate SAM on Au has been used to direct subsequent metallization procedures.[62] In this way, a smooth junction between adjacent Ni and Au metal layers was formed on the surface of the SiO_2.

The interactions between the thiol alligator clips of methanethiol or benzenethiol and an Au nanoparticle containing 13 atoms, Au_{13}, was simulated using DFT.[63] Calculations showed significant structural modifications occurred when the thiolate formed a bond to the Au cluster. Carrying out the calculations with the limited-size cluster allowed simplification of the problem. The type of carbon atom to which the S atom was bound did not make much difference in the bonding of the S atom to the Au_{13} cluster.

5.3.5. Self-Assembly of Quantum Islands and Quantum Dots

Quantum dots (QDs, small conducting regions less than 1 μm in size that can contain from one to a few thousand electrons and have a variety of geometries and dimensions) are a component of the quantum cellular automata (QCA) approach to molecular computing[64] while quantum islands (QIs) are wider versions of QDs that have similar characteristics. If one places four QDs in a square array to form a QCA, such that the electrons can tunnel betweens QDs but cannot leave the array, then Coulomb repulsion will force the electrons to occupy QDs on opposite corners of the array. By aligning groups of QCAs, then one could transmit potentials from one end of the line to the other, essentially doing computations.

InAs QIs have been grown on an InP{001} substrate via a self-assembly process using solid source molecular-beam epitaxy (SSMBE) in the Stranski-Krastonov growth mode.[65] A high growth temperature and low As pressure and growth rate gave the narrowest QI size distribution. Photoluminescence and photoluminescence excitation spectroscopy indicated good confinement and/or spatial localization of the carriers. Similarly, GaInNAs/GaAs QDs were grown by self-assembly methods using SSMBE in a radio frequency N_2 plasma.[66] The QDs were 30 nm in size with a density on the surface of $\sim 10^{10}$ cm^{-2}.

While the use of QDs in the demonstration of QCA is a good first step in reduction to practice, the ultimate goal is to use individual molecules to hold the electrons and pass electrostatic potentials down QCA wires. We have synthesized molecules that have been shown by *ab initio* computational methods to have the capability of transferring information from one molecule to another through electrostatic potential.[67] The molecules synthesized included three-terminal molecular junctions, switches, and molecular logic gates. Unfortunately, the methods for depositing individual molecules at precisely defined positions on a QCA circuit have not been developed. Additionally, the degradation of only one molecule in the QCA array can cause failure of the entire circuit. Research continues in the development of QCA circuits using molecules. In the mean time, QDs have been found to be useful as tags in biological systems.[68]

5.3.6. Self-Assembled Optoelectronics

Multilayer lattices of self-assembled monolayers derived from the molecule shown in Figure 5.12, sandwiched between polysiloxane layers formed from octachlorotrisiloxane, have been produced for organic light emitting diode (OLED) and electrooptical (EO) modulator applications.[69] OLEDs can be used in displays such as instrument and computer screens, and the use of EO modulaters could lead to increased date transfer rates, resulting in better network capacity, speed, and bandwidth in data networking and telecommunications.[70]

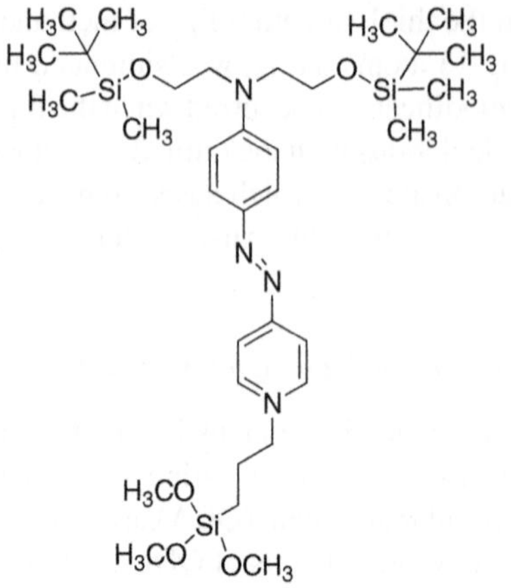

FIGURE 5.12. Iterative deposition of this molecule leads to the construction of optically active superlattices.

The thermally stable superlattices have been built on glass, silicon, and indium tin oxide-coated glass substrates. They exhibit very large EO responses with a second order non-linear optical susceptibility, $\chi^{(2)}$, ~220 pm/V (the known stilbazolium-based multilayers exhibit $\chi^{(2)} = 150\text{–}200$ pm/V but are laborious to produce). In the present case, the chromophore layer is deposited on the surface via reaction of the trimethoxylsilane-end group with reactive hydroxyl sites. Treatment of the SAM with NH_4F removes at least one of the *tert*-butyldimethylsilyl (TBDMS) protecting groups from each molecule but does not affect the Si-O bond on the surface. Analytical data indicated that about 50% of the TBDMS groups were removed. It is likely that the two TBDMS groups on each molecule are not in the same environment, i.e. one may be on the surface of the monolayer while the other may be buried in the monolayer, preventing its removal by the deprotection reagent. The SAM was then capped by depositing a polysiloxane layer on top. The SAM formation and capping process was repeated five more times to produce a six bi-layer lattice of chromophores.

OLEDs using small organic molecules are typically constructed of complex multilayed structures of hole transport layers (HTLs), emissive layers (EMLs), and electron transport layers (ETLs), all thought to be essential for high luminescence and quantum efficiency. Recent work has shown that the HTL component might not be necessary.[71] SAMs were formed using the saturated *n*-butylsiloxane shown in Figure 5.13 instead of the typical

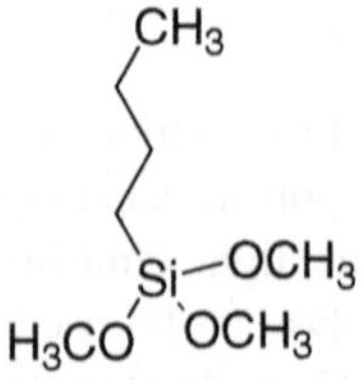

FIGURE 5.13. The *n*-butylsiloxane molecule used to form SAMs of an OLED device that did not require a hole transport layer (HTL).

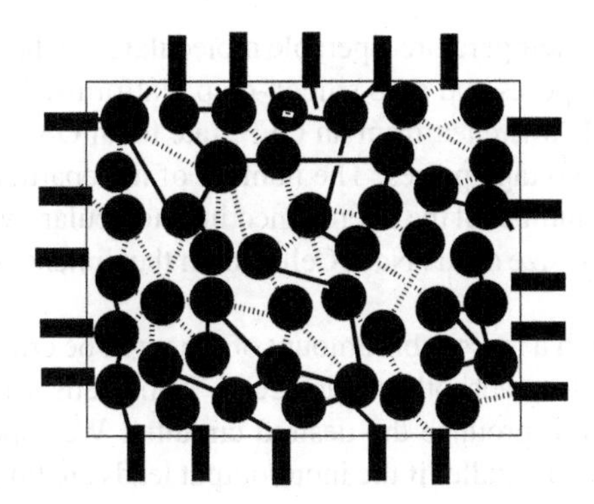

FIGURE 5.14. An illustration of the NanoCell concept for building a molecular-based computer.

triarylamine HTL component. The resulting devices exhibited the same or greater efficiency and luminance compared to compositions containing the HTL components.

5.3.7. Self-Assembly in the NanoCell Approach to Molecular Computing

For the last two years our lab has focused on the NanoCell approach to molecular computing. We have shown via computer simulation that it is possible to program an assembled NanoCell.[72] The NanoCell is illustrated in Figure 5.14.

This NanoCell architecture involves an approach where molecular switches are not specifically directed to a precise location and the internal topology is generally disordered. A NanoCell is a two-dimensional (three-dimensional models could also be considered) network of self-assembled metallic particles or islands connected by molecules that show reprogrammable (can be turned ON or OFF) NDR (or other switching and/or memory properties although these are not initially addressed). The NanoCell is surrounded by a small number of lithographically defined access leads at the edges of the NanoCell. Unlike typical chip fabrication, the NanoCell is not constructed as a specific logic gate and the internal topology is, for the most part, disordered. Logic is created in the NanoCell by training it post-fabrication, similar in some respects to a field-programmable gate array (FPGA). Even if this process is only a few percent efficient in the use molecular devices, very high logic densities will be possible. Moreover, the NanoCell has the potential to be reprogrammed throughout a computational process via changes in the ON and OFF states of the molecules, thereby creating a real-time dynamic reconfigurable hard-wired logic. The CPU of the computer would be comprised of arrays of NanoCells wherein each NanoCell would have the functionality of many transistors working in concert. A regular array of NanoCells is assumed to manage complexity, and ultimately, a few NanoCells, once programmed, should be capable of programming their neighboring NanoCells through bootstrapping heuristics. Alternatively, arrays could be programmed one NanoCell at a time via an underlying CMOS platform.

Once the physical topology of the self-assembly is formed in the NanoCell, it remains static; there is no molecule or nanoparticle dynamic character (other than bond rotations or vibrations) to the highly crosslinked network. The only changeable behavior is in the molecular states: conducting ON or non-conducting OFF, as set by voltage pulses from the periphery of the cell, or as defined by the search algorithms in these simulations.

Several types of room temperature-operable molecular switches have been synthesized and demonstrated in nanopores and atop silicon-chip platforms. The functional molecular switches can be reversibly switched from an OFF state to an ON state, and/or the reverse, based on stimuli such as voltage pulses. The number of nanoparticles (usually metallic or semiconducting) and the number of the interconnecting molecular switches can vary dramatically based on the chosen size of the NanoCell and on the dimensions of the nanoparticles and molecules chosen.[10]

Training a NanoCell in a reasonable amount of time will be critical. Eventually, trained NanoCells will be used to teach other NanoCells. NanoCells will be tiled together on traditional silicon wafers to produce the desired circuitry. We expect to be able to make future NanoCells 0.1 μm^2 or smaller if the input/output leads are limited in number, i.e. one on each side of a square.

In our initial approach to a molecular computer based on the NanoCell, a small 1 μm^2 feature is etched into the surface of a silicon wafer. Using standard lithography techniques, 10 to 20 Au electrodes are formed around the edges of the NanoCell. The Au leads are exposed only as they protrude into the NanoCell's core; all other gold surfaces are nitride-coated. The silicon surface at the center of the NanoCell (the molehole—the location of "moleware" assembly) is functionalized with $HS(CH_2)_3SiO_x$. A two dimensional array of Au nanoparticles, about 30–60 nm in diameter, is deposited onto the thiol groups in the molehole. The Au leads (initially protected by alkane thiols) are then deprotected using UV/O_3 and the molecular switches are deposited from solution into the molehole, where they insert themselves between the Au nanoparticles and link the Au nanoparticles around the perimeter with the Au electrodes.

We have had problems with this approach to the NanoCell due to troubles getting the nanoparticles anchored in the molehole. In an alternative process, we fabricated a two-dimensional unit of juxtaposed Au electrodes atop a Si/SiO_2 substrate. A discontinuous Au film was vapor deposited onto the SiO_2 in the central region. Electrical measurements confirmed the absence of DC conduction paths across the discontinuous Au film between the pairs of $\sim$5 μm-spaced electrodes ($<$1 picoamp up to 30 V). Each pair can serve as an independent memory bit address system.

The assembly of OPE molecules and Au nanowires (instead of nanoparticles) in the central portion of the NanoCell was then carried out, under N_2, to provide a current pathway across the NanoCell. The OPE used is shown in Figure 5.15. The nanowires were first covered with SAMs of the OPE, with the thiol end bonding to the surface of the nanowires while the protected thioacetyl group projected away from the nanowire. After covering the nanowires with the OPE, the acetyl protecting groups were removed with dilute NH_4OH, followed by exposing a chip containing 10 NanoCells to the solution of coated nanowires. The coated nanowires then assembled on the discontinuous Au film, producing a completed circuit. Remarkably, the NanoCells thus prepared exhibit reproducible room temperature NDR switching behavior with excellent PVRs, peak currents in the milliamp range and

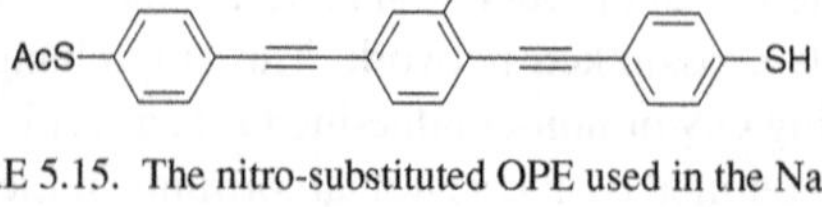

FIGURE 5.15. The nitro-substituted OPE used in the NanoCell.

reprogrammable memory states that are stable for more than a week with substantial 0:1 bit level ratios.[73] However, the conduction and switching pathway might be Au filament in nature rather than molecular electronic.[74,75] Work continues in our lab to differentiate between the two mechanisms.

5.4. CONCLUSION

The use of self-assembly techniques in molecular electronics has proven to be useful, as shown by the many publications cited. We expect the field to continue to develop and mature as researchers fine-tune their procedures and new methods are developed. Processes refined for the molecular electronics field will find applications in other nanotechnology areas; the reverse will also be true. Thus, as it will be beneficial for those in the solid-state microelectronics field to look toward molecular electronics for solutions to their problems, it will also be beneficial for those in the field of self-assembled molecular electronics to look outside that narrow range of technology for potential solutions to their problems. The coming years will surely see many exciting developments.

ACKNOWLEDGMENT

The authors gratefully acknowledge our collaborators and colleagues on the cited papers that originated from our labs; we also acknowledge funding from DARPA, ONR, and ARO.

REFERENCES

1. G. E. Moore, Cramming more components onto integrated circuits, *Electronics* **38**(8) 114–117 (1965).
2. J. M. Tour, Molecular electronics. Synthesis and testing of components, *Acc. Chem. Res.* **33**, 791–804 (2000).
3. M. A. Reed and J. M. Tour, The birth of molecular electronics, *Sci. Am.* **June**, 69–75 (2000).
4. J. M. Tour, D. K. James, Molecular electronic computing architectures, in *Handbook of Nanoscience, Engineering, and Technology*, edited by W. A. Goddard, D. W. Brenner, S. E. Lyshevski, and G. J. Iafrate (CRC Press, New York, 2002), 4-1–4-28.
5. L. Peters, Industry confronts sub-100 nm challenges, *Semiconductor Int.* **January**, 42–48 (2003).
6. IBM claims world's smallest silicon transistor, *Electronic News*, 9 December, 2002.
7. Intel to convert Arizona factory to 300 mm technology, Intel press release, 18 Feb. 2003, accessed on the web 11 April 2003; http://www.intel.com/pressroom/archive/releases/20030218corp_a.htm.
8. G. M. Whitesides, Self-assembling materials, *Sci. American* **273**(3) 146 (1995).
9. G. M. Whitesides, J. P. Mathias, and C. T. Seto, Molecular self-assembly and nanochemistry: a chemical strategy for the synthesis of nanostructures, *Science* **254**, 11312–1319 (1991).
10. J. M. Tour, *Molecular Electronics: Commercial Insights, Chemistry, Devices, Architecture and Programming* (World Scientific, New Jersey, 2003).
11. H. H. Jian and J. M. Tour, En route to surface-bound electric field-driven molecular motors, *J. Org. Chem.* **68**, 5091–5103 (2003).
12. J. Chen, M. A. Reed, S. M. Dirk, D. W. Price, A. M. Rawlett, J. M. Tour, D. S. Grubisha, and D. W. Bennett, Molecular Electronic Devices, in *Molecular Electronics: Bio-sensors and Bio-computers*, (Kluwer, Netherlands, 2003) pp. 59–195.
13. R. L. Carroll and C. B. Gorman, The genesis of molecular electronics, *Angew. Chem. Int. Ed.* **41**, 4378–4400 (2002).

14. Intentional Technology Roadmap for Semiconductors web pages, http://public.itrs.net/Files/2002Update/2002Update.pdf (accessed April 2003).
15. O. Tairo, Primitive protein models, *Seibutsu Butsuri* **9**, 101–109 (1969).
16. K. M. Ulmer, Biological assembly of molecular ultracircuits, *NRL Memo. Rep.*, 167–177 (1981).
17. J. S. Lindsey, Self-assembly in synthetic routes to molecular devices. Biological principles and chemical perspectives: a review, *New J. Chem.* **15**, 153–180 (1991).
18. J. M. Tour, L. Jones, D. L. Pearson, J. J. S. Lamba, T. P. Burgin, G. M. Whiteseides, D. L. Allara, A. N. Pariky, and S. V. Atre, Self-assembled Monolayers and multilayers of conjugated thiols, α,ω-tithiols, and thioacetyl-containing adsorbates–understanding attachments between potential molecular wires and gold surfaces, *J. Am. Chem. Soc.* **117**(37), 9529–9534 (1995).
19. L. Jones, II, J. S. Schumm, and J. M. Tour, Rapid solution and solid phase syntheses of oligo(1,4-phenylene ethynylene)s with thioester termini: molecular scale wires with alligator clips. Derivation of iterative reaction efficiencies on a polymer support, *J. Org. Chem.* **62**(5), 1388–1410 (1997).
20. D. L. Pearson and J. M. Tour, Rapid syntheses of oligo(2,5-thiophene ethynylene)s with thioester termini: potential molecular scale wires with alligator clips, *J. Org. Chem.* **62**(5) 1376–1387 (1997).
21. J. M. Seminario, A. G. Zacarias, and P. A. Derosa, Theoretical analysis of complementary molecular memory devices, *J. Phys. Chem. A*, **105**(5), 791–795 (2001).
22. P. A. Derosa, and J. M. Seminario, Electron transport through single molecules: scattering treatment using density functional and green function theories, *J. Phys. Chem. B* **105**, 471–481 (2001).
23. J. S. Schumm, D. L. Pearson, L., II Jones, R. Hara, and J. M. Tour, Potential molecular wires and molecular alligator clips, *Nanotechnology* **7**, 430–433 (1996).
24. J. M. Seminario, A. G. Zacarias, and J. M. Tour, Molecular alligator clips for single molecule electronics. Studies of group 16 and isonitriles interfaced with Au contacts, *J. Am. Chem. Soc.* **121**(2), 411–416 (1999).
25. L. Patrone, S. Palacin, J. P. Bourgoin, J. Laoute, T. Zambelli, and S. Gauthier, Direct comparison of the electronic coupling efficiency of sulfur and selenium anchoring groups for molecules adsorbed onto gold electrodes, *Chem. Phys.* **281**, 325–332 (2002).
26. M. A. Reed, C. Zhou, C. J. Muller, T. P. Burgin, and J. M. Tour, Conductance of a molecular junction, *Science* **278**, 252–254 (1997).
27. A. Nitzan and M. A. Ratner, Electron transport in molecular wire junctions, *Science* **300**, 1384–1389 (2003).
28. J. Chen, M. A. Reed, A. M. Rawlett, and J. M. Tour, Large on-off ratios and negative differential resistance in a molecular electronic device, *Science* **286**, 1550–1552 (1999).
29. J. Chen, W. Wang, J. Klemic, M. A. Reed, B. W. Axelrod, D. M. Kaschak, A. M. Rawlett, D. W. Price, S. M. Dirk, J. M. Tour, D. S. Grubisha, and D. W. Bennett, Molecular wires, switches, and memories, *Ann. N. Y. Acad. Sci.* **960**, 69–99 (2002).
30. R. F. Service, Molecules get wired, *Science* **294**, 2442 (2001).
31. S. M. Dirk, D. W. Price, Jr., S. Chanteau, D. V. Kosynkin, and J. M. Tour, Accoutrements of a molecular computer: switches, memory components, and alligator clips, *Tetrahedron* **57**, 5109–5121 (2001).
32. J. M. Tour, A. M. Rawlett, M. Kozaki, Y. Yao, R. C. Jagessar, S. M. Dirk, D. W.; Price, M. A. Reed, C.-W. Zhou, J. Chen, W. Wang, and I. Campbell, Synthesis and preliminary testing of molecular wires and devices, *Chem. Eur.* **5**, 5118–5134 (2001).
33. S. H. Chanteau and J. M. Tour, Synthesis of potential molecular electronic devices containing pyridine units, *Tetrahedron Lett.* **42**, 3057–3060 (2001).
34. D. V. Kosynkin and J. M. Tour. "Phenylene ethynylene diazonium salts as potential self-assembling molecular wires, *Org. Lett.* **3**, 993–995 (2001).
35. D. W. Price, Jr. and J. M. Tour, Biphenyl- and Fluorenyl-Based Potential Molecular Electronic Devices, *Tetrahedron*, **59**, 3131–3156 (2003).
36. S. M. Dirk and J. M. Tour, Synthesis of Nitrile Terminated Potential Molecular Electronic Devices, *Tetrahedron* **59**, 287–293 (2003).
37. D. W. Price, Jr., S. M. Dirk, F. Maya, and J. M. Tour, Improved and New Syntheses of Potential Molecular Electronic Devices, *Tetrahedron* **59**, 2497–2518 (2003).
38. J.-J. Hwang and J. M. Tour, Combinatorial synthesis of oligo(phenylene ethynylene)s, *Tetrahedron* **58**, 10387–10405 (2002).
39. S. M. Dirk, A. K. Flatt, and J. M. Tour, to be submitted.

40. L. Cai, Y. Yao, J. Yang, D. W. Price, Jr., and J. M. Tour, Chemical and potential assisted assembly of thioacetyl-terminated oligo(phenylene ethynylene)s on gold surfaces, *Chem. Mater.* **14**, 2905–2909 (2002).
41. A. M. Rawlett, T. J. Hopson, L. A. Nagahara, R. K. Tsui, G. K. Ramachandran, Lindsay, Electrical measurements of a dithiolated electronic molecule via S. M. conductiong atomic force microscopy, *Appl. Phys. Lett.* **81**, 30443–3045 (2002).
42. D. J. Wold and C. D. Frisbie, Fabrication and characterization of metal-molecule-metal junctions by conducting probe atomic force microscopy, *J. Am. Chem. Soc.* **123**, 5549–5556 (2001).
43. B. De Boer, H. Meng, D. F. Perepichka, J. Zheng, M. M. Frank, Y. J. Chabal, and Z. Bao, Synthesis and characterization of conjugated mono- and dithiol oliogmers and charaterization of their self-assembled monolayers, *Langmuir* **20**, 1539–1542 (2004).
44. Z. J. Donhauser, B. A. Mantooth, K. F. Kelly, L. A. Bumm, J. D. Monnell, J. J. Stapleton, D. W. Price, Jr. A. M. Rawlett, D. L. Allara, J. M. Tour, and P. S. Weiss, Conductance switching in single molecules through conformational changes, *Science* **292**, 2303–2307 (2001).
45. F.-R. F. Fan, Y. Yang, L. D. W. CaiPrice, Jr. S. M. Dirk, D. V. Kosynkin, Y. Yao, A. M. Rawlett, J. M. Tour, and A. J. Bard, Charge transport through self-assembled monolayers of compounds of interest in molecular electronics, *J. Am. Chem. Soc.* **124**, 5550–5560 (2002).
46. J. G. Kushmerick, D. B. Holt, S. K. Pollack, M. A. Ratner, J. C. Yang, T. L. Schull, J. Naciri, M. H. Moore, and R. Shashidhar, Effect of bond-length alternation in molecular wires, *J. Am. Chem. Soc.* **124**, 10654–10655 (2002).
47. M. R. Bryce and M. C. Petty, Electrically conductive Langmuir-Blodgett films of charge-transfer materials, *Nature* **374**, 771–776 (1995).
48. M. Matsumoto, H. Tachibana, and T. Nakamura, *Applied Phys.*, (4 Organic Conductors), 759–790 (1994).
49. J. R. Heath, P. J. Kuekes, G. S. Snider, and R. S. A Williams, Defect-tolerant computer architecture: opportunities for Nanotechnology, *Science* **280**, 1716–1721 (1998).
50. J. R. Heath, Wires, switches, and wiring. A route toward a chemically assembled electronic nanocomputer, *Pure Appl. Chem.* **72**, 11–20 (2000).
51. U. Luo, C. P. Collier, J. O. Jeppesen, K. A. Nielsen, E. Delonno, G. Ho, J. Perkins, H.-R. Tseng, T. Yamamoto, J. F. Stoddart, and J. R. Heath, Two-dimensional molecular electronics circuits, *Chemphyschem* **3**, 519–525 (2002).
52. N. A. Melosh, A. Boukai, F. Diana, B. Grardot, A. Badolato, P. M. Petroff, and J. R. Heath, Ultrahigh-density nanowire lattices and circuits, *Science* **300**, 112–115 (2003).
53. X. Duan, Y. Huang, and C. M. Lieber, Nonvolatile memory and programmable logic from molecule-gated nanowires, *Nano Lett.* **2**, 487–290 (2002).
54. Y. Cui, Z. Zhong, D. Wang, W. U. Wang, and C. M. Lieber, High performance silicon nanowire field effect transistors, *Nano Lett.* **3**, 149–152 (2003).
55. L. Cai, J. L. Bahr, Y. Yao, and J. M. Tour, Ozonation of single-walled carbon nanotubes and their assemblies on rigid self-assembled monolayers, *Chem. Mater.* **14**, 4235–4241 (2002).
56. S. Sortino, S. Petralia, S. Conoci, and S. Di Bella, Novel self-assembled Monolayers of dipolar ruthenium(III/II) pentaammine(4,4′-bipyridinium) complexes on ultrathin platinum films as redox molecular switches, *J. Am. Chem. Soc.* **125**, 1122–1123 (2003).
57. C. Hortholoary, F. Minc, C. Coudret, J. Bonvoisin, and J.-P. Launay, "A new redox site as an alternative to ferrocene to study electron transfer in self-assembled Monolayers" *Chem. Commun.* 1932–1933 (2002).
58. G. M. Credo, A. K. Boal, K. Das, T. H. Galow, V. M. Rotello, D. L. Feldheim, and C. B. Gorman, Supramolecular assembly on surfaces: manipulating conductance in noncovalently modified mesoscale structures, *J. Am. Chem. Soc.* **124**, 9036–9037 (2002).
59. G. L. Fisher, A. E. Hooper, R. L. Opila, D. L. Allara, and N. Winograd, The interaction of vapor-deposited Al atoms with CO2H groups at the surface of a self-assembled alkanethiolate monolayer on Au, *J. Phys. Chem. B* **104**, 3267–3273 (2000).
60. G. L. Fisher, A. V. Walker, A. E. Hooper, T. B. Tighe, K. B. Bahnck, H. T. Skriba, M. D. Reinard, B. C. Haynie, R. L. Opila, N. Winograd, and D. L. Allara, Bond insertion, complexation, and penetration pathways of vapor-deposited aluminum atoms with HO- and CH_3O-terminated organic monolayers, *J. Am. Chem. Soc.* **124**, 5528–5541 (2002).

61. S. P. Chenakin, Sputtering of organic molecular Monolayers self-assembled onto Ag(111) and Au(111) surfaces, *Vacuum*, **66**, 157–166 (2002).
62. P. K.-H. Ho, R. W. Filas, D. Abusch-Magder, and Z. Bao, Orthogonal self-aligned electroless metallization by molecular self-assembly, *Langmuir* **18**, 9625–9628 (2002).
63. J. A. Larsson, M. Nolan, and J. C. Greer, Interations between thiol molecular linkers and the Au13 nanoparticle, *J. Phys. Chem. B* **106**, 5931–5837 (2002).
64. G. L. Snider, A. O. Orlov, I. Amlani, X. Zuo, G. H. Bernstein, C. S. Lent, J. L. Merz, and W. Porod, Quantum-dot cellular automata: review and recent experiments (invited), *J. Appl. Phys.* **85**, 4283–4283 (1999).
65. B. Salem, T. Benyattou, G. Guillot, C. Bru-Chevallier, and G. Bremond, Strong carrier confinement and evidence for excited states in self-assembled InAs quantum islands grown on InP(001), *Phys. Rev. B* **66**, 193305 (2002).
66. Z. Z. Sun, S. F. Yoon, K. C. Yew, W. K. Loke, S. Z. Wang, and T. K. Ng, Self-assembled GaInNAs/GaAs quantum dots grown by solid-source molecular beam epitaxy, J. Crystal Growth **242**, 109–115 (2002).
67. J. M. Tour, M. Kozaki, and J. M. Seminario, Molecular scale electronics: a synthetic/computational approach to digital computing, *J. Am. Chem. Soc.* **120**, 8486 (1998).
68. R. Salvador, D. Mitra, and S. A. Michael, Acoustic modes in free and embedded quantum dots, *J. Appl. Phys.* **93**, 2900–2905 (2003), and references therein.
69. M. E. Van der Boom, A. G. Richter, J. E. Malinsky, P. A. Lee, N. R. Armstrong, P. Dutta, and T. J. Marks, Single reactor route to polar superlattices. Layer-by-layer self-assembly of large-response molecular electrooptic materials by protection-deprotection, *Chem. Mater.* **13**, 15–17 (2001).
70. L. R., Dalton, *et al.*, From molecules to opto-chips: organic electro-optic materials, *J. Mater. Chem.* **9**, 1905–1920 (1999).
71. Q. Huang, J. Cui, H.; Yan, J. G. C. Veinot, and T. J. Marks, Small molecule organic light-emitting diodes can exhibit high performance without conventional hole transport layers, *Appl. Phys. Lett.* **81**, 3528–3530 (2002).
72. J. M. Tour, W. L. Van Zandt, C. P Husband, S. M. Husband, L. S. Wilson, P. D. Franzon and D. P. Nackashi, Nanocell logic gates for molecular computing, *IEEE Transactions on Nanotechnology* **1**, 100–109 (2002).
73. J. M. Tour, L. Cheng, D. P. Nackashi, Y. Yao, A. K. Flatt, S. K. St. Angelo, T. E. Mallouk, and P. D. Franzon, NanoCell molecular electronic memories, *J. Am. Chem. Soc.* **123**, 13279–13283 (2003).
74. R. E. Thurstans and D. P. Oxley, The electroformed metal-insulator-metal structure: a comprehensive model, *J. Phys. D: Appl. Phys.* **35**, 802–809 (2002).
75. D. B. Neal, M. I. Newton, and G. McHale, Negative differential resistance in thin metal films with a cadmium arachidate overlayer, *Int. J. Electronics* **76**, 771–775 (1994).

6

Multivalent Ligand-Receptor Interactions on Planar Supported Membranes

An On-Chip Approach

Seung-Yong Jung, Edward T. Castellana, Matthew A. Holden, Tinglu Yang, and Paul S. Cremer

6.1. INTRODUCTION

Since their initial fabrication two decades ago by McConnell and coworkers, fluid supported phospholipid bilayers (SLBs) have played a key role in the development of nanoscale assemblies of biological materials on artificial supports.[1,2] The reason for this is quite straightforward. SLBs can serve as biomimetics for chemical and biological processes which occur in cell membranes. A thin aqueous layer (approximately 1 nm thick) is trapped between the bilayer and the underlying support (Figure 6.1). This water layer acts as a lubricant allowing both leaflets of the bilayer to remain fluid.[3–10] Consequently, planar supported membranes retain many of the physical properties of free vesicles or even native cell surfaces when the appropriate recognition components are present.[4] Specifically, SLBs are capable of undergoing lateral rearrangements to accommodate binding by aqueous proteins, viruses, toxins, and even cells.[11] As substrate supported entities, they are convenient to study by a host of interface-sensitive techniques and are far less fragile than either unsupported membranes or full-blown cellular systems.

SLBs can be formed by either Langmuir Blodgett methods or through the fusion of small unilamellar vesicles to a planar solid substrate.[2,8] Either way it is relatively straightforward to incorporate ligands into these membrane models. Indeed, any numbers of species including

Department of Chemistry, Texas A&M University, College Station, TX 77843. tel: 979 862-1200 cremer@mail.chem.edu

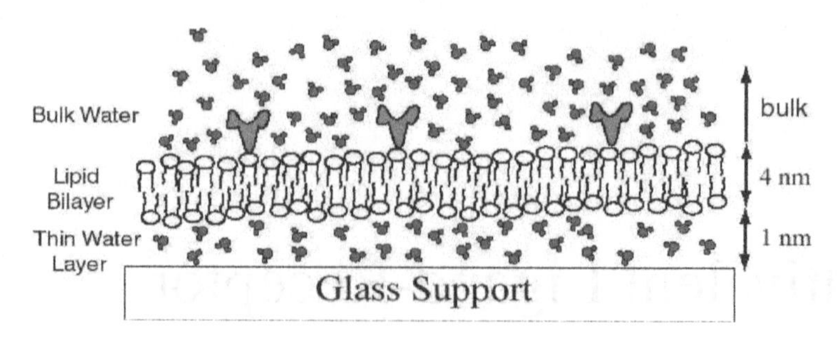

FIGURE 6.1. Schematic diagram of a supported phospholipid bilayer membrane containing a covalently attached ligand molecule.

peptides, peptide labeled lipids, biotinylated lipids, channel forming proteins, and antigenic ligands have been presented in supported bilayers.[4]

Herein we will focus on research performed in our laboratory with supported phospholipid bilayers on glass supports.[10,12–18] The work described below concerns the creation of two-dimensional membrane arrays and microfluidic devices for the investigation of multivalent ligand-receptor interactions. Probing these events at a fluid interface is fundamentally different than with ligands attached to polymer backbones or on the surface of other scaffolds[19–23] because each ligand on the bilayer is free to rearrange its position on the surface to maximize its interactions with an incoming multivalent protein, virus, bacterium, or toxin.[24–26] For example, in the case of anti-2,4 dinitrophenyl IgG antibodies interacting with a supported bilayer containing N-dinitrophenylaminocaproyl phosphatidylethanolamine (DNP-Cap PE), the overall interaction should be bivalent with both the bound IgG and ligands able to rearrange to accommodate proper binding (Figure 6.2). In this case, both binding sites are identical. The use of SLBs allows the ligand-receptor binding process to be facilely probed as a function of membrane chemistry. For example, the amount of cholesterol in the membrane, the charge on the membrane, the ligand density, the types of lipids present, as well as the presence of additional peptides or glycosylation can be varied on a single chip. Conducting such experiments is vital because cells carefully control the type of lipids and ligands present in each leaflet of each membrane. Establishing and maintaining such fine control is a significant metabolic burden to the cell.[27] Yet until now there has been only limited understanding as to the purpose of this lipid differentiation. Part of the reason for this paucity of understanding stems from the fact that testing ligand-receptor binding as a function of cell membrane chemistry is difficult due to the lack of high throughput platforms for probing multivalent interactions in a fluid membrane environment. This problem is overcome by combining SLBs with microfluidic and array based techniques. Therefore, we will first describe the formation of two-dimensional bilayer arrays and microfluidic platforms for making high throughput thermodynamic measurements. Then we demonstrate the use of these techniques for probing the effects of ligand density on bivalent antibody binding.

6.2. CREATION OF SPATIALLY ADDRESSED ARRAYS

The use of planar supports for presenting large arrays of spatially addressed molecules is one of the most powerful and versatile methods for creating combinatorial libraries.[19,28–29]

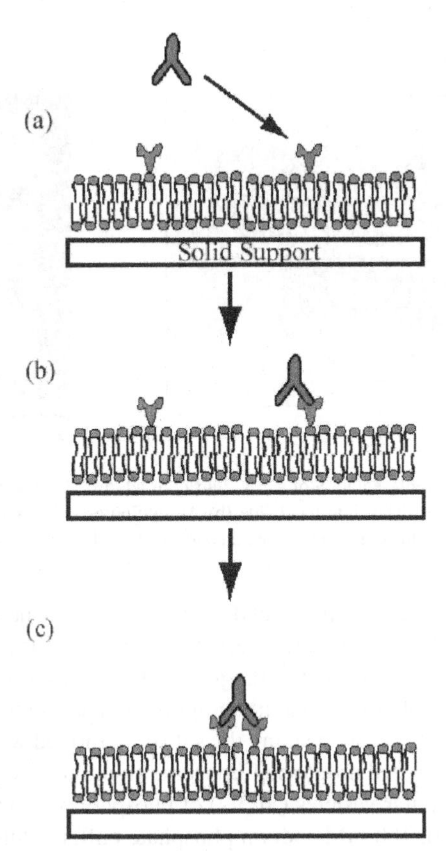

FIGURE 6.2. (a) An IgG antibody approaches a phospholipid membrane containing fluid ligands. (b) The species first binds to one ligand and (c) then diffuses laterally to bind to a second ligand.

Extending this approach to supported phospholipid bilayer membranes is an especially valuable goal because of the ability of these systems to mimic many of the properties of native cell surfaces as suggested above (Figure 6.3a).[4] Addressing biomembrane mimics on planar supports, however, presents unique challenges, as the two-dimensional fluidity of the biomembrane must be preserved in many cases for it to function properly.[30–32] The bilayer deposition process must take place in an aqueous environment and the entire system must continue to remain submerged under water to preserve the planar supported structure. Because of this physical constraint as well as the inherent complexities of biomembrane materials, traditional technologies such as light-directed synthesis for addressing peptide or DNA sequences onto solid supports are inherently difficult to apply.[19] We have, therefore, employed an alternate approach[12] based upon depositing mesoscopic quantities of aqueous solution onto lithographically patterned hydrophilic surface[33] well plates, followed by the immersion of the entire substrate into buffer. This is a general and flexible method for directing chemically distinct phospholipid membranes into individually addressable surface sectors.

Previous investigators showed that patterned surfaces allow partitioning of one fluid lipid bilayer from the next.[34–35] Molecules within an individual membrane are free to move

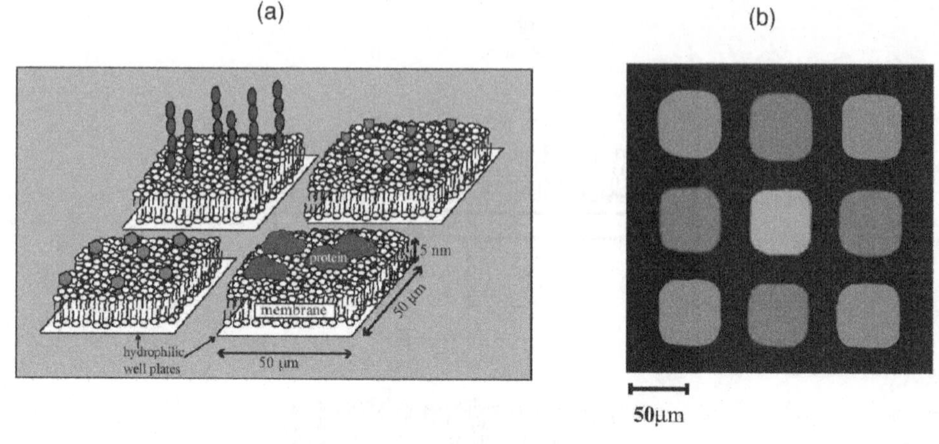

FIGURE 6.3. (a) A schematic representation of a 2×2 array of addressed supported fluid biomembranes with various components in each box. The right-hand side (b) An epifluorescence image of a 3×3 array of fluid biomembranes that have been addressed using the spatial addressing techniques described above.

within the confines of a single partition, but do not crossover to a neighboring region. In our laboratory, planar borosilicate substrates were partitioned into arrays of micrometer sized hydrophilic boxes using standard photolithography. Patterning was achieved by exposing the surface to ultraviolet light through a lithographic mask consisting of an array of square boxes. Developing the pattern and cleaning the substrate formed well plates of hydrophilic glass onto which picoliter-sized droplets of liposome solution were placed. The liposomes, which were small unilamellar vesicles (SUVs) of phospholipids, were present at 1 mg/ml concentration in a pH 7.0, 100 mM sodium phosphate buffer solution. Figure 6.3b shows the epifluorescence image of nine $50\,\mu m \times 50\,\mu m$ well plates that have been addressed with three chemically distinct types of supported phospholipid bilayers. The boxes at the middle of each side appear red in color and contain 1 mol% Texas Red DHPE fluorescent probes while the boxes at the corners appear green and contain 3 mol% NBD-DHPE probes. The center box, which appears dark yellow, contains both kinds of fluorophores. Fluorescence recovery after photobleaching (FRAP)[36–37] demonstrated that the lipids were free to move throughout each two-dimensional box, but were otherwise completely confined.

With spatially addressed membrane arrays in hand, we wished to investigate the effect of ligand density and cholesterol content on the binding of anti-2,4 dinitrophenyl IgG antibodies with DNP-Cap PE lipids in phosphatidylcholine membranes. The antibodies were labeled with Texas Red dye so that they could be visualized at the bilayer interface. Experiments were conducted on a 4×4 membrane array with cholesterol content varying from 0 to 20 mol% and DNP-Cap PE concentrations ranging from 0 to 5 mol% (Figure 6.4).[15] The results indicated that binding was dependent on both ligand density and cholesterol content.

It has been previously suggested that the addition of cholesterol to the phospholipid membranes increases the availability of the DNP ligand to antibodies in the bulk solution.[38] Significantly, we see here that the effect is much larger at lower ligand concentrations then it is at high ligand concentrations.

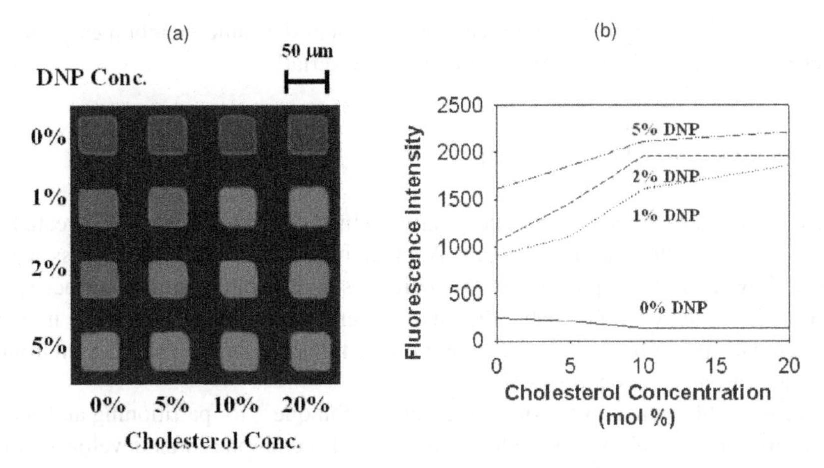

FIGURE 6.4. (a) An epifluorescence image of a 4 × 4 membrane array. (b) A quantitative plot of fluorescence intensity vs. cholesterol content at each DNP ligand concentration.

Although the techniques described above allow addressed arrays of bilayers to be formed, it is a time consuming process. This is because each bilayer has to be injected separately from a clean microcapillary tube. To improve the efficiency of this technique, the bilayer materials need to be deposited in parallel. One way to achieve this would be to employ a patterned master and then transfer the array of phospholipid vesicle chemistries to the membrane biochip simultaneously using a field of tips (Figure 6.5a). It should be noted, however, that it is still necessary to keep the vesicle solutions properly hydrated. Another important consideration is that the tips used for transfer are on the order of 1 mm apart. This creates a 10 × 10 array of membranes over approximately a one cm^2 area. To view such a large area, very low magnification objectives with reasonably high numerical apertures are needed. Typical epifluorescence microscopes are not usually equipped to handle such objectives. We therefore designed and built an epifluorescence macroscope for this purpose. The instrument was capable of taking ×1 images of the surface with an N.A. of 0.4. An example of a supported bilayer array taken under the macroscope is shown in Figure 6.5b.

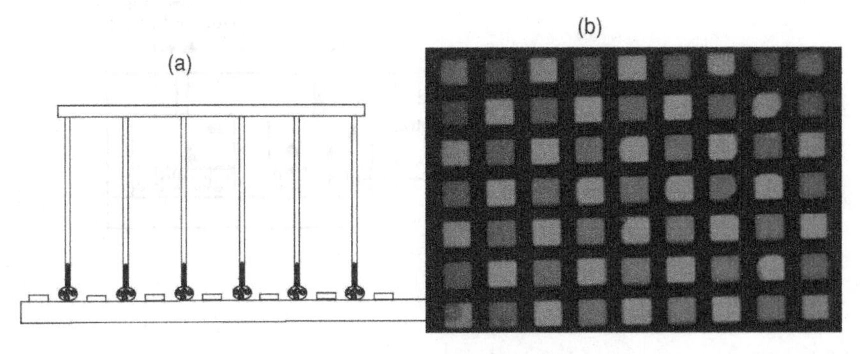

FIGURE 6.5. (a) An array of tips is used to transfer droplets of vesicle materials from a patterned master onto a biochip containing the same registry. (b) TIR image of a 7 × 9 array of lipid bilayers taken with an epifluorescence macroscope. The center to center spacing of the bilayers was 800 μm.

The boxes are alternatingly red and green and represent the same membranes chemistries as described for Figure 6.3; the deposition was done serially.

6.3. MICROCONTACT DISPLACEMENT

In the work described above, chemically distinct bilayer arrays were created with common aqueous solutions above them. Another important goal is to address aqueous solutions above an array of planar supported bilayers. In combination with surface specific detection, this strategy would enable the rapid screening of a library of soluble molecules for their efficacy in inhibiting ligand-receptor interactions in a fluid membrane environment that is similar to *in vivo* conditions.[13]

To achieve this goal a novel soft lithographic technique[39] for partitioning and addressing aqueous solutions above supported phospholipid membranes was developed (Figure 6.6).[13] This methodology affords the ability to create a large number of aqueous compartments consisting of various chemistries, pH values and ionic strengths, as well as ligand and inhibitor concentrations above patterned arrays of lipid membranes on a single planar support. Such methods may ultimately serve as screens for the efficacy of libraries of small molecule inhibitors to prevent multivalent ligand-receptor attachment to bilayer surfaces.

The basic procedure for creating addressed aqueous compartments above individual bilayers is outline in Figure 6.6. In the first step, a lithographically patterned PDMS mold is brought into direct contact with a planar supported lipid bilayer containing covalently attached ligands, to create hermetically sealed compartments by microcontact displacement

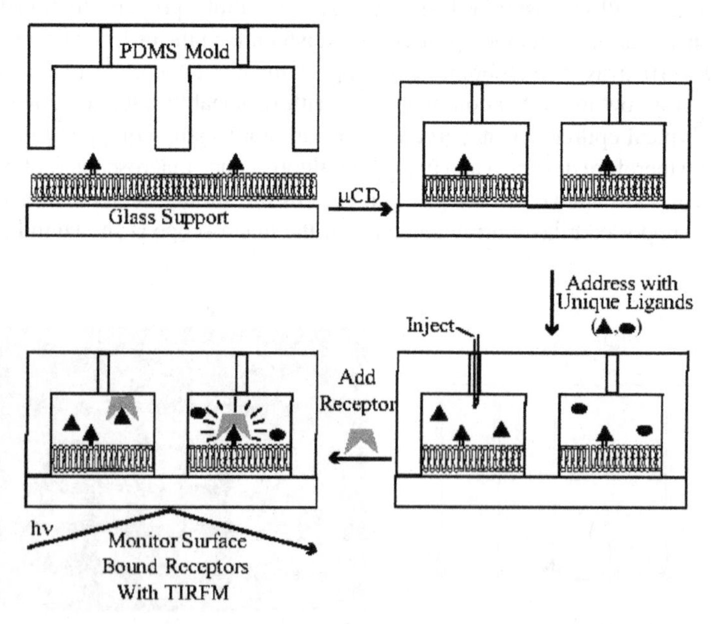

FIGURE 6.6. Schematic depiction of the screening assay for monitoring ligand-receptor interactions on a supported fluid lipid bilayer in the presence of a library of soluble inhibitors. Surface specific observation is achieved using TIRFM.

(μCD) of the phospholipid membrane from the substrate. Leaving the molds in place creates sealed and addressable aqueous compartments above the membrane array. Each compartment can be separately addressed by reaming small holes into the elastomer above a confined compartment. Specifically, circular holes were reamed with a syringe needle that were 100 μm in diameter.[13,40] Pulled microcapillary tips mounted on an XYZ translation stage were then employed to transfer solutions containing small molecule inhibitors into the holes under a microscope. Finally, a fluorescently labeled protein containing receptor sites for binding to the phospholipid bilayer can be introduced into all the compartments. TIRFM is then employed to monitor the extent of surface binding (see section 5).[41–55]

It is straightforward to demonstrate the utility of this partitioning procedure for screening libraries of small soluble molecules for their ability to inhibit binding between surface bound ligands and aqueous phase receptors. To do this a supported phospholipid bilayer was formed which contained 2 mol% biotinylated lipids.[56] Next, μCD was performed with a PDMS mold containing a 2 × 2 patterned array. Eight nanoliter droplets of 1mM N-2,4-DNP-glycine, 1 mM bovine serum albumin, and 1 mM biotin PEO-amine were added to the aqueous compartments in boxes 1, 2, and 3 respectively.[57] Compartment 4 was left unaltered as a control. Upon addition of 8 nL droplets of a 10 μM concentration of fluorescently labeled streptavidin to all four of the compartments, it could be seen that the fluorescence from the aqueous phase[58] above each of the bilayers looked identical (Figure 6.7a). Using the evanescence wave generated from the 594 nm line of a HeNe laser to excite only the surface bound, Texas Red labeled streptavidin reveals that a high concentration of streptavidin is bound in boxes 1, 2, and 4 while the surface of box 3 remained essentially protein free (Figure 6.7b).

The experiments described above demonstrate the ability to deliver soluble analytes to bilayer arrays and to evaluate the ability of these species to prevent surface absorption of proteins. Therefore, we have demonstrated methods of creating spatially addressed arrays of aqueous solutions above phospholipid membranes as well as arrays of phospholipid membranes with unique chemistry in each bilayer.[13] These two concepts were carried out in separate assays. To be able to control both surface chemistry and aqueous chemistry

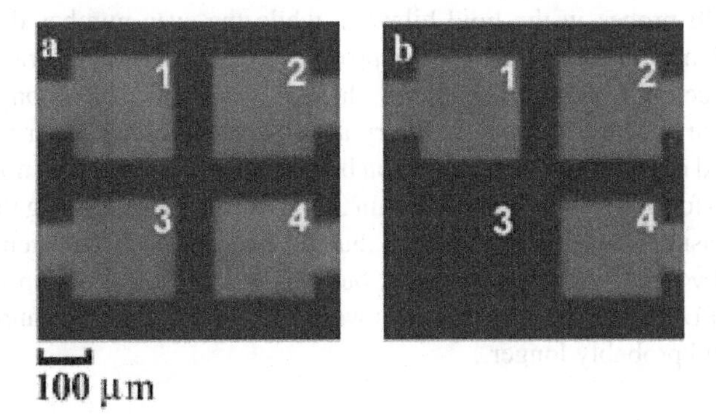

FIGURE 6.7. Epifluorescent images of (a) Texas Red labeled streptavidin confined in aqueous compartments above solid supported phospholipid bilayers containing 2 mol% biotinylated lipids. Each box contains a different aqueous phase small molecule as described in the text. (b) Same system as in (a), but fluorescence is excited by total internal reflection revealing that streptavidin is surface bound in boxes 1, 2, and 4, while the surface in box 3 remains protein free.

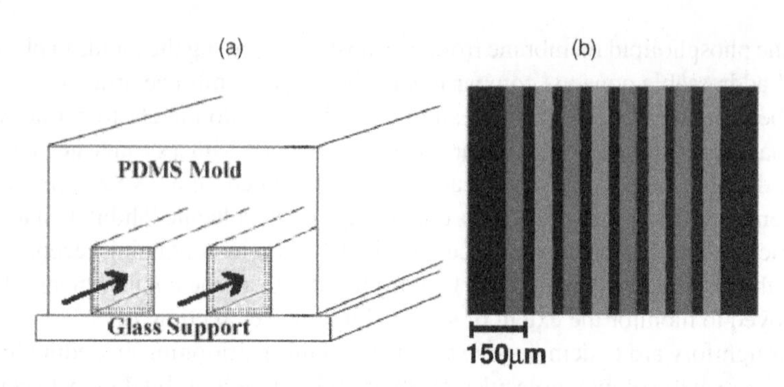

FIGURE 6.8. (a) Schematic diagram of bilayer coated PDMS microchannels on a planar glass substrate. The bilayer coats both the glass and polymer surfaces. Protein solutions can then be injected into the channels as indicated by the arrows. (b) Epifluorescence image of a spatially addressed array of eight Egg PC bilayer coated microchannels. The channels are alternately coated with bilayers containing 1 mol% Texas Red and 3 mol% fluorescein labeled lipids.

simultaneously, we switched to supported bilayer microfluidics.[14] As will be shown below, the use of microchannels for studying bilayers has the added advantage that thermodynamic data can be abstracted.

6.4. SUPPORTED BILAYER MICROFLUIDICS

Exploiting fluid bilayer-based platforms for sensor design, biocompatibility studies, or fundamental investigations of lipid membranes could be made even more powerful by incorporating them into microfluidic networks for lab-on-a-chip assays.[14,59–70] A schematic diagram of the concept is shown in Figure 6.8a, while 6.8b shows the epifluorescence image of an array of eight microchannels coated with fluorescently labeled supported lipid bilayers. Each microchannel was addressed individually by injecting SUVs in a 10 mM PBS buffer at pH 7.2 into the channel inlet ports. The odd numbered microchannels contained 1 mol% Texas Red DHPE probes in the lipid bilayers, while the even numbered channels were prepared with 3 mol% fluorescein DHPE probes (contrast is lost in gray scale image). Vesicle fusion occurred on both the PDMS channel walls[71] as well as on the underlying borosilicate substrate. Fluorescence recovery after photobleaching experiments indicated that the supported membranes were mobile on both materials. It should be noted that vesicle injection was performed in all channels within 3–4 minutes after exposing the PDMS mold to the oxygen plasma. This not only ensured that the channels were sufficiently hydrophilic to induce flow by positive capillary action, but also led to the formation of high quality bilayers on the PDMS surface. The bilayers were stable on the microchannels for at least several weeks and probably longer.

6.5. CREATION OF IMMUNOASSAYS

One of the most important potential uses for bilayer microfluidics is the creation of quantitative on-chip immunoassays. Using parallel arrays of microchannels allows an entire binding curve to be obtained in one-shot simply by coating each channel with the same ligand-containing membrane and flowing different IgG concentrations over each. Such a methodology for performing heterogeneous immunoassays not only produces rapid results, but also requires much less protein than traditional procedures and eliminates some standard sources of experimental error.[14]

To discriminate between surface bound protein molecules and those in bulk solution, total internal reflection fluorescence microscopy (TIRFM)[41–55] was employed. TIRFM creates an evanescence wave that decays as a function of distance from the surface as:

$$I(z) = I_0 e^{-z/d} \quad \text{and} \quad d = \frac{\lambda_0}{4\pi}\left[n_1^2 \sin^2\theta - n_2^2\right]^{-1/2} \tag{6.1}$$

where I is the intensity of the incident light beam, z is the distance from the interface measured in the normal direction, n_1 is the index of refraction of medium 1 (high index medium from which the light is incident) and n_2 is the index of refraction of medium 2 (low index medium into which the evanescent field is propagating), θ is the angle which the radiation makes with the surface normal and λ_0 is the wavelength. The substrate beneath the biochip on which the bilayers resided was made of borosilicate float glass, which has an index of refraction of about 1.52 (Figure 6.9). The aqueous solution above the surface has an index of refraction close to 1.33. Since the laser beam was incident on this surface at approximately 79°, the intensity of the evanescence field fell to 37% ($1/e$) of its initial value at a distance of 70 nm above the liquid/solid interface for the 594 nm radiation employed.

The DNP/anti-DNP system was chosen as a test case for immunoassay fabrication. In order to obtain sufficient data for quantitative measurements of ligand-receptor binding as a function of antibody concentration, measurements were made over nearly two orders of magnitude in protein concentration. Twelve microchannels were arrayed on a single chip. Each channel was injected with a solution of small unilamellar vesicles composed of 92 mol% egg PC, 5 mol% DNP-PE and 3 mol% fluorescein DHPE in a 10 mM PBS buffer at pH 7.2. After flushing out excess vesicles, various concentrations of Alexa dye labeled anti-DNP were injected into the channels, with the highest protein concentration on the left side and the lowest on the right (Figure 6.10a).

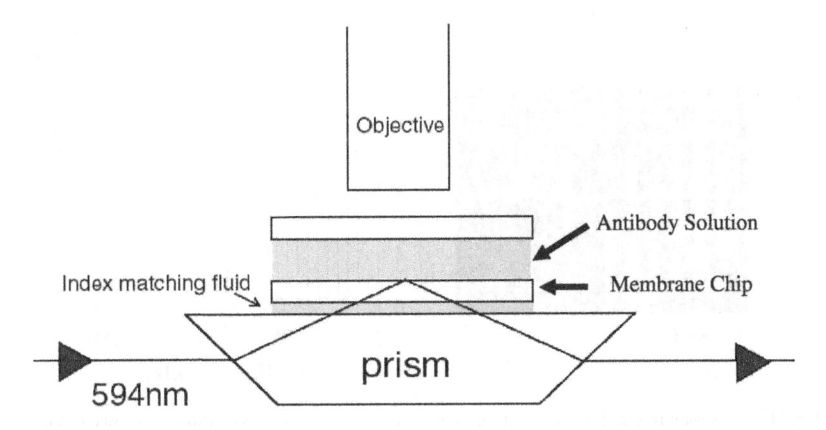

FIGURE 6.9. TIRFM setup for imaging antibody-antigen interactions on bilayer coated chips.

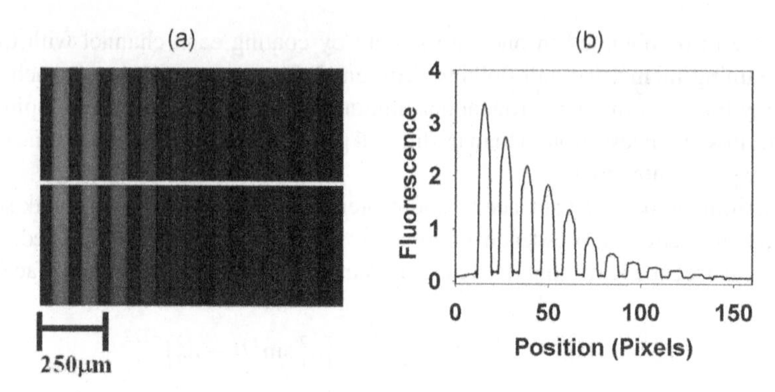

FIGURE 6.10. (a) Bulk phase epifluorescence image of Alexa 594 dye-labeled anti-DNP inside bilayer coated microchannels. Starting from the left-hand-side, the antibody concentrations are 13.2 μM, 8.80 μM, 5.87 μM, 3.91 μM, 2.61 μM, 1.74 μM, 1.16 μM, 0.77 μM, 0.52 μM, 0.34 μM, 0.23 μM, and 0.15 μM. A line scan of fluorescence intensity (dotted line) across the microchannels is plotted in (b).

Because surface binding causes bulk concentration depletion, protein solution was flowed continuously through the channels until the bulk concentration became stabile. This required an aliquot roughly equal to 4 or 5 times the channel volume in the microchannels with the lowest protein concentration, while the highest concentration channels required considerably less flow, as expected. A line profile generated from the epifluorescence image of these channels is shown in Fig. 6.10b. The majority of the signal emanated from the bulk.

Figure 6.11a shows the same system as 10a, but now imaged by TIRFM. As can be seen from the line profile (Figure 6.11b), the fluorescence intensity no longer showed a logarithmic decay with respect to bulk concentration. Instead, the signal represented a combination of specifically bound antibody, nonspecifically bound antibody, and near surface antibody in the bulk solution. To account for signal arising from the latter two effects, the experiment was repeated under the identical conditions, but in the absence of DNP-PE lipids in the SLBs. In this case, the fluorescence intensity from the total internal reflection experiments was dramatically reduced (data not shown). A binding curve for coverage vs. bulk protein concentration was obtained for the DNP/anti-DNP system by

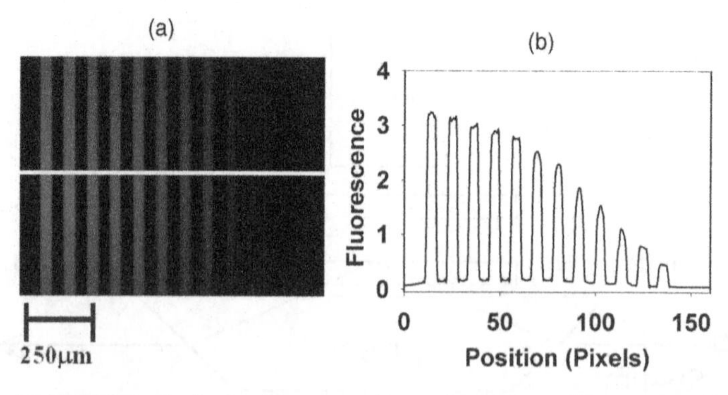

FIGURE 6.11. (a) A total internal reflection fluorescence image of the same conditions as shown in Figure 6.10a. (b) Line scan of fluorescence intensity (dotted line) across the microchannels.

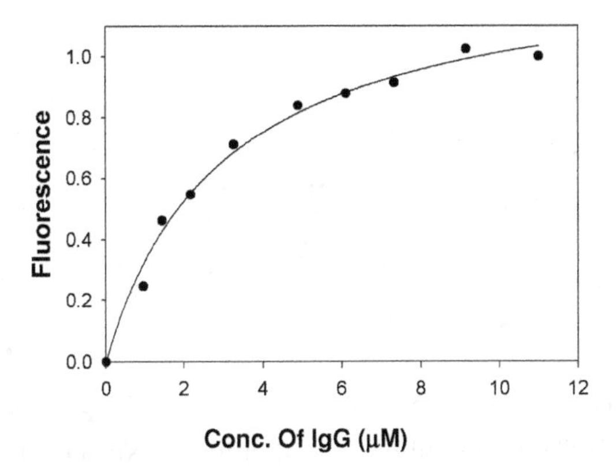

FIGURE 6.12. Binding curve for the DNP/anti-DNP system on a fluid phospholipid membrane.

subtracting the background from the data in Fig. 6.11b. This is plotted in Figure 6.12. The half coverage point on the curve occurs at a concentration of 1.8 μM and is in good agreement with data from previous experiments.[42,72]

6.6. MULTIVALENT BINDING AS A FUNCTION OF LIGAND DENSITY

Microfluidics provides a high throughput means of obtaining binding information as a function of membrane content for proteins binding multivalently. One of the key parameters to consider for multivalent binding is ligand density.[18] Very few studies have looked at the relationship between hapten density and apparent binding constant. This is presumably because of the tedious, time consuming nature and great expense of experiments performed without the benefit of a high throughput, low sample volume assay. In fact, to the best of our knowledge, previous binding experiments on fluid membranes have been done with no more than three different ligand densities[43] and very few investigators have ever considered the separate dissociation constants.[73] This is unfortunate because the most recent calculations for multivalent ligand-receptor binding now allow the effects of large proteins binding with small molecules at the membrane interface to be taken into account.[74] This would make it possible to devise simple models for determining both the first and second binding constants for antibody-antigen interactions at the biomembrane surface provided that binding curves are available for a sufficient number of antigen concentrations.

Using the microfluidic technique described above, binding curves were obtained on supported bilayers as a function of hapten density for eleven concentrations ranging from 0.1 to 5.0 mol% DNP-cap-PE in Egg PC membranes. In addition to the hapten-conjugated lipid, the membranes contained 1 mol% fluorescein DHPE and the remainder was egg PC. All data were repeated at least three times and the results were averaged. Five representative binding curves are shown in Figure 6.13a. Data were taken at a total of eleven different ligand concentrations. The point at which the surface is half saturated with protein (K_{Dapp}) is plotted as a function of ligand density (Figure 6.13b).

Previous efforts have been made to model the surface association in bivalent systems.[42,75–84] Beginning with this notation, two equilibria can be written corresponding

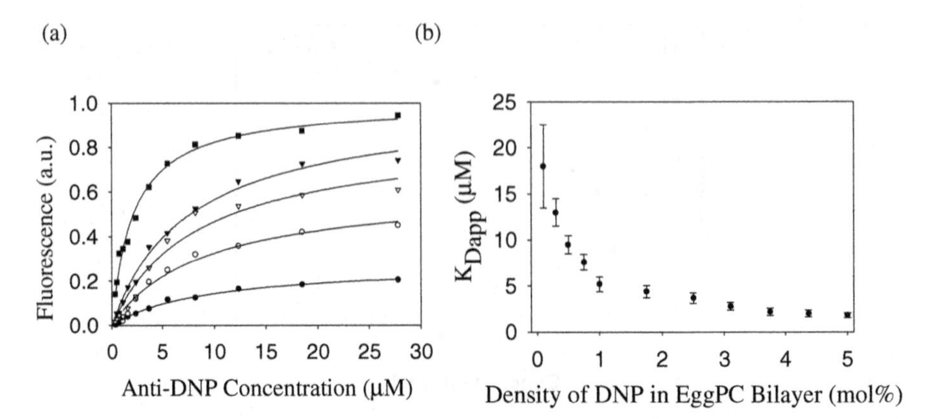

FIGURE 6.13. (a) Normalized binding curves for antibody binding to lipid bilayers containing 0.1 (solid circles, •), 0.50 (open circles, ○), 0.75 (solid triangles, σ), 1.75 (open triangles ▼) and 3.75 (solid squares ■) mol% DNP-cap PE. (b) K_{Dapp} vs. mol% of DNP-cap PE in the lipid bilayers for anti-DNP binding.

to the two sequential binding events:

$$K_{A1} = \frac{[BL]_s}{[B][L]_s} \quad \text{and} \quad K_{A2} = \frac{[BL_2]_s}{[BL]_s[L]_s} \tag{6.2}$$

The protein is now represented as, B, for bivalent. The units of K_{A1} are in M^{-1}, while K_{A2} has units of dm^2/moles (1/number density).[85] The total surface site density $[S]_s$ can be expressed as:

$$[S]_s = [BL_2]_s + \frac{1}{2}[BL]_s + \frac{1}{2}[L]_s \tag{6.3}$$

The surface normalized fluorescence,

$$\frac{F([B])}{F([\infty])} = \frac{[BL]_s + [BL_2]_s}{2[S]_s} \tag{6.4}$$

after substitution of Eqs. 6.2 and 6.3 into Eq. 6.4, can be written as

$$\frac{F([B])}{F([\infty])} = \frac{\alpha K_A[B]}{1 + K_A[B]}, \tag{6.5}$$

where:

$$K_A = K_{A1} + 2K_{A1}K_{A2}[L]_s \tag{6.6}$$

and

$$\alpha = \frac{1 + K_{A2}[L]_s}{1 + 2K_{A2}[L]_s} \tag{6.7}$$

In this notation, the bivalent binding model (Eq. 6.5) takes on the same general form as a simple Langmuir isotherm. The term K_A for bivalent binding is interpreted as an association or affinity parameter for the bivalent process. K_A is not a constant, as it varies with the ligand density in the membrane, which is being changed in these experiments. It is, however, constant at any fixed ligand density and, hence, does not perturb the shape of the Langmuir isotherms. The parameter, α, which can vary between 0.5 and 1.0, also depends on ligand density, but like K_A is constant at each fixed density value. The value of α should reach 1.0 when all protein binds monovalently to the surface and fall to 0.5 when binding is entirely bivalent.

Experimental data are typically expressed in terms of dissociation constants where $K_D = 1/K_A$. Therefore, if the individual dissociation constants are written as K_{D1} and K_{D2}, the apparent experimental parameter, K_{Dapp}, extracted from the fit to the isotherm can be written as follows:

$$K_{Dapp} = \frac{K_{D1} K_{D2}}{K_{D2} + 2[L]_s} \tag{6.8}$$

Both K_{D1} and K_{D2} can be determined by fitting this equation to the data in Figure 6.13b provided that $[L]_s$ is known. Apparent values for the number density of ligands can be calculated by noting that the area occupied by an Egg PC headgroup is ~70 Å^2/molecule[86] and further assuming that DNP-Cap PE is roughly the same size. Such values can, however, be perturbed if the binding protein is sufficiently large.[74] In the present case each IgG antibody covers an area of about 60 nm^2 on the membrane surface.[87–88] Therefore, some portion of the unbound antigens should be covered by proteins at any given instance. This perturbation to ligand availability becomes greater at higher ligand densities because the average distance between ligand moieties decreases. This effect can be accounted for using an appropriate model for the surface binding of large proteins.[74] The results of such a calculation show that almost all the ligands (99.0%) are available at 0.1 mol%, but only 56.8% at 5.0 mol%. The corrected values of $[L]_s$ require a recalibration of the x-axis for the data in Figure 6.13b. Using rescaled data and fitting it to Eq. 6.8 yields the following dissociation constants:

$$K_{D1} = 2.46 \times 10^{-5}\ \mathrm{M}$$
$$K_{D2} = 1.37 \times 10^{-10}\ \mathrm{moles/dm^2}$$

These numbers can be utilized to calculate the corresponding free energy changes for the individual ligand-receptor binding steps.[85] Invoking $\Delta G = -RT\ln K_A$ under the conditions of this experiment ($T = 295$ K) yields:

$$\Delta G_1 = -26.0 \pm 0.4\ \mathrm{kJ/mol}$$
$$\Delta G_2 = -55.7 \pm 0.7\ \mathrm{kJ/mol}$$

The decrease in free energy of the second binding event is, therefore, approximately a factor of two greater than for the first. It should be noted that the activity coefficients are assumed to be unity for obtaining these thermodynamic values.

Once K_{D2} was determined, the α value could be easily calculated for each ligand density using Eq. 6.7. The mole fraction that is bivalently bound is equal to $2(1-\alpha)$. These values are plotted in Figure 6.14.

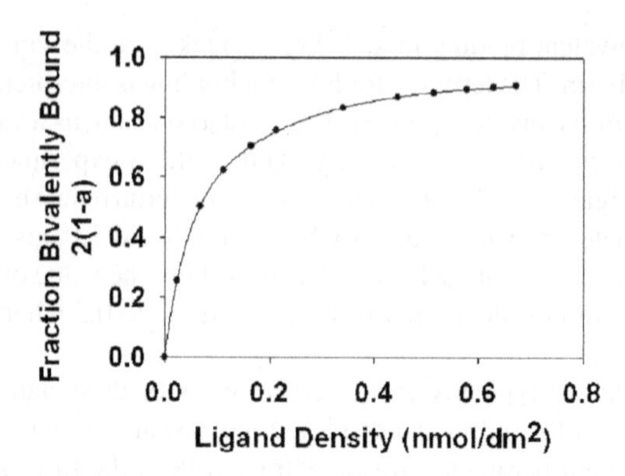

FIGURE 6.14. Plot of the extent of ligand density vs. the extent of bivalently bound anti-DNP on a phospholipid membrane.

As can be seen, nearly all protein is bivalently bound at the highest hapten density (91%), but at the lowest hapten density only 26% of the IgG molecules are bivalently bound. Moreover, the percentage of bivalently bound protein rose steeply at first with one in two proteins already bivalently bound at 0.3 mol% hapten density and about three out of four proteins bivalently bound at 1.00 mol% hapten density.

The binding of IgG antibodies to hapten containing membranes is not a directly cooperative process. In other words, no allosteric mechanism is at work and the intrinsic affinity of the second binding event should not be affected by the first. This has been verified by homogeneous phase binding measurements where K_{D2} was actually found to be slightly larger (weaker) than K_{D1}.[89–90] This was the case because only one binding site is available once a first ligand has bound and there are two choices for unbinding. In other words, entropic considerations lead to $K_{D1} < K_{D2}$ in the homogeneous phase.

For sequential binding at membrane surfaces, there is also an entropic difference between the first and second binding event. In this case the result is the opposite and of much greater magnitude than in the homogeneous phase. The difference in entropy at the interface for the two sequential binding events can be calculated by simply assuming that the enthalpy of both events is the same. The free energy of each event can be written as follows:

$$\Delta G_1 = \Delta H_1 - T\Delta S_1 \text{ (first binding event)}$$
$$\Delta G_2 = \Delta H_2 - T\Delta S_2 \text{ (second binding event)}$$

A difference in free energy, $\Delta G_2 - \Delta G_1 = T(\Delta S_1 - \Delta S_2)$, results if the two enthalpy values are assumed to be identical. ΔS is therefore 100.7 J/mol·K higher for the second event than for the first.[85] This difference presumably stems from the fact that the entropic cost of bringing the protein to the surface and properly orienting it is almost entirely born by the first binding step. Moreover, the second binding site on the antibody is already at the interface and at least partially aligned and positioned to undergo the subsequent binding step in an environment which is now much richer in hapten than was the corresponding environment in three dimensions. ΔS_2, therefore, predominantly contains a lateral diffusion component and the local entropic changes associated with binding (*e.g.* the displacement of water molecules from the binding pocket upon interaction with the ligand). The entropic factors which affect the initial binding constant are almost certainly lowering the on rate, k_{on}, while leaving the off rate, k_{off}, unaffected. This argument is bolstered by previous measurements of the off rate for Fab fragments from lipid-linked 2,4 dinitrophenyl moieties in supported membranes which gave values of $k_{off} \approx 1\ sec^{-1}$, a quite standard value.[45]

6.7. CONCLUSIONS AND FUTURE OUTLOOK

In the work described above, high throughput microfluidic and array based techniques were designed to study protein-ligand interactions at fluid membrane interfaces. Usually, array-based techniques are used for either materials discovery purposes or as part of a rapid screening protocol. We believe, however, that combinatorial and high throughput methods will gain increasing importance for studying the molecular level details of mechanistic process in physical/biophysical chemistry. Indeed, as the systems under investigation become ever more complex, this may become a necessity. For example, the methods described above can now be exploited to studying higher order multivalent processes such as the binding of cholera toxin to GM1 in membrane rafts. In such cases it will be imperative to use rapid assays to begin to appreciate the physical and chemical variables that contribute to tight binding.

ACKNOWLEDGMENTS

This work was funded by an ONR-YIP Award (NOOO14-00-1-0664) and by ARO (DAAD19-01-1-0346). PSC gratefully acknowledges the receipt of a Beckman Young Investigator Award, a Nontenured Faculty Award from 3M Corporation, an Alfred P. Sloan Fellowship, an ACS PRF grant, and a Research Innovation Award from Research Corporation.

REFERENCES

1. Brian and H. M. McConnell, Allogeneic stimulation of cyto-toxic T-cells by supported planar membranes, *Proc. Natl. Acad. Sci.* **81**, 6159–6163 (1984).
2. L. K. Tamm and H. M. McConnell, Supported phospholipid bilayers, *Biophys. J.* **47**, 105–113 (1985).
3. S. J. Johnson, T. M. Bayerl, D. C. McDermott, G. W. Adam, A. R. Rennie, R. K. Thomas, and E. Sackmann, Structure of an adsorbed dimyristoylphosphatidycholine bilayer measured with specular reflection of neutrons, *Biophys. J.* **59**, 289–294 (1991).
4. E. Sackmann, Supported membranes: Scientific and practical applications, *Science* **271**, 43–48 (1996).
5. L. M. Williams, S. D. Evans, T. M. Flynn, A. Marsh, P. F. Knowles, R. J. Bushby, and N. Boden, Kinetics of the unrolling of small unilamellar phospholipid vesicles onto self-assembled monolayers, *Langmuir* **13**, 751–757 (1997).
6. M. Stelzle, R. Miehlich, and E. Sackmann, 2-Dimensional microelectrophoresis in supported lipid bilayers, *Biophys. J.* **63**, 1346–1354 (1992).
7. Y. Cheng, N. Boden, R. J. Bushby, S. Clarkson, S. D. Evans, P. F. Knowles, A. Marsh, and R. E. Miles, Attenuated total reflection Fourier transform infrared spectroscopic characterization of fluid lipid bilayers tethered to solid supports, *Langmuir* **14**, 893–844 (1998).
8. E. Kalb, S. Frey, and L. K. Tamm, Formation of supported planar bilayers by fusion of vesicles to supported phospholipid monolayers, *Biochim. Biophys. Acta* **1103**, 307–316 (1992).

9. P. S. Cremer and S. G., Boxer, Formation and spreading of lipid bilayers on planar glass surfaces, *J. Phys. Chem. B* **103**, 2554–2559 (1999).
10. J. Kim, G. Kim, and P. S. Cremer, Investigations of water structure at the solid/liquid interface in the presence of supported lipid bilayers by vibrational sum frequency spectroscopy, *Langmuir* **17**, 7255–7260 (2001).
11. Heldin, Dimerization of cell surface receptors in signal, *Cell* **80**, 213–223 (1995).
12. P. S. Cremer and T. Yang, Creating spatially addressed arrays of planar supported fluid phospholipid membranes, *J. Am. Chem. Soc.* **121**, 8130–8131 (1999).
13. T. Yang, E. E. Simanek, and P. S. Cremer, Creating addressable aqueous microcompartments above solid supported phospholipid bilayers using lithographically patterned poly(dimethylsiloxane) molds, *Anal. Chem.* **72**, 2587–2589 (2000).
14. T. Yang, S. Y. Jung, H. Mao, and P. S. Cremer, Fabrication of phospholipid bilayer coated microchannels for on chip immunoassays, *Anal. Chem.* **73**, 165–169 (2001).
15. H. Bayley and P. S. Cremer, Stochastic sensors inspired by biology, *Nature* **413**, 226–230 (2001).
16. H. Mao, T. Yang, and P. S. Cremer, Design and characterization of immobilized enzymes in microfluidic systems, *Anal. Chem.* **74**, 379–385 (2002).
17. H. Mao, T. Yang, and P. S. Cremer, A microfluidic device with a linear temperature gradient for parallel and combinatorial measurements, *J. Am. Chem. Soc.* **124**, 4432–4435 (2002).
18. T. Yang, O. K. Baryshnikova, H. Mao, M. A. Holden, and P. S. Cremer, Investigation of bivalent antibody binding on fluid supported phospholipid membranes: the effect of hapten density, *J. Am. Chem. Soc.* **125**, 4779–4784 (2003).
19. S. P. Fodor, J. L. Read, M. C. Pirrung, L. Stryer, A. T. Lu, and D. Solas, Light-directed, spatially addressable parallel chemical synthesis, *Science* **251**, 767–773 (1991).
20. J. M. Brockman, A. G. Frutos, and R. M. Corn, A multistep chemical modification procedure to create DNA arrays on gold surfaces for the study of protein-DNA interactions with surface plasmon resonance imaging, *J. Am. Chem. Soc.* **121**, 8044–8051 (1999).
21. G. MacBeath, A. N. Koehler, and S. L. Schreiber, Printing small molecules as microarrays and detecting protein-ligand interactions en masse, *J. Am. Chem. Soc.* **121**, 7967–7968 (1999).
22. M. Mrksich and G. M. Whitesides, Patterning self-assembled monolayers using microcontact printing-a new technology for biosensors, *Trends Biotechnol.* **13**, 228–235 (1995).
23. M. Mrksich and G. M., Whitesides, Using self-assembled monolayers to understand the interactions of man-made surfaces with proteins and cells, *Ann. Rev. Biomol. Struct.* **25**, 55–78 (1996).
24. A. Jans, *The mobile receptor hypothesis: The role of membrane receptor lateral movement in signal transduction* (Chapman & Hall, Austin, TX, 1997).
25. M. Mammen, S. K. Choi, and G. M. Whitesides, Polyvalent interactions in biological systems: Implications for design and use of multivalent ligands and inhibitors, *Angew. Chem. Int. Ed.* **37**, 2754–2794 (1998).
26. L. L. Kiessling and N. L. Pohl, Strength in numbers: Non-natural polyvalent carbohydrate derivatives, *Chem. Biol.* **3**, 71–77 (1996).
27. L. Finegold, *Cholesterol in Membrane Models* (CRC Press, Boca Raton, FL, 1993).
28. X. D. Xiang, X. Sun, G. Briceno, Y. Lou, K. A. Wang, H. Chang, W. G. Wallace-Freedman, S. W. Chen, P. G. Schultz, A combinatorial approach to materials discovery *Science* **268**, 1738–1740 (1995).
29. Reddington, A. Sapienza, B. Gurau, R. Viswanathan, S. Sarangapani, E. S. Smotkin, and T. E. Mallouk, Combinatorial electrochemistry: A highly parallel, optical screening method for discovery of better electrocatalysts, *Science* **280**, 1735–1737 (1998).
30. H. M. McConnell, T. H. Watts, R. M. Weis, and A. A. Brian, Supported planar membranes in studies of cell-cell recognition in the immune-system, *Biochim. Biophys. Acta* **864**, 95–106 (1986).
31. T. H. Watts, H. Gaub, and H. M. McConnell, T-cell-mediated association of peptide antigen and major histocompatibility complex protein detected by energy-transfer in an evanescent wave-field, *Nature* **320**, 179 –181 (1986).
32. Tozeren, P. K.-L. Sung, L. A. Sung, M. L. Dustin, P. Y. Chan, T. A. Springer, and S. Chien, Micromanipulation of adhesion of a jurkat cell to a planar bilayer-membrane containing lymphocyte function-associated antigen 3 molecules, *J. Cell. Biol.* **116**, 997–1066 (1992).
33. Kumar and G. M. Whitesides, Patterned condensation figures as optical diffraction gratings, *Science* **263**, 60–62 (1994).

34. J. T. Groves, N. Ulman, and S. G. Boxer, Micropatterning fluid lipid bilayers on solid supports *Science* **275**, 651–653 (1997).
35. J. T. Groves, N. Ulman, P. S. Cremer, and S. G. Boxer, Substrate-membrane interactions: Mechanisms for imposing patterns on a fluid bilayer membrane, *Langmuir* **14**, 3347–3350 (1998).
36. Axelrod, D. E. Koppel, J. Schlessinger, E. Elson, and W.W. Web, Mobility Measurement by Analysis of Flurescence Photobleaching Recovery Kinetics, *Biophys. J.* **16**, 1055–1069 (1976).
37. D. M. Soumpasis, Theorectical-analysis of fluorescence photobleaching recovery experiments, *Biophys. J.* **41**, 95–97 (1983).
38. K. Balakrishnan, S. Q. Mehdi, and H. M. McConnell, Availability of dinitrophenylated lipid haptens for specific antibody binding depends on the physical properties of host bilayer membranes, *J. Biol. Chem.* **257**, 6434–6439 (1982).
39. Y. Xia and G. M. Whitesides, Soft lithography, *Angew. Chem. Int. Ed.* **37**, 550–575 (1998).
40. D. C. Duffy, J. C. McDonald, O. J. A. Schueller, and G. M. Whitesides, Rapid prototyping of microfluidic systems in poly(dimethylsiloxane), *Anal. Chem.* **70**, 4974–4984 (1998).
41. D. Axelrod, T. P. Burghardt, and N. L. Thompson, Total internal-reflection fluorescence, *Ann. Rev. Biophys. Bioeng.* **13**, 247–268 (1984).
42. M. L. Pisarchick and N. L. Thompson, Binding of a monoclonal antibody and its Fab fragment to supported phospholipid monolayers measured by total internal reflection fluorescence microscopy, *Biophys. J.* **58**, 1235–1249 (1990).
43. Kalb, J. Engel, and L. K. Tamm, Binding of proteins to specific target sites in membranes measureed by total internal reflection fluorescence microscopy, *Biochemistry* **29**, 1607–1613 (1990).
44. M. M. Timbs, C. L. Poglitsch, M. L. Pisarchick, M. T. Sumner, and N. L. Thompson, Binding and mobility of anti-dinitrophenyl monoclonal-antibodies on fluid-like, langmuir-blodgett phospholipid monolayers containing dinitrophenyl-conjugated phospholipids, *Biochim. Biophys. Acta* **1064**, 219–228 (1991).
45. M. L. Pisarchick, D. Gesty, and N. L. Thompson, Binding-kinetics of an antidinitrophenyl monoclonal Fab on supported phospholipid monolayers measured by total internal reflection with fluorescence photobleaching recovery, *Biophys. J.* **63**, 216–223 (1992).
46. N. L. Thompson, C. L. Poglitsch, M. M. Timbs, and M. L. Pisarchick, Dynamics of antibodies on planar model membranes, *Acc. Chem. Res.* **26**, 567–573 (1993).
47. N. L. Thompson, K. H. Pearce, and H. V. Hsieh, Total internal reflection fluorescence microscopy: Application to substrate-supported planar membranes, *Eur. Biophys. J.* **22**, 367–378 (1993).
48. V. Hsieh and N. L. Thompson, Theory for measuring bivalent surface binding-kinetics using total internal-reflection with fluorescence photobleaching recovery, *Biophys. J.* **66**, 898–911 (1994).
49. W. V. Broek and N. L. Thompson, When bivalent proteins might walk across cell surfaces, *J. Phys. Chem.* **100**, 11471–11479 (1996).
50. Z. P. Huang and N. L. Thompson, Imaging fluorescence correlation spectroscopy: Nonuniform IgE distributions on planar membranes, *Biophys. J.* **70**, 2001–2007 (1996).
51. N. L. Thompson and B. C. Lagerholm, Total internal reflection fluorescence: Applications in cellular biophysics, *Curr. Opin. Biotech.* **8**, 58–64 (1997).
52. N. L. Thompson, A. W. Drake, L. Chen, and W. V. Broek, Equilibrium, kinetics, diffusion and self-association of proteins at membrane surfaces: Measurement by total internal reflection fluorescence microscopy, *Photochem. Photobiol.* **65**, 39–46 (1997).
53. C. Lagerholm and N. L. Thompson, Temporal dependence of ligand dissociation and rebinding at planar surfaces, *J. Phys. Chem. B* **104**, 863–868 (2000).
54. C. Lagerholm, T. E. Starr, Z. N. Volovyk, and N. L. Thompson, Rebinding of IgE Fabs at haptenated planar membranes: Measurement by total internal reflection with fluorescence photobleaching recovery, *Biochemistry* **39**, 2042–2051 (2000).
55. T. E. Starr and N. L. Thompson, Total internal reflection with fluorescence correlation spectroscopy: Combined surface reaction and solution diffusion, *Biophys. J.* **80**, 1575–1584 (2001).
56. T. L. Calvert and D. Leckband, Two-dimensional protein crystallization at solid-liquid interfaces, *Langmuir* **13**, 6737–6745 (1997).
57. The final concentrations of these species are attenuated by approximately a factor of 20 through dilution as the aqueous container volumes were roughly 150 nL.

58. Some streptavidin was observed to adsorb to the walls of the PDMS mold. This process could be suppressed by preadsorbing bovine serum albumin.
59. Figeys and D. Pinto, Lab-on-a-chip: A revolution in biological and medical sciences, *Anal. Chem.* **72**, 330A–335A (2000).
60. J. Harrison and A. van den Berg (Eds.), *Micro Total Analysis Systems 98* (Kluwer Academic Publishers, Dordrecht, 1998).
61. van den Berg and P. Bergveld (Eds.) *Micro Total Analysis Systems* (Kluwer Academic Publishers: Dordrecht, 1995).
62. S. C. Jacobson, T. E. McKnight, and J. M. Ramsey, Microfluidic devices for electrokinetically driven parallel and serial mixing, *Anal. Chem.* **71**, 4455–4459 (1999).
63. K. Hosokawa, T. Fujii, and I. Endo, Handling of picoliter liquid samples in a poly(dimethylsiloxane)-based microfluidic device, *Anal. Chem.* **71**, 4781–4785 (1999).
64. P. J. A. Kenis, R. F. Ismagilov, and G. M. Whitesides, Microfabrication inside capillaries using multiphase laminar flow patterning, *Science* **285**, 83–85 (1999).
65. L. Colyer, T. Tang, N. Chiem, and D. J. Harrison, Clinical potential of microchip capillary electrophoresis systems, *Electrophoresis* **18**, 1733–1741(1997).
66. N. H. Chiem and D. J. Harrison, Monoclonal antibody binding affinity determined by microchip-based capillary electrophoresis, *Electrophoresis* **19**, 3040–3044 (1998).
67. C. Duffy, H. L. Gillis, J. Lin, N. F. J. Sheppard, and G. J. Kellogg, Microfabricated centrifugal microfluidic systems: Characterization and multiple enzymatic assays, *Anal. Chem.* **71**, 4669–4678 (1999).
68. Delamarche, A. Bernard, H. Schmid, B. Michel, and H. Biebuych, Patterned delivery of immunoglobulins to surfaces using microfluidic networks, *Science* **276**, 779–781 (1997).
69. R. B. M. Schasfoort, S. Schlautmann, J. Hendrikse, and A. van den Berg, Field Effect Flow Control for Microfabricated Fluidic Networks, *Science* **286**, 942–945 (1999).
70. S. Kunneke and A. Janshoff, Visualization of molecular recognition events on microstructured lipid-membrane compartments by in situ scanning force microscopy, *Angew. Chem. Int. Ed.* **41**, 314–316 (2002).
71. J. S. Hovis and S. G. Boxer, Patterning barriers to lateral diffusion in supported lipid bilayer membranes by blotting and stamping, *Langmuir* **16**, 894–897 (2000).
72. M. Mammem, F. A. Gomez, and G. M. Whitesides, Determination of the binding of ligands containing the N-2,4 dinitrophenyl group to bivalent monoclonal rat anti-DNP antibody using affinity capillary electrophoresis, *Anal. Chem.* **67**, 3526–3535 (1995).
73. M. L. Dustin, L. M. Ferguson, P. Chan, T. A. Springer, and D. E. Golan, Visualization of CD2 interaction with LFA-3 and Determination of Two-dimensional dissociation constant for adhesion receptors in a contact area, *J. Cell Biol.* **132**, 465–474 (1996).
74. W. S. Hlavacek, R. G. Posner, and A. S. Perelson, Steric effects on Multivalent Ligan-Recaeptor Binding: Exclusion of Ligand Sites by Bound Cell Surface Receptors, *Biophys. J.* **76**, 3031–3043 (1999).
75. Delisi, The biophysics of ligand-receptor interactions, *Q. Rev. Biophys. J.* **13**, 201–230 (1980).
76. Shoup, G. Lipari, and A. Szabo, Diffusion-controlled bimolecular reaction rates- the effect of rotational diffusion and orientation, *Biophys. J.* **36**, 697–714 (1981).
77. Delisi and F. W. Wiegel, Effect of nonspecific forces and finite receptor number on rate constants of ligand cell-bound receptor interactions, *Proc. Nat. Acad. Sci.* **78**, 5569–5572 (1981).
78. Shoup and A. Szabo, Role of diffusion in ligand-binding to macromolecules and cell-bound receptors, *Biophys. J.* **40**, 33–39 (1982).
79. O. G. Berg and P. H. Vonhippel, Diffusion-controlled macromolecular interactions, *Annu. Rev. Biophys. Biophys. Chem.* **14**, 131–160 (1985).
80. R. Zwanzig, Diffusion-controlled ligand-binding to spheres partially covered by receptors- an effective medium treatment, *Proc. Nat. Acad. Sci.* **87**, 5856–5857 (1990).
81. A Lauffenburger and J. J. Linderman, *Receptors: Models for Binding, Trafficking, and Signaling* (Oxford University Press, New York, 1993).
82. D. Axelrod and M. D. Wang, Reduction of dimensionality kinetics at reaction limited cell surface receptors, *Biophys. J.* **66**, 588–600 (1994).
83. Balgi, D. E. Leckband, and J. M. Nitsche, Transport effects on the kinetics of protein surface binding, *Biophys. J.* **68**, 2251–2260 (1995).

84. B. Goldstein and M. Dembo, Approximating the effects of diffusion on reversible reactions at the cell surface ligand receptor kinetics, *Biophys. J.* **68**, 1222–1230 (1995).
85. The use of mol/dm^2 for K_{D2}, which is two-dimensional, corresponds well to mol/L or mol/dm^3 that is used for the first binding constant, which is three-dimensional. The use of these units accounts for the difference in the $\Delta\Delta S$ and ΔG_2 values reported here and those reported in ref. 18.
86. S. H. White and G. I. King, Molecular packing and area compressibility of lipid bilayers, *Proc. Nat. Acad. Sci.* **82**, 6532–6536 (1985).
87. L. K. Tamm and I. Bartoldus, Antibody-binding to lipid model membranes - the large-ligand effect, *Biochemistry* **27**, 7453–7458 (1988).
88. P. M. Colman, J. Deisenhofer, R. Huber, and W. Palm, Structure of human antibody molecule Kol (immunoglobulin-G1)-electron density map at 5 A resolution, *J. Mol. Biol.* **100**, 257–282 (1976).
89. M. Mammen, F. A. Gomez, and G. M.Whitesides, Determination of the binding of ligands containing the N-2,4 dinitrophenyl group to bivalent monoclonal rat anti-DNP antibody using affinity capillary electrophoresis, *Anal. Chem.* **67**, 3526–3535 (1995).
90. J. M. Gargano, T. Ngo, J. Y. Kim, D. W. K. Acheson, and W. J. Lees, Multivalent inhibition of AB(5) toxins, *J. Am. Chem. Soc.* **123**, 12909–12910 (2001).

7

Aggregation of Amphiphiles as a Tool to Create Novel Functional Nano-Objects

K. Velonia, J. J. L. M. Cornelissen, M. C. Feiters, A. E. Rowan, and R. J. M. Nolte

7.1. INTRODUCTION

Nature often uses the self-assembly of amphiphilic building blocks as a tool for the structuring of matter. The most representative example of the functionality that can be achieved through the interplay between different, structurally simple, monomeric units is the cell membrane.[1,2] In this natural supramolecular structure, the organization is achieved by the self-assembly of different types of individual functional molecules (phospholipids, glycolipids, glycoproteins, membrane spanning peptide helices, the cytoskeleton, etc). The resulting cell membrane combines functionality, compartmentalization, order and mobility, characteristics all essential for life. The realization of the importance of self-assembly in Nature has led scientists to extensively explore its basic principles in the last decades. The designed self-assembly of individual molecules has led to macromolecular structures of one, two or three-dimensional nature. These supramolecular structures can contain between 10^1 and 10^5 molecules and thus resemble synthetic and biological polymers in molecular mass. As G. M. Whitesides stated; "*Nature's mechanisms are much more complex and much more highly evolved than the self-assembly currently used in laboratories. The ultimate goal is to look at these mechanisms, abstract the principles from them, and then embed those principles in non-biological systems to make functional, very sophisticated small machines.*" Though significant progress has been achieved, it still is a challenge to try to understand in full detail the principles that govern the self-organization of individual molecules leading to the formation of nanometer sized assemblies, and furthermore to be able to manipulate these nanometer scale functionalities and enhance their properties.

Nijmegen Science Research Institute for Materials (NSRIM), Department of Organic Chemistry, University of Nijmegen, Toernooiveld 1, 6525 ED Nijmegen. nolte@sci.kun.nl

7.1.1. Amphiphiles

The word ***amphiphile*** is derived from the Greek words "*ἀμφι-*" meaning "*dual*" and "*φίλος*" meaning "*loving*" and refers to a molecule that -due to this dual structure- has a strong attraction towards both polar solvents (like a *hydrophile*) and non-polar solvents (like a *hydrophobe*). Amphiphilic molecules contain a polar head group and an apolar tail covalently connected through a -rigid- linker. The water-soluble, hydrophilic head group can be neutral, ionic or zwitterionic, while the hydrophobic tail generally consists of one or more saturated or unsaturated alkyl chains. Because of this ambivalent character, amphiphiles tend to aggregate, upon dispersion in water, into various superstructures following either a ***programmed***[3] or a ***synkinetic*** way.[4,5] Utilizing the same mechanisms, assemblies can be also formed in organic solvents and ionic liquids. The programmed aggregation process, self-assembly, is the process by which molecules spontaneously form supramolecular architectures.[3,6] This process relies on the recognition between simple subunits and can be compared to computer programming: the structure of a single molecule can be viewed as the algorithm giving the set of instructions that defines its aggregation into a particular superstructure. The formation of helical self-assemblies for example, is predefined by a "helical molecular programming"[7] process dictated by the structure of the molecular amphiphile (algorithm). In the synkinetic aggregation route the monomeric building blocks "synkinons" are held together by non-covalent interactions, which follow the synkinetic plans of the chemist rather than a self-organization process.

The driving force for the self-assembling behavior of amphiphiles in water is considered to be the ***hydrophobic effect***, which is caused by the disruption of the strong attractive forces between water molecules upon dissolving a solute.[8,9] The concept of hydrophobic interactions was already alluded to by Kauzmann in 1959.[10] In his work he discussed the entropy gain upon interaction of apolar compounds in aqueous environment by the release of structured water molecules during a destructive overlap of hydrophobic hydration shells. Since then, the hydrophobic effect has been discussed in many publications and opinions have evolved while recently it has been redefined by considering the thermodynamics of solvation as the favorable overlap of the hydration shells of hydrophobic groups.[11,12] Above a certain concentration (called ***critical aggregation concentration***), the formation of the assembly is entropically favoured over a solution of individual molecules. Upon dispersal into water, both the natural and synthetic surfactants organize in such a way that the polar head groups become oriented towards the water, while at the same time the hydrophobic tails cluster together. This can lead to various superstructures such as micelles, vesicles, multilayers and lyotropic liquid crystalline phases (at high concentrations).[13,14] More exotic aggregation morphologies have also been observed and involve the formation of chiral superstructures such as "cigars",[15] twisted ribbons,[16] helices,[17] tubes,[18] braids,[19] boomerangs,[20] and superhelices.[21,22] The specific superstructure formed by the amphiphiles is determined by the combination of three terms of free energy:

1. A favorable hydrophobic contribution caused by the clustering of the hydrophobic tails within the interior of the aggregates,
2. A surface term that reflects the opposing tendencies of the molecules to crowd closely together minimizing the unfavorable hydrocarbon-water interactions and to spread apart as a result of the electrostatic (head group) repulsion, hydration and steric hindrance,

3. A packing term that requires that the hydrophobic core of the aggregate excludes water and polar head groups and which, as a consequence, limits the possible geometries of the aggregates.

The aggregation process also makes use of molecular information such as the shape, rigidity and flexibility and dispersity of the monomeric unit. Furthermore, surface properties, charge, polarizability, mass and magnetic dipole also influence the aggregation process. The interactions between the individual molecules within the aggregates can be hydrogen-bonding, (chiral) dipole-dipole interactions, π-stacking, non-specific van der Waals, repulsive steric forces or electrostatic.

A theoretical model predicting the shape-structure relationship between the monomeric units and their aggregates was developed by Israelachvili and was based on statistical mechanics of phospholipids.[23] This model predicts the type of the aggregate formed on the basis of the packing parameter (P), which relates the volume of the molecule (V) to its length (l) and to the mean cross-sectional (effective) head group surface area (a):

$$P = \frac{V}{l \cdot a} \tag{7.1}$$

From this formula it can be deduced that cone-shaped amphiphiles, where $0 < P < 1/3$ form micelles in aqueous solutions (Table 7.1). Wormlike micelles are formed if $< P < 1/2$, whereas vesicles are formed from cylindrical surfactants ($1/2 < P < 1$). Finally, inverted structures are formed when the shape of the amphiphile is dominated by the size of the head group, $P > 1$.[24] The predictions of this model are in agreement with most of the experimental results for phospholipids, and small amphiphiles with conventional aliphatic chains,[25] as well as for several more complicated molecules such as diblock copolymers (*e.g.* amphiphiles consisting of a dendrimer acting as polar head group connected to a polystyrene tail).[21,26,27] However, several authors have criticized this packing parameter approach for predicting the aggregate morphology.[5,28] Deviations of the model have been reported for surfactants containing rigid segments (such as diphenylazomethine,[22,29] biphenyl[30] or azobenzene moieties) or multiple hydrogen bonding units (like urea and acylurea).[31] It is postulated that the rigid aromatic segments can obtain specific orientation with respect to each other (due to interactions such as π-stacking that are manifested as red or blue shifts in the UV-vis spectra of these molecules) and therefore align at the air-water interface to form bilayers. As an instructive example of this concept it is of interest to refer to the *anthocyans*–a group of biological flower pigments- that owe their characteristic colours to the formation of specific assemblies formed by a modulation of π-stacking of their aromatic groups by carbohydrate substituents.[32] This modulation is delightfully expressed by changes in the characteristic wavelengths of the monomers and other UV and CD changes and consequently expressed in to the rainbow of colours seen in plants.

7.2. MOLECULAR ASSEMBLY

In the rapidly evolving fields of nanosciences, which sit at the interface of supramolecular chemistry, surface chemistry and the soft matter chemistry, one of the main goals is to be able to design functional systems possessing architectures with a well-defined, programmed

TABLE 7.1 Relationship between the shape of the surfactant monomer and the predicted aggregate morphology.

Aggregation Morphologies

Effective Shape of the Surfactant	*Packing Parameter ($P = v/a\,l$)*	*Aggregate Morphology*
cone	<⅓	spherical micelles
truncated cone	⅓ - ½	wormlike micelles
cylinder	½ - 1	bilayers, vesicles
inverted (truncated) cone	>1	inverted micelles

Effective Shape of the Surfactant	Packing Parameter	Aggregate Morphology
Cone	$<{}^{1}/_{3}\ P = \frac{V}{a \cdot l}$	Spherical micelles
Truncated Cone	${}^{1}/_{3} - {}^{1}/_{2}$	Wormlike micelles
Cylinder	${}^{1}/_{2} - 1$	Bilayers Vesicles
Inverted (truncated) cone	>1	Inverted micelles

geometry. Using Mother Nature as a guide, it can be foreseen that the primary tool toward this goal is self-assembly. It is therefore of great importance to understand the processes of assembly in great detail in order to engineer through molecular design, and ultimately manipulate molecular systems into supramolecular functional nanostructures.[33,34]

The colloids and simple amphiphiles have been extensively studied in this respect for more than 150 years. Only in the last 20 to 30 years have we reached some understanding on the behavior of super-amphiphilic block copolymer systems. In the above-mentioned systems though, it is still clear that as soon as supramolecular interactions are involved, there are still no definitive rules predicting the resulting structures.

In this chapter we will discuss, almost chronologically and through increasing complexity, the advances in the understanding of the relation between the structure of individual amphiphilic molecules and the superstructures of the resulting aggregates. Successful methodologies have yielded a large variety of non-covalently bonded structures in both solutions and crystals. However, the possibility of regulating size and shape of nanostructures in relation to function remains a current challenge of exceptional interest.[35,36,37,38,39,40,41]

It is our goal to show the scientific knowledge concerning the creation of superstructures has advanced (evolved) from the single level architectures (lipids, small amphiphiles), to systems where the molecular chirality is expressed (phospholipids, aminoacid amphiphiles,

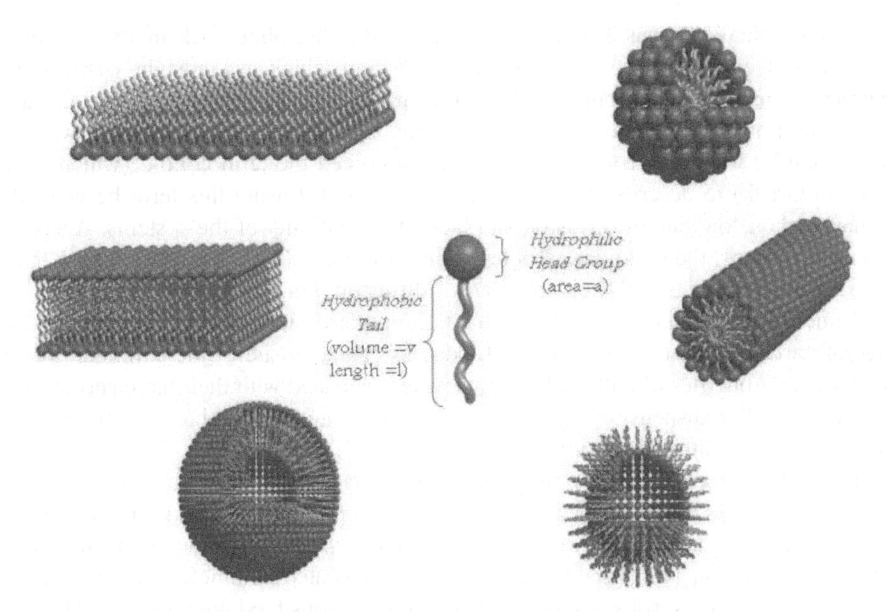

FIGURE 7.1. Aggregate morphologies of single-chain amphiphiles.

gluconamides), via systems which contain different levels of hierarchy (gluconamides, amyloid fibers), to finally programmed systems which form finite structures (*e.g*. fullerene and peptide amphiphiles) (Figure 7.1).

While Nature uses primarily small amphiphilic building blocks, chemists have access to a more vast array of amphiphiles ranging from the natural analogues through to block copolymer "*super amphiphiles*" and more recently the biohybrids and "*giant amphiphiles*". The application of this range of amphiphiles in the construction of nanosized assemblies, some with unique properties, is clearly going to be at the forefront of the future developments in the nanosciences.

7.2.1. Lipids/Colloids

The use of colloids as functional materials dates back to the very early records of civilization; stabilized colloidal pigments were found in stone-age paintings and in writings on papyrus by ancient Egyptians. Soap-making for example has been known to mankind since 2800 BC, as proven by the soap-like substances found in clay cylinders during the excavation of ancient Babylon. Inscriptions on these cylinders state that fats were boiled with ashes but do not refer to the purpose of this first soap. The first known written mention of soap was on Sumerian clay tablets (which are interestingly surfactants themselves, 2500 BC) found in the area of Tigris and Euphrates rivers. The tablets refer to the use of this soap for wool-washing. Another Sumerian tablet (2200 BC) describes a "soap" formulation from water, alkali and cassia oil. The Ebers papyrus, a medical document from 1500 BC shows evidence that the Egyptians combined animal and vegetable oils with alkaline salts to create a substance used for washing and treating skin diseases. In fact, many of the very early technological processes (such as pottery, papermaking and fabrication of soap, cosmetics and dyes) evolved around the manipulation of amphiphilic systems.

Colloid chemistry was established as a scientific discipline back in 1845, with the description of colloidal particles by Francesco Selmi, which in hindsight were the first example of functional nano-objects. He defined the common properties of what he called "pseudosolutions" (suspensions of silver chloride, sulphur and prussian blue in water). It was not until 1861 though that Thomas Graham coined the term colloid (which means "*glue*" in Greek) to describe Selmi's pseudosolutions. By using this term he wanted to emphasize their low rate of diffusion and lack of crystallinity of the systems. According to his descriptions, the colloidal particles were fairly large (with diameters bigger than 1 nm) a fact which explained their low rate of diffusion, but still had an upper size limit of ~1 micrometer, which accounted for the failure of sedimentation of the particles. This range of particle sizes is still considered today as characteristic of the colloidal domain. The intrinsic properties of colloidal systems were associated with their large surface areas (a typical micellar suspension containing 0.1 M of an amphiphile has ca. 40,000 m^3 of interfacial area per litre of solution).

Today, colloid science studies the suspensions of one substance into another. Numerous fundamental studies have lead to a better understanding of the colloidal properties encouraging the extensive use of colloids in a variety of systems of scientific and technological importance (including paints and other coatings, sophisticated ceramics, cosmetics, agricultural sprays, detergents, soils, biological cells and numerous food preparations). However, the increasing complexity of these developing systems exceeds the ability of theoreticians to explain and predict their properties and to a large extent the colloid chemistry remains an empirically derived science. The description of the field by Hedges as "*To some the word colloidal conjures up visions of things indefinite in shape, indefinite in chemical composition and physical properties, fickle in chemical development, things infilterable and generally unmanageable*" remains therefore contemporary.

7.2.2. Natural Surfactants—Biological Lipids.

Nature utilizes surfactants for a variety of additional roles. A natural surfactant, using a strict definition, is a surfactant taken directly from a natural source (isolated by a separation procedure from either a plant or an animal origin).[42] Lecithin, obtained either from soybean or from egg yolk, is probably the best example of a truly natural surfactant. Other natural originated surfactants are the various soap-like surfactants for the removal of fatty/oily substances. These compounds produce a rich lather when dispersed into water and are found in various natural systems (such as chestnuts, in leaves and seeds of *Saponaria Officinalis* (soapwort), in the bark of the South American soaptree *Quillaja saponaria Molina* and in the fruits of *Acacia Auriculiformis* (Figure 7.2).[43,44]

Amphiphilic molecules possessing specific structural characteristics are also one of the main building blocks of membranes.[45] These membrane amphiphiles consist of two hydrophobic aliphatic chains connected through a rigid connector to a hydrophilic head group (Scheme 7.1). The structural characteristics present in these molecules contain all the necessary information, which dictates the aggregation of these amphiphiles into the well-defined superstructures that define both the shape and the properties of the biological membranes (compartmentation, permeability, functionality). The first proof for the hypothesis that the structure of the individual monomer defines the aggregation profile was given by

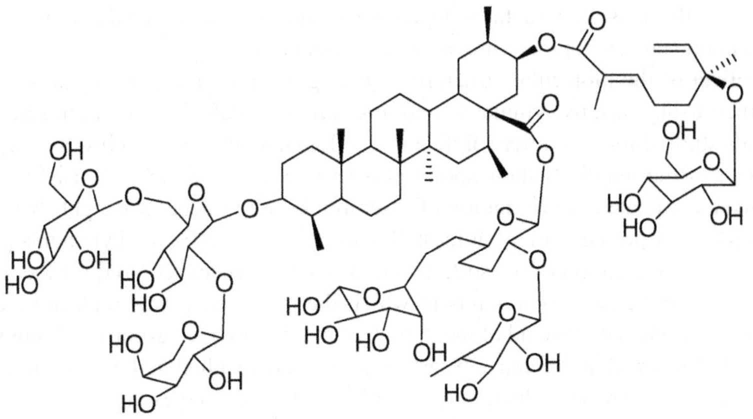

FIGURE 7.2. Chemical structure of a saponin from ***Acacia Auriculiformis***.

the demonstration that isolated biological amphiphiles spontaneously form bilayers when re-dispersed in water.[46]

The first synthetic amphiphiles found to self-organize into bilayers, were quaternary ammonium salts bearing two long alkyl chains **1**.[13,47,48,49] It is interesting to note that these molecules did not contain a connector moiety between the polar and the apolar part, as in the case of the biolipids. While the physicochemical properties of these bilayers were found to be comparable to those of the biological membranes, the synthetic lipids were found to

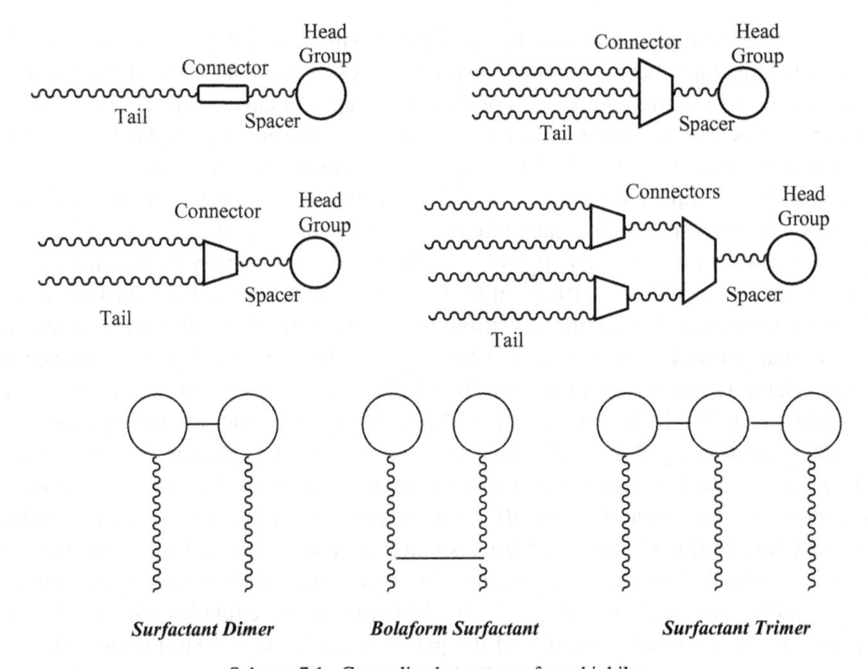

Scheme 7.1. Generalized structure of amphiphiles.

aggregate in differing superstructures depending on the specific molecular structure and the physical circumstances (temperature, concentration, etc.).[50,51]

The effect of the molecular structure on the aggregation morphology was systematically studied using various symmetric and unsymmetric dialkyl ammonium salts. All the compounds studied formed, after sonification colloidal solutions, which were surprisingly independent of the length of their apolar tails (n = 12, 14, 16, 18, 22). Light scattering measurements indicated the formation of assemblies with molecular weights between 10^6 and 10^7 Dalton. A prerequisite of at least 10 carbon atoms on both alkyl chains proved to be essential for the creation of the sufficient hydrophobic forces required for the formation of vesicles and lamellae. Compounds in which the head group was replaced (sulfonate, phosphate or carboxylate instead of the dimethylammonium head group),[52,53] but still possessed alkyl chains with more than ten carbon atoms, also resulted in the formation of stable bilayer membranes. The introduction of a connector moiety to these molecules was found to promote the alignment of the alkyl chains,[54,55,56] while the addition of an extra spacer unit was observed to have a significant effect on the bilayer orientation. Finally the incorporation of fluorocarbon chains within these structures has also proven to have an effect on the balance between the solvophobic and solvophilic parts of the molecules in both organic solvents[57,58,59] and water[60,61,62] leading to much more complex structures.

It is of interest to note that in spite of the fact that biological lipids are often chiral, their chirality is very rarely expressed in the resulting functional assemblies. One of the few examples of this expression of chirality has been observed in lecithin dispersions in water, where the formation of helical intermediates was monitored with polarizing microscopy.[63,64]

7.2.3. Synthetic Phospholipids

Monomeric lecithin analogues functionalized with a diacetylenic function in their fatty acid chains, have been shown to form liposomes in aqueous dispersions and above the phase transition temperature (Figure 7.3). One representative example is the formation of liposomes from 1,2-bis(tricosa-10,12-diynoyl)-*sn*-glycero-3-phosphocholine (**2** with m = 8, n = 9; $DC_{8,9}PC$) above the T_c of 43 °C.[65] A gradual lowering of the temperature until a few degrees under the transition (38 °C), resulted in the transformation of all the liposomes structures into hollow nanotubules,[66] with diameters ranging from 0.4 to 1 μm with aspect ratios of 10-100. The liposome walls were found to vary in thickness from 2 to approximately 10 bilayers (10–50 nm). The effect of the molecular structure and the structural prerequisites for tubule formation were comparatively studied using a series of $DC_{m,n}PC$ analogues (**2**, m = 4–15, n = 17–6).[67,68,69] More specifically, it was proven that the diacetylenic moiety is a prerequisite for tubule formation but that the position of the diacetylene group had little or no effect on the structure of the tubules and that symmetry in the alkyl chain (m equal or nearly equal to n) was preferred. *Using these simple building blocks it was clearly demonstrated that the architectures can be readily tuned by varying the head-group, the ionic strength or the pH of the solutions.* The effects of the alternative head groups and the addition of additives were also investigated. When the trimethylammonium group was substituted by a hydroxyl group, (**3**), tubules were only formed in the presence of certain metal ions, such as Cu^{2+}.[70,71] The formation of these tubules was also found to be dependent on the ionic strength and the pH of the solution.[72] Extreme pH values led to a reduction of the diameter of the tubules and to lipid degradation while addition of

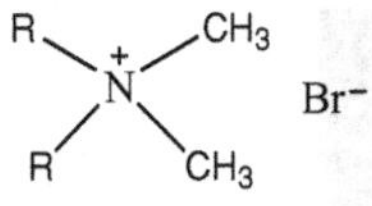

1

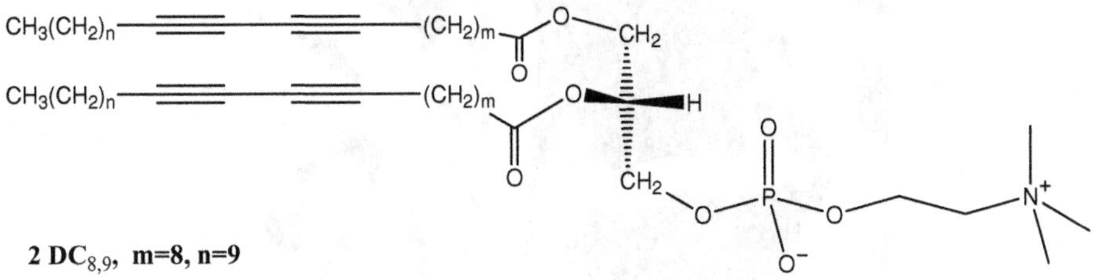

2 $\mathbf{DC_{8,9}}$**, m=8, n=9**

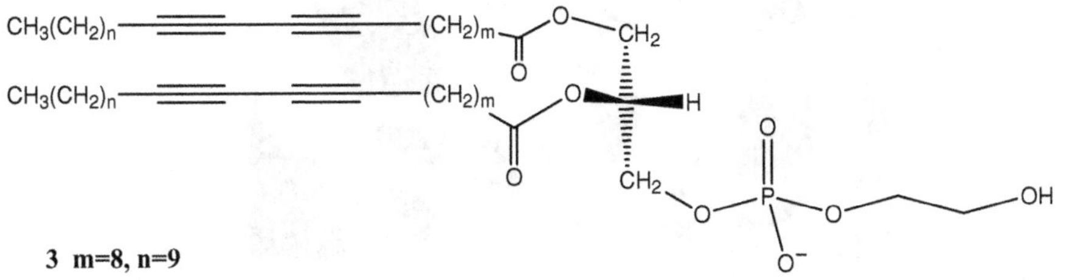

3 m=8, n=9

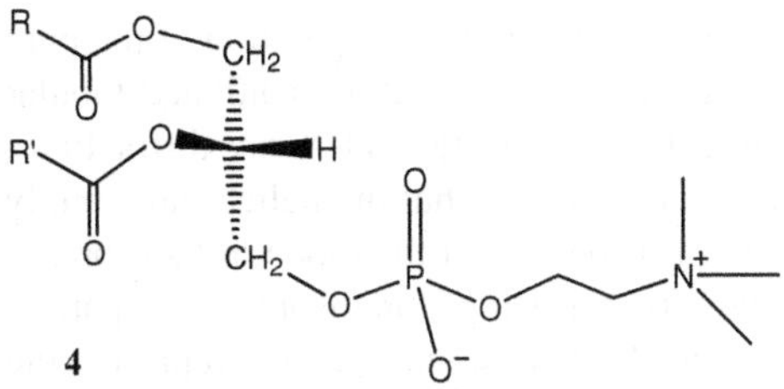

4

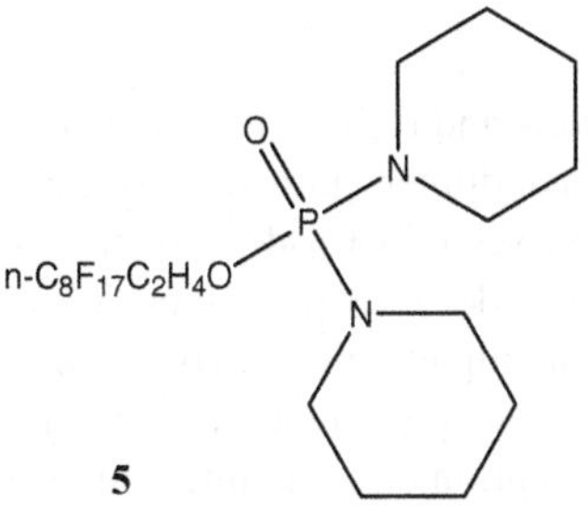

5

FIGURE 7.3. Electron micrograph of helical structures with diameters approx. 5 μm formed from the racemic mixture of $DC_{8,9}PC$ (m = 8, n = 9) Reproduced from Singh *et al.*, *Chem. Phys. Lipids* **47**, 135 (1988) with permission of Elsevier Science.

NaCl or $CaCl_2$ in 1M concentration resulted in shorter tubules. Precipitation of $DC_{8,9}PC$ in ethanol upon the addition of water also led to the formation of tubules, which highlighted their thermodynamic stability and in addition accompanying helical structures were also observed.[72] One very important outcome of this study was the expression of the chirality of the monomer synthetic phospholipids in the resulting helical aggregates. More specifically when the L-amphiphile was dispersed, only right handed helices were observed while the ratio between the tubules and the precipitated helices, as well as their respective dimensions (length, pitch, diameter) were found to be dependent on the ratio solvent/non-solvent and the cooling rate.[73] Interestingly, when racemic mixtures of $DC_{8,9}PC$ were studied both

FIGURE 7.4. Some examples of functionalized lipids.

left- and right-handed helices were observed, due to a lateral phase-separation of the pure enantiomers into their respective aggregates (Figure 7.4).[74] The aforementioned tubular phases of $DC_{8,9}PC$ are reasonably ordered (as confirmed by IR and X-ray diffraction). In order to explain these observations, the authors considered that the helices are simply incomplete tubules. According to this assumption irrespective of the mode of their preparation (precipitation or cooling), the tubules must be formed by one isomer of a pair of enantiomers. The interactions between the chiral head groups do not seem to represent the factor determining the thermodynamics of the gel-to-liquid phase transition.[74] This was clearly demonstrated by the fact that corresponding amphiphilic molecules lacking the diacetylenic function, such as the phosphatidyl choline DPPC (**4**, $R{=}R'{=}n\text{-}C_5H_{31}$) do not exhibit any resolution or lateral phase separation.[75,76] It was therefore concluded that the improved "pseudo-chirality" of the unsaturated fatty acid chains caused by the presence of the diacetylenic group is responsible for the improved ordering of the molecules in the observed superstructures.[74] This group imposes steric restraints forcing the hydrophibic chains to orientate in a more energetically favoured way, parallel to one another and with either a clockwise or a counter-clockwise twist in the resulting lipid bilayer.[74,77]

In general, it has been argued that tubular morphology is an expression of the chirality of the monomeric species, though there are a number of examples of non-chiral surfactants (such as the dimorpholinophospho-amidate **5**) that have been shown to self-assemble into such structures.[78,79] In some contradictory reports it is even argued that chirality is a requirement and that cylinders are in fact sheet-like structures that are curled.[69]

It is of interest for the understanding of the factors governing the aggregation morphology of amphiphiles to also consider the mechanisms that rule this process. In the case of $DC_{8,9}PC$ there is a number of different theories explaining the mechanisms that rule the formation of tubules from liposomes and bilayers.[80,81] The most generally accepted theory for the formation of helical (tape-like) structures from chiral bilayer membranes, includes the bending elasticity of bilayers.[82] According to this theory, the bending energy is influenced by the competition between two basic components, the spontaneous torsion of the bilayer edges caused by the chirality of the molecules and the bending stiffness of the bilayer. Though the parameters for a specific bilayer are difficult to calculate, it has been possible to estimate that the bending energy is minimal when the gradient angle is 45°, a result which is in agreement with most of the experimental observations.

A more extended model,[83,84] leading also to a 45° gradient helix angle, considered also the anisotropic bending force caused by the tilting of chiral surfactants in C_2 symmetrical layers (such as bilayers and S_c^* smectic layers). In the case of crystalline membranes no tilt requirement for anisotropy was found to be necessary since calculations demonstrated that the minimization of elastic energy occurs through the formation of tubes and tubules possessing a circular cross-section. During the consideration of cholesterol aggregates, the anisotropy factor was also worked out to account for helices with other gradient angles. In further extensions of this theory, the effect of thermal fluctuations[85,86] and the formation of helical substructures[87] were also explained. All calculations were consistent with experimental results pointing out that the tubule diameter is independent of the ionic strength and the length of the tubule. *In conclusion, though the predictive value of the theory is not yet fully explored, it predicts that the formation of the tubules is driven by the chirality of the bilayer, the magnitude of which also affects the tubule diameter. Tilting of the molecules with respect to the bilayer is necessary and its alteration results in variation of the tubule diameter.*

Important information concerning the factors affecting the tubule formation and indicating ways to manipulate it and form predesigned nanotudes, was obtained from experimental CD measurements. The significantly enhanced molar ellipticity measured for tubules in comparison to that measured for solutions and vesicles, indicated that the mechanism of tubule formation involves initially a curving of a membrane of chiral molecules into wound ribbons (because of the favoured twist between these molecules) and fusing at a later stage into the final cylindrical tubules (Figure 7.5).[88] Further experimental studies also pointed out an odd-even effect. When a series of $DC_{m,n}PC$ lipids with $m + n = 21$ (1 mg/ml) was studied in methanol/water (80:20), the compounds with even m values exhibited an increasing multilayer character as the value of m increased, while in the ones with odd m the diameter of the tubules was not affected. Accordingly, lipids with even m melt at higher temperatures that the ones with odd m values both in methanol/water and in water.[89]

As mentioned in the previous paragraphs, the striking relationship between the chirality of the individual molecular components and the corresponding helical handedness of the $DC_{8,9}PC$ tubules, has led scientists to believe that the tubule formation is driven by the molecular chirality. However, according to more recent studies the process of tubule formation is more complex than previously thought. It initially involves the formation of enantiopure L_α-phase vesicles which are then transformed to $L_{\beta'}$-phase helical ribbons composed of a nearly racemic mixture of left and right handed helices.[90] In the few minutes following the sphere-to-tubule transition, monomeric lipids from the saturated

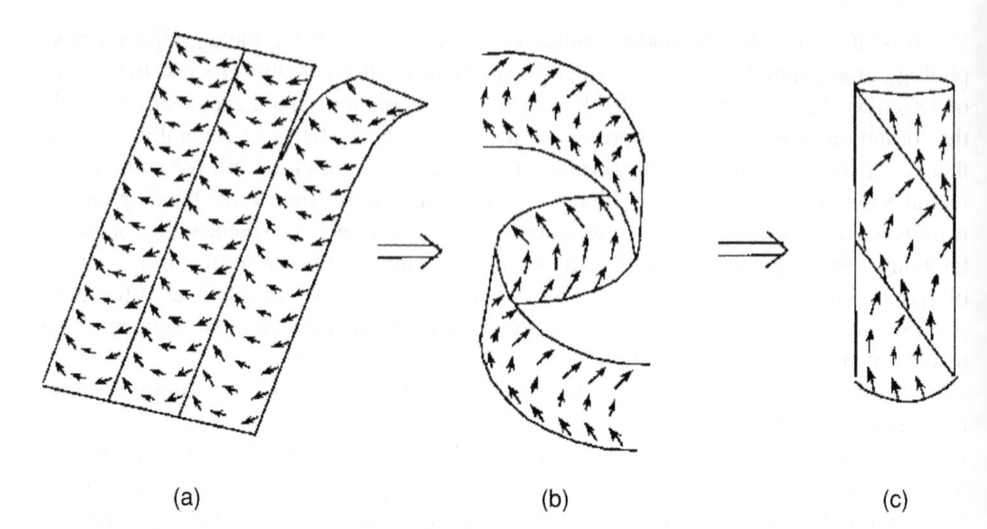

FIGURE 7.5. Scenario for the kinetic evolution of flat membranes into tubules. (a) When a membrane is cooled into a tilted phase, it develops stripes in the tilt direction and then breaks up along the domain walls to form ribbons. (b) Each ribbon twists in solution to form a helix. (c) A helical ribbon may remain stable or may grow wider to form a tubule. Reproduced from Jonathan V. Selinger, Mark S. Spector, and Joel M. Schnur *J. Phys. Chem. B*, **105**(30) (2001).

$DC_{8,9}PC$ solution completely ensheath the helices through the coaxial helical growth of a second outer cylinder. In contrast to the racemic core, the outer cylinders have a uniform handedness which corresponds to the chirality of the monomeric unit. These results have been further strengthened with studies performed on $DC_{8,9}PC$ analogues lacking the phosphoryl oxygen or possessing a methylene ($-CH_2-$) unit instead.[91,92] The fact that all these resembling molecules behave in the same manner suggests that this process is apparently a random membrane chiralization involving a chiral symmetry-breaking mechanism as a consequence of the L_α-to-$L_{\beta'}$-phase transition during which the helices are formed.

Studies on the tubule formation by $DC_{8,9}PC$ racemates are also supporting this new theory. The previously proposed packing theories predicted that tubules formed from non-enantiopure $DC_{8,9}PC$ preparations would exhibit increased tubule diameters, while racemates would only form flat sheets (tubule diameter was predicted to be infinite in the latter case). However, the experimental results did not verify the predictions, showing tubules with unchanged diameters and both helical senses of handedness in the $DC_{8,9}PC$ racemate.[93,94,95]

An alternative interpretation of the tubule formation in the $DC_{8,9}PC$ racemates can be given by new chiral symmetry breaking models not depending on the molecular chirality.[95,96] These models predict equal amounts of left and right-handed helices. In the most qualitative of these models[95] the symmetry breaking process strongly depends on the $DC_{8,9}PC$ tail diynes but is also strongly biased by the chirality of the head group. Very recent studies[97] are now focusing on resolving the role of the chirality of the $DC_{8,9}PC$ in the formation of tubules by studying the achiral $DC_{8,9}PC$ isomer, β-TFL (**6**). In this molecule the lack of a chiral centre eliminates the possibility of enantiomeric microphase separation leading to chiral microdomains as a result of any chiral symmetry model. Surprisingly, the β-TFL preparations produced equal numbers of left and right-handed tubules identical to those of the chiral $DC_{8,9}PC$, suggesting that the molecular chirality cannot be essential for the tubule formation and that a chiral symmetry-breaking process must be the driving force. CD experiments of β-TFL and doped with R-$DC_{8,9}PC$, β-TFL solutions indicated a cooperative "*sergeants-and-soldiers*" chiralization, similar to that observed in achiral polyisocyanate polymer systems[98,99] and achiral bow-shaped smectic liquid crystals.[100] This co-operativity effect suggests that the Ising-based collective chiral breaking models developed for polymer systems[101] might also be adaptable to the $DC_{8,9}PC$ system too. Thus, the latest experimental results indicate that *a satisfactory theory explaining tubule formation should abandon the assumption that the helical wound core and the ribbons are equilibrium structures and locate the genesis of tubule chirality in the hydrocarbon tail diynes.* The challenge for theoretical studies still remains how to explain the formation of nanotubules and nanometer-sized helical ribbons based upon symmetry considerations. This theory should have some symmetrical dependent parameters which would predict the behaviour of any particular system and enable nanoscientists to control the dimensions of the aggregates through subtle changes in the molecular structure or processing conditions.[80]

7.2.4. Phospholipid Analogues

Phospholipid analogues exhibit also very rich aggregation and hence potential nanoarchitecture chemistry and are therefore worth reviewing (Figure 7.6). A representative group of compounds are the phosphate containing amphiphiles (**7**) which are based on C4 sugars, contain phosphates at C1 and C4 and possess stearic acid ester derivatized C2 and C3 hydroxyl groups. The D-threitol derived, S,S-amphiphile (S,S-**7**) forms stacked bilayers at very high concentrations (80% in 2 mM 1,4-piperazinediethanesulfonic acid (PIPES) buffer, pH 7) and platelets at low concentrations (0.1%). In these aggregates, the volume defining head group cross-sectional area is compensated for by a tilting of the hydrophobic tails.[102] This cross sectional area becomes even larger because of hydration causing intercalation of the alkyl chains and formation of other bilayer aggregates. At increased concentration (25% in 0.1 M ammonium formate buffer, pH 8) and after incubation at $-18\,°C$ for long periods of time ($\sim$2 months), the formation of tubules was also observed. These tubules are probably rolled-up aggregates and were found to further assemble in to thread-like structures with diameters of approximately 500 nm).

The comparison of the effects of Ca^{2+} addition to the S,S- and its *meso* stereoisomer (*meso* **7**) lead to interesting observations.[103] The dispersions of 0.1% (w/w) of both compounds revealed the formation of vesicles with diameters of 15–25 nm for the chiral compound and 50–100 nm for its *meso* analogue. These nano-assemblies can be readily manipulated by addition of Ca^{2+} which results in fusion of the chiral compound derived vesicles (diameter 50–100 nm) and in breaking of the vesicles from the *meso*-surfactant (diameter 10–25 nm). The different effect of Ca^{2+} complexation was accounted for by the difference in the resulting structures using a model in which the molecules have their alkyl tails aligned in a preferential conformation. The conformation of the *meso* surfactant allows the intermolecular Ca^{2+} chelation while the more or less parallel alignment of the alkyl chains is maintained resulting in an overall decrease in the size of the head group (through

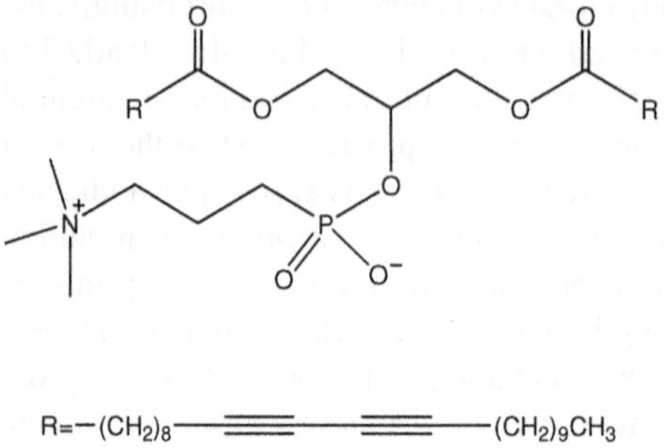

6

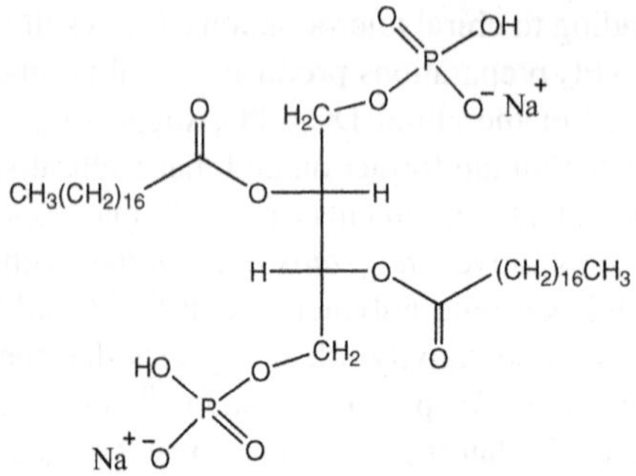

7

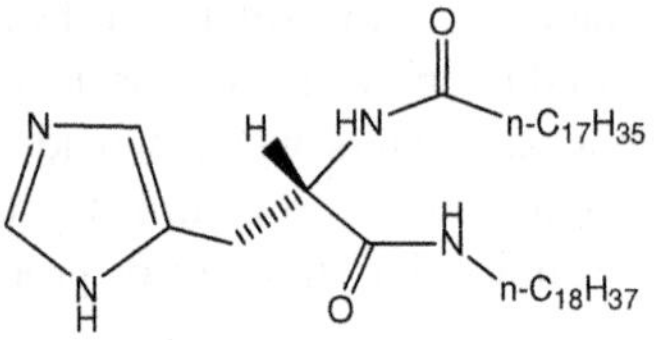

8

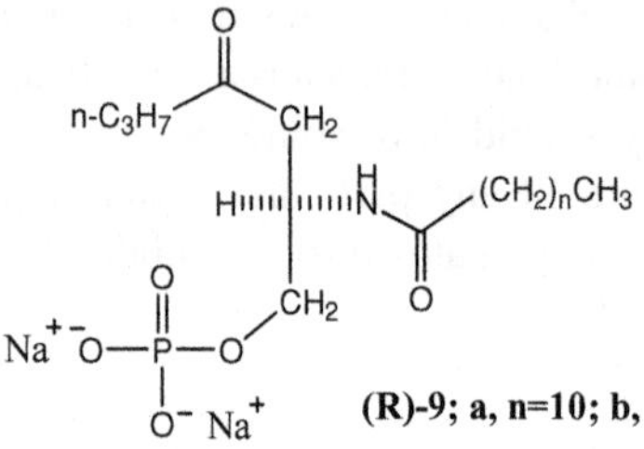

(R)-9; a, n=10; b, n=16

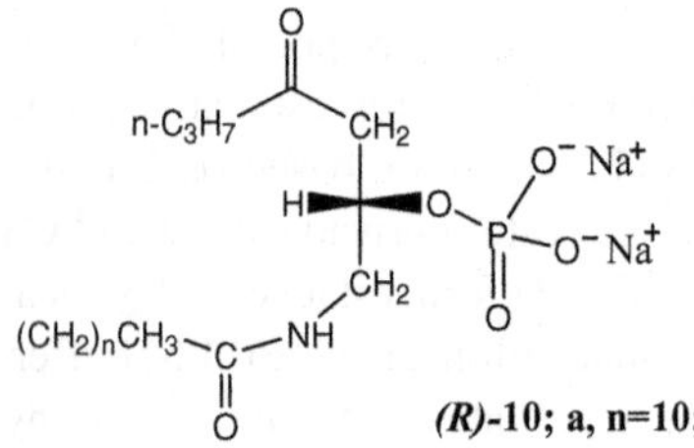

(R)-10; a, n=10; b, n=16

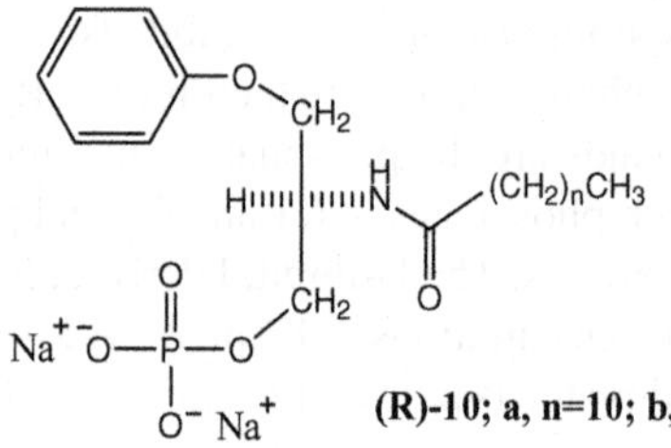

(R)-10; a, n=10; b, n=16

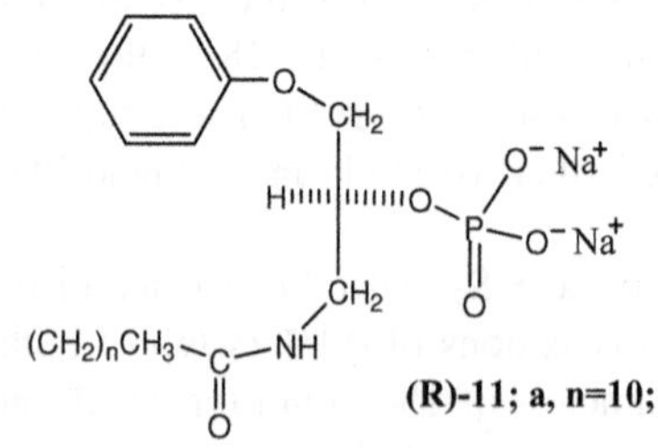

(R)-11; a, n=10; b, n=16

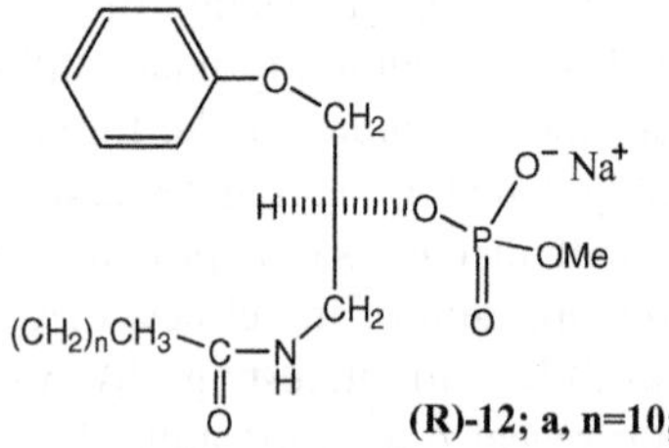

(R)-12; a, n=10; b, n=16

FIGURE 7.6. Examples of several phopholipid analogues.

dehydration) which destabilizes the intercalation of the alkyl chains and forced the bilayer to form smaller vesicles of a higher curvature. The conformation of the chiral compound on the other hand, allows only intermolecular Ca^{2+} complexation resulting in stabilization of the aggregates and allowing the formation of larger vesicles.

An interesting effect on the superstructures of various C4-phospholipid isomers (**7**) was observed upon the addition of an histidine based surfactant (**8**).[104] This histidine based monomer forms at pH 2.5 long, thin fibers which tend to further curl up into right handed, twisted structures. Addition of copper triflate (Cu:**8** = 1:4) results in the formation of boomerang-like scrolls.[105] At pH 6.5 the interactions of **8** with the **7** surfactants proved to be stereoselective.[106] In the case of the *meso* isomer, increase of the **8** molar percentage in the preparation, resulted to the change of the vesicular structures formed in low concentration to extended bilayers and multilayers. Irregular rod-like structures were observed in the doping of the R,R-isomer with **8**, while the S,S-isomer formed helical aggregates in the presence of 30% of the histidine compound suggesting thus that a 1:2 complex of **7**:**8** is the building block of the observed nanostructures.

In an effort to express the chirality of the monomeric phospholipid analogues into the resulting self-assembled nanostructures, new families of analogues incorporated strategically positioned functionalities having the purpose of inducing extra interactions. For example amino groups were introduced in a series of phospholipids based on C3 sugars with the aim to express molecular chirality on the supramolecular level with the aid of hydrogen bonding.[106] Indeed the molecular geometry proved to strongly influence the morphology of the superstructures.[107,108] The position of the phosphate group for example, proved to be crucial to the self-assembling process. The regioisomer with the phosphate group in 1-position (**9**) formed plate-like structures, while the one bearing the phosphate in 2-position (**10**) formed left-handed helical strands (Figure 7.7). The difference in the hydrogen-bonding network arises from the more or less linear conformation of the 2-phosphate **10** amphiphile, which therefore packs more tightly than the **9** allowing intermolecular hydrogen-bonding network formation and hence express the molecular chirality information on to the next level of supramolecular packing.

The left-handed helical morphology observed in 2-(R)-phosphate **10** dispersions proved to be sensitive in small changes in pH and ionic strength which were found to induce a coiling up and the formation of large right-handed superhelices[108] in a process analogous to that of the formation of supercoiled DNA.[109] When the (S)-enantiomer was studied, the aggregation process proved to be enantiomorphic giving rise to right-handed helices. In the case of the (S)-amphiphile **10** the aggregation process was also followed in time. The kinetically favoured vesicles (with diameters between 25 and 100 nm), fused into ribbons (width 50 nm) shortly after their formation while the thermodynamically stable right-handed helices (diameter 20–40 nm, pitch 85 nm) were not observed until 24 hours later. In a further attempt to influence aggregation and the expression of chirality on the supramolecular level, the phenoxy groups of the phospholipids analogues were substituted with butyryloxy groups.[108] Dispersion of the regioisomer 1-phosphate **11** at pH 6.5 revealed the formation of non-chiral vesicles and ribbons while its 2-phosphate analogue **12** formed non-chiral fibers. Upon lowering the pH to 2.5 the ribbons of **11** twisted to afford left-handed helices and ultimately tubular structures. DSC (differential scanning calorimetry) experiments showed a new transition during the formation of tubules, verifying an increase in the molecular organization and in the thickness of the bilayers (from 34Å to 45Å). Helical structures were also observed in dispersions of the monomethylated 1-phosphate analogue **13** in the presence of $CaCl_2$ (Figure 7.8).[110] *The studies with the above mentioned phospholipid analogues proved therefore that the helicity of the supramolecular nanostructures is not*

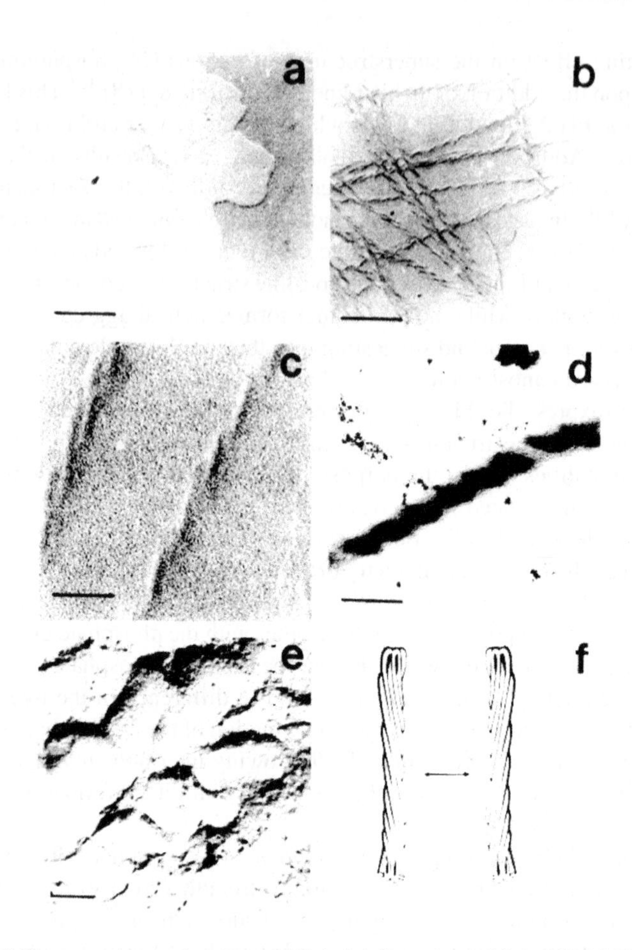

FIGURE 7.7. (a–f) Electron micrographs taken of 2% (w/w) dispersions of **9** and **10**. (a) Planar structures of **37** (Pt shadowing, bar 250 nm). (b–c) Left-handed helices of **10** (Pt shadowing, bar (b) 500 and (c) 100 nm), (d–e) right-handed super helix of **10** ((d) non-stained, bar 500 nm, (e) freeze fracture, bar 125 nm). (f) Schematic representations of the model proposed for the chiral packing of DNA molecules. Reproduced with permission of the American Chemical Society.

only the result of the structure and chirality of individual molecules, but also expresses the complementarity of intermolecular interactions.

7.2.5. Amino Acid and Peptide Amphiphiles.

Early studies on amino acid/peptide derived amphiphiles revealed that their ability to assemble into various architectures was controlled by surprisingly small structural differences. The hydrophobic residues in the peptide amphiphiles have shown to shield themselves from water and to self-assemble in a manner similar to the protein folding. The similarity of their self-assembling properties to those of proteins, initiated numerous studies for the determination of the relationship between the structure of an individual amino acid or a specific peptide sequence and the resulting architecture.

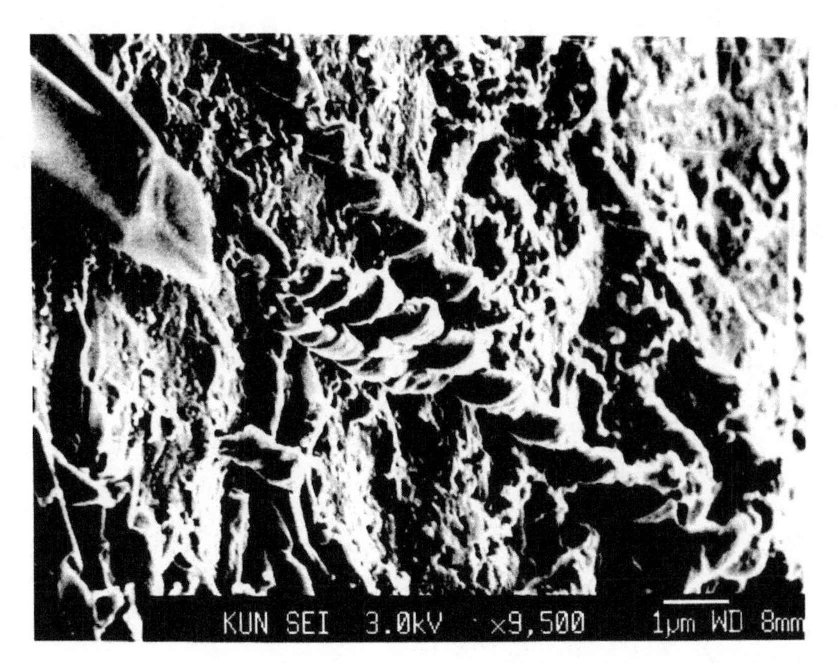

FIGURE 7.8. Scanning electron micrograph of a 0.1 w/v dispersion of **12** in 10 mM $CaCl_2$. (Courtesy of Dr. P. J. J. A. Buynsters).

All the characteristics expressed in the biological lipids and phospholipids are also expressed in peptide amphiphiles. A variety of alkyl and dialkyl oligoglutamates will be initially discussed as representative of such molecules, which helped in the determination of the structure/assembly relationship (Figure 7.9). The latest advances that demonstrate how using this knowledge, peptide-derived materials were tailored for the construction of functional nanostructures amphiphiles will also be discussed.

A variety of different akyl and dialkyl oligoglutamates were studied in this aspect and will be used here as a representative example.[111,112] Dialkyl oligoglutamates **13**[112] of the general formula $2C_mGlu_n$($m = 12$ and 16, $n = 14$), were found (TEM, molybdate staining) to contain lamellar structures with widths dependent on the length of the alkyl chain.[111,112] An L-Glu didodecylamide amphiphile possessing poly(L-Asp) as head group, (3 residues, **14**, n = 3) formed fine fibrous assemblies upon incubation in water. These assemblies gradually transformed into right-handed helical superstructures.[113,114] The lengths of these structures were observed to shrink in the presence of $BaCl_2$, $CaCl_2$, $MgCl_2$, and KCl whilst NaCl, NH_4Cl, and $FeCl_3$ had no effect.[115] The length reduction due to the complexation with the cations was also accompanied with a better packing of the surfactants and, as a result, reduction of the membrane fluidity. Furthermore this reduction in length was found to decrease in the same manner as the selectivity of cation binding by the oligopeptides ($Ba^{2+} \gg Ca^{2+} = Mg^{2+} = K^+ > Na^+ = NH_4^+ = Fe^{3+}$).[116] Addition of excess of Ba^{2+} resulted in the formation of crystallization nuclei and, as a consequence, in shorter crystals. In a different study[117] the left-handed helical ribbons formed by an amphiphilic Pro-trimer **14** (n = 3), were shown to be connected to tubular structures. Even a very a small extension of the peptide chain to 4 Pro residues **14** (n = 4) led to very stable vesicles.

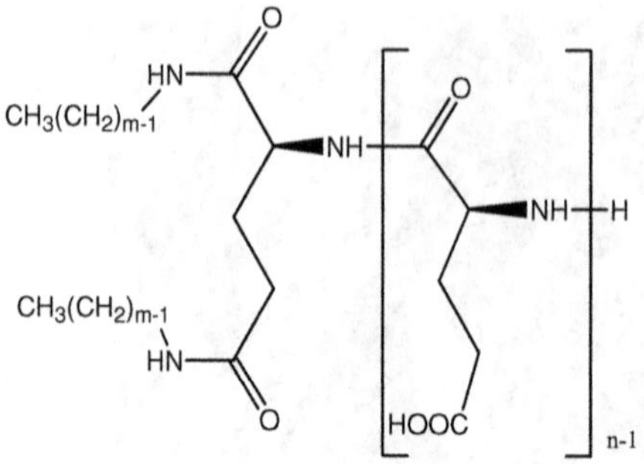

13a, m=12, n=14
13b, m=16, n=14

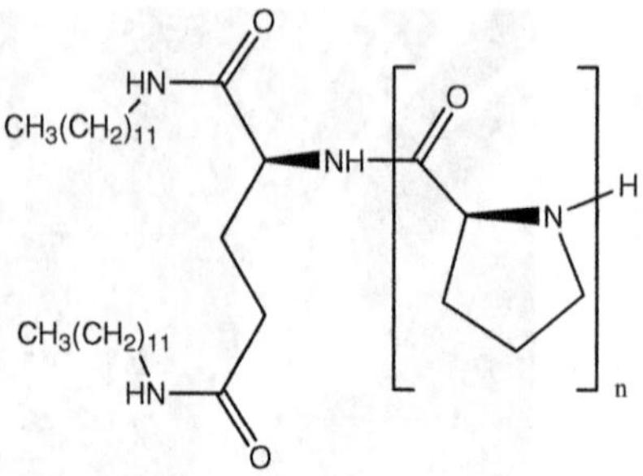

14

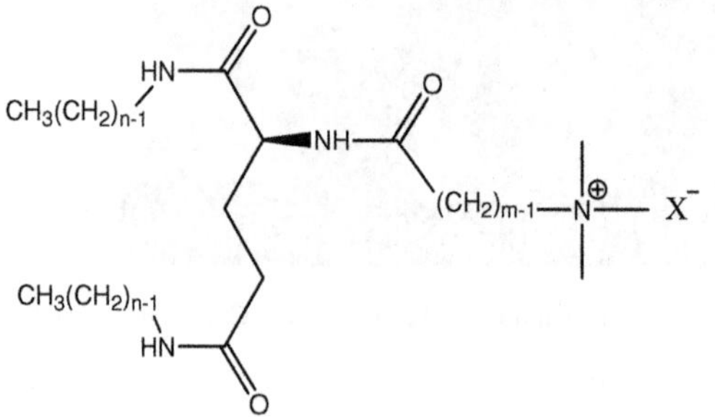

15, $2C_n$-*L*-Glu-C_mN^+
a, m=2, X=Cl,
b, m=11, X=Br

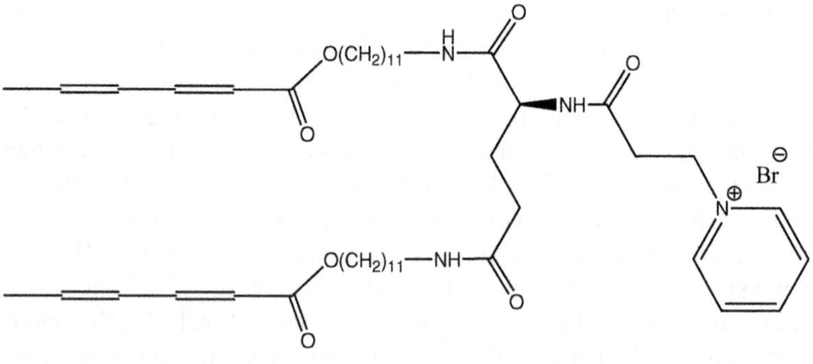

16

FIGURE 7.9. Examples of amino acid and peptide analogues.

It was generally found that double chain ammonium glutamic acid amphiphiles of the general formula $2C_n$-Glu-C_mN^+ (**15**) typically form typical bilayer vesicles above their phase transition temperature.[118] It is interesting to refer to the compound $2C_{12}$-Glu-$C_{11}N^+$ (**15b**, n = 12)[111,119] in which the predominant superstructures were flexible filaments with

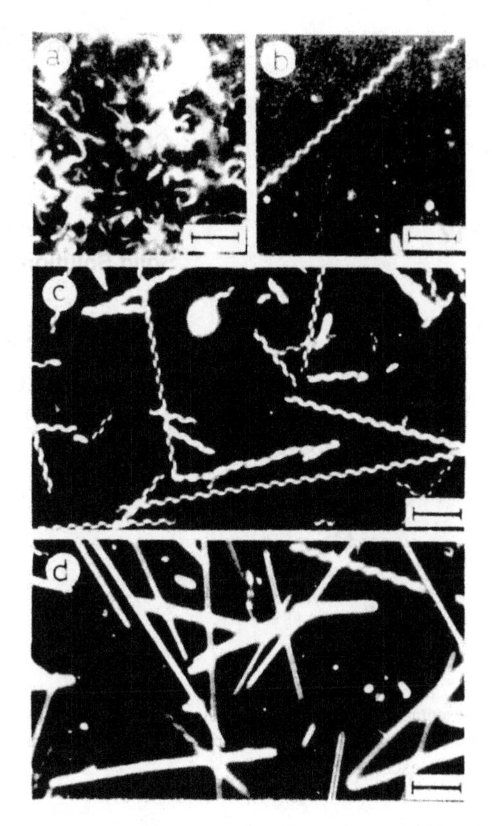

FIGURE 7.10. Dark-field optical micrographs of aqueous dispersions of 2C12-L-Glu-C11N+ (65b, n = 12) (10–3 M). Aging condition: (a) 20 °C, several hours; (b) 15–20 °C, 1 day; (c) 5–6 hours after (b); (d) after 1 month at 15–20 °C. Bars, 10 μm. Reproduced with permission of the American Chemical Society.

small lengths (length 5–50 μm) accompanied by a small percentage of vesicles (average diameter of 1–10 μm). Aging of these samples (24 hours at room temperature) resulted in a twisting of the filaments in to helices, while even further aging resulted in rod-like structures (Figure 7.10).[111,119,120] Interestingly the helical pitch of the twisted tapes or the helical ribbons remained constant during slow gradual growth, suggesting that the transformation occurs through a broadening of the tape. In these studies an expression of the chirality of the monomer unit was also observed as separate enantiomers were found to result in defined helices of opposite handedness upon aging. The L-amphiphile derived helix was found to be right-handed while that of the D-amphiphiles left-handed. The racemic $2C_{12}$-(*DL*)Glu-$C_{11}N^+$ (**15b**, n = 12) on the other hand was found to form elastic fibres instead of helices. The helices of the separate enantiomers were found to be stable below the phase transition temperature of the respective bilayers or membranes,[111] but to transform instantaneously to spherical vesicles when heated. It was proven that at temperatures close to T_c, the pitch of the helix becomes smaller causing initially the formation of tubes, which in turn break up to form vesicles. The chemical structure of the individual amphiphiles also strongly affects the resulting aggregates apparently, linking alkyl spacers (high *m* values) are crucial for the formation of helical structures.[111] In all the examples of glutamic acid

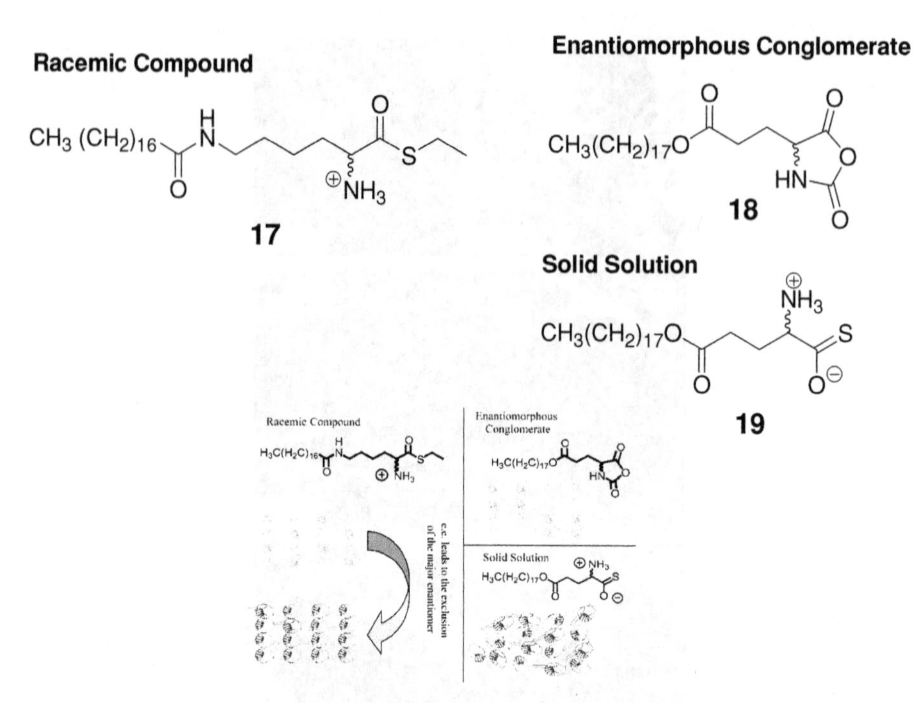

FIGURE 7.11. Schematic representation of the tree types of crystals formed by the two-dimensional assembly of the enantiomeric structures shown above at the air water interface. Each hand represents an enantiomer within the crystal, e.e. stands for enantiomeric excess.

amphiphiles mentioned above the dimensions, the enantiomorphism and the presence of helices as intermediates to the final cylindrical structures are all features analogous to those observed in synthetic lecithins.[72]

Another interesting category within these glutamate surfactants are amphiphiles with a single unsaturation or a polymerizable group in their alkyl chain.[121] For example, the glutamate amphiphile (**16**) possessing a pyridinium head group and a photoreactive 2,3–hexadienoyl (sorboyl) hydrophobic tail resulted in the formation of helical aggregates (4–6 nm thick, diameter across the helix 25–30 nm). In these aggregates, CD indicated that the sorboyl groups were stacked below T_c and as a result reacted 25 times faster than in temperatures above T_c resulting into nanometer-sized chiral tubular aggregates (as compared to the twisted fibers formed above T_c).[122,123,124]

The formation of two-dimensional nanocrystals, by peptide amphiphiles is also influenced by the chirality of the peptide building block.[125] Three types of two-dimensional crystals were observed after the assembly of the functionalized peptide amphiphiles **17–19** shown in Figure 7.11 (above) at the air-water interface, which was followed by polymerization. These two dimensional crystals include: (*i*) a racemic compound, in which each enantiomer is packed with its mirror image in a crystalline order, (*ii*) enantiomorphous conglomerates, in which each enantiomer is segregated into small domains, and (*iii*) a solid solution, in which all molecules are randomly distributed without crystalline order. Interestingly, in the case of the two-dimensional system arising from the racemic compound,

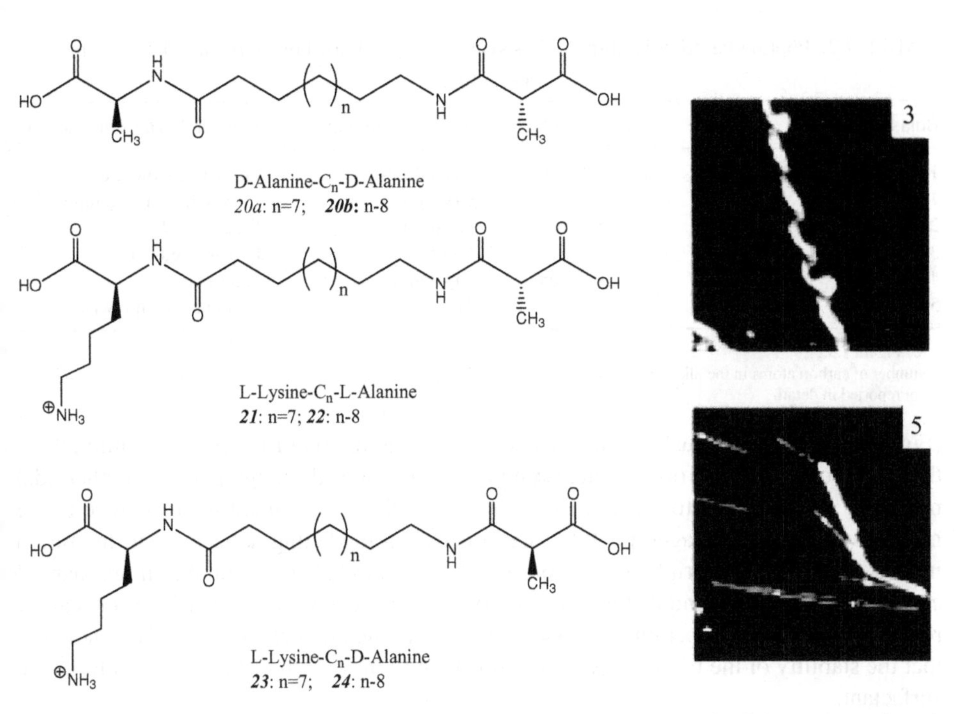

FIGURE 7.12. Structures of the bola-amphiphiles 20–24. Scanning force micrographs of the aggregates formed by 22 and 24.

the self-assembly leads to the exclusion of the enantiomeric excess of the pure form which makes this result potentially useful for the resolution of chiral polymeric amphiphiles.

Furhrhop *et al.*, constructed a series of bola-amphiphiles formed by tethering two amino acids with an alkyl chain and found that they formed crystallites, nanofibers and aqueous gels.[126] They observed that very small structural changes (such as the addition of one methylene unit to the alkyl chains) led to pronounced changes in the resulting superstructures (Figure 7.12). For example, when 1% aqueous solutions of the bola-amphiphiles **20–24** (pH 2.0 to 2.5) were slowly cooled down, highly organized nano superstructures were formed with morphologies directly dependent upon the structure of the monomer unit (Table 7.2). An odd-even effect was also observed in the different superstructures (crystallites vs. fibers) or the different morphologies expressing the difference in the packing constraints of odd and even chain lengths. Interestingly, a single inversion in the chirality of one stereochemical center (**22** vs. **24**) led to a dramatic change on the macroscopic morphology and symmetry of the aggregated structures[127] presumably by influencing the hydrogen-bonding networks within the aggregates. Using this bottom-up approach it was demonstrated for the first time that architecture of nanostructures can be readily tuned.

The progress in understanding the properties ruling the aggregation of molecular amphiphiles has allowed the *rational design* of peptide based amphiphilic molecules that self-assemble into predictable aggregated structures. Inspired by the structure of biological scaffolds, Tirrell *et al.* created systems that mimic these structural motifs.[128] They demonstrated that an amphiphile tethered with collagen-like peptide sequences

TABLE 7.2. Peptide-based bola-amphiphiles synthesized by Furhrhop *et al.*, and their assembling properties.

Bola-amphiphile[a]	n[b]	Amino acid	Type of Assembly	Structural details of assembly
1a	11	D-Ala, D-Ala	Crystallites	Sheet-like structure
1b	12		Crystallites	Curved helical assembly
2	11	L-Lys, L-Ala	Crystallites	Precipitates[c]
3	12		Fibers	Left-handed helices
4	11	L-Lys, D-ala	Crystallites	Precipitates[c]
5	12		Fibers	Ribbons with no twist

[a] See Figure 7.3.
[b] Number of carbon atoms in the alkyl chain of the bola-amphiphiles.
[c] Not reported in detail.

(Figure 7.13) spontaneously forms a triple-helix in the region of the peptide—mimicking the corresponding motifs of biological systems- and that it further aggregates into spheroidal micelles.[128] The ability to manipulate these triple helices is of great interest since in the natural systems the collagen peptides are effective in mediating adhesion and migration in cells only when this triple helix is present. For an alkyl chain containing more than 17 carbon atoms, it was found that the triple helix is disrupted at room temperatures but can be restored at elevated temperatures in which the crystalline alkyl chains melt. This indicates that the stability of the triple helix is very sensitive to the length of the alkyl chains of the surfactant.

Since the peptide amphiphiles are biologically-inspired systems, the structure of the peptide region which is incorporated into the amphiphile can be responsible for the introduction of a wide range of functions. Using such an approach, Chmielewski *et al.* have used the peptide based amphiphile **26** (Figure 7.14) in drug design.[129, 130] This amphiphile which consisted of two peptides mimicking the interface of HIV-1 protease, linked by an aliphatic chain (Figure 7.14b) showed a great potential in inhibiting the protease. The possibility of optimization of this potency by varying the residues on the amphiphile was also explored by

(Gly-Pro-Hyp)$_4$ - IVH1 - NH$_2$

25

IVH1 = Gly-Val-Lys-Gly-Asp-Lys-Gly-Asn-Pro-Gly-Trp-Pro-Gly-Ala-Pro

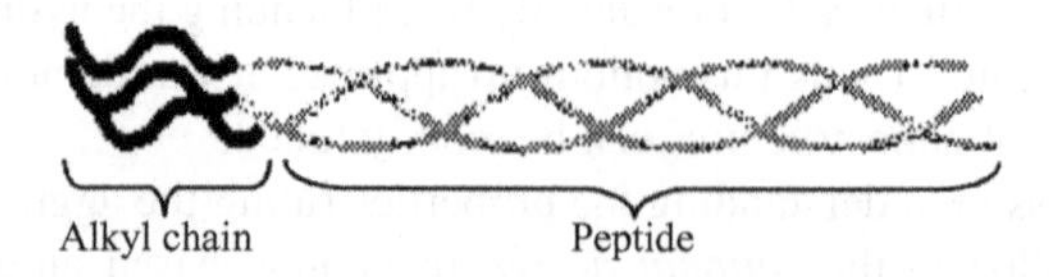

FIGURE 7.13. Schematic representation of the triple-helix formed by double tail amphiphiles that contain the peptide sequence of type IV collagen. The structure and sequence of peptides in the amphiphile is shown below the illustration.

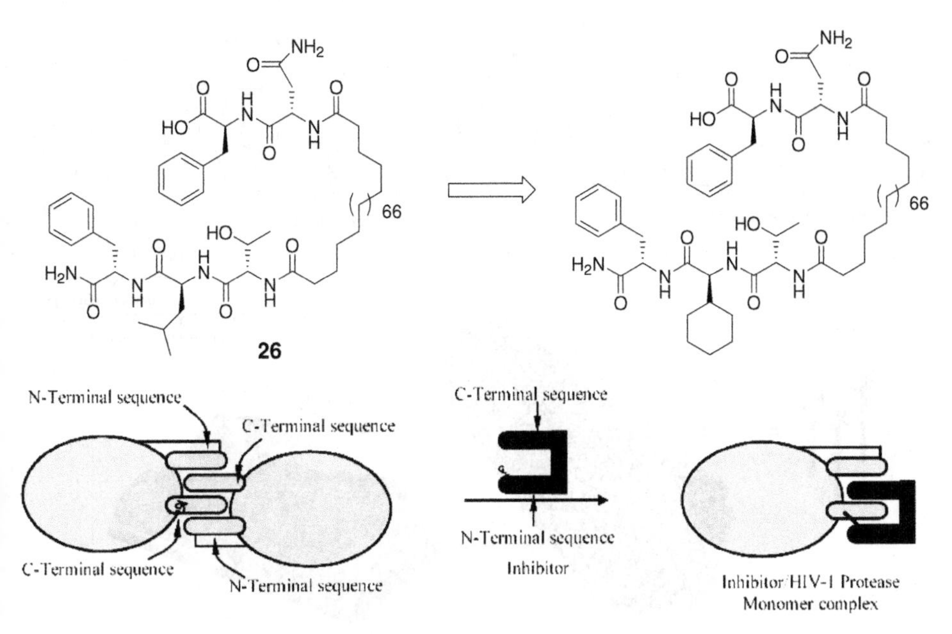

FIGURE 7.14. Schematic illustration of the inhibition of HIV-1 protease homodimer by amphiphiles that tether two peptides mimicking the interface of the protease. Optimization of the potency of the inhibitor is achieved by varying the residues on the amphiphiles, as shown below the illustration.

the same group by the synthesis of a second generation of amphiphiles which were 10-fold more active than their precursor. Interestingly, the different configurational degrees of freedom at the binding site present at amphiphiles with varying lengths of the hydrophobic chain influenced significantly the inhibitory action of the surfactant. Since the amphiphilic nature of the molecules is strongly dependent on the chain length, it is postulated that the process of self-organization of the molecule (and thus the process of self-association) determines its inhibitory potency.[131] This postulation remains to be investigated.

In an effort to synthesize artificial bone materials arising from peptide based structures, Stupp *et al.* synthesized a series of surfactants that reversibly assemble into nanometer-sized fibers (Figure 7.15).[132,133] One of the more important aspects of this work is probably the fact that Stupp was able to rationally design these molecules. The sequence of peptides within the amphiphile comprised a hydrophilic domain of 11 amino acids and an alkyl chain of 16 or more carbon atoms. The peptide domain itself, combined four different subgroups, each designed to introduce a different function. More specifically (Figure 7.15), the four consecutive cystein residues were aimed to polymerize the aggregates (thiols can be reversibly transformed to disulfides), the three glycines to provide conformational flexibility within the head group, the phosphorylated serine residue to introduce mineralization (through its strong interaction with divalent ions) and the RGD ligand to mediate cell adhesion. Surprisingly, all different domains were able to carry out their preprogrammed tasks since it was shown that the assembly of the peptide amphiphile **27** could be triggered by the addition of acid, divalent ions and oxidizing agents (leading to the formation of disulfide bonds). Furthermore, the assembled bundles of fibers were successfully used for templating the mineralization of hydroxyapatite ($Ca_{10}(PO_4)_6(OH)_2$) from $CaCl_2$ and Na_2HPO_4.

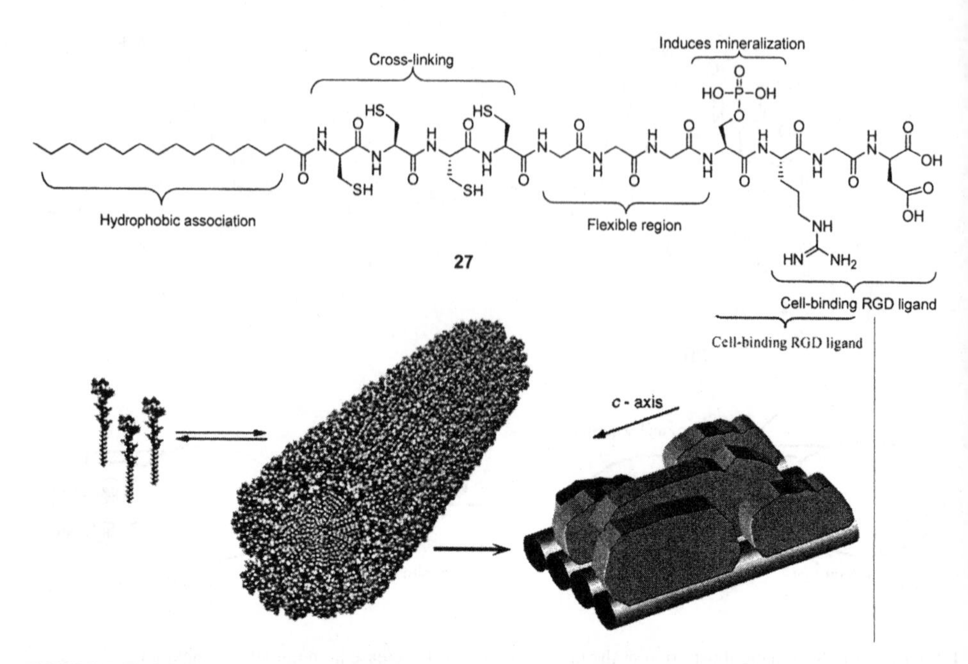

FIGURE 7.15. Molecular structure of a peptide amphiphile that self-assembles into a cylindrical aggregate. The five key structural elements of the peptide are highlighted. The cylindrical aggregate assembles into fibers that template the mineralization of the hydroxapatite crystal. The *c* axis of the crystal is aligned with the long axis of the fiber.

Many amphiphilic peptides, consisting exclusively of natural L-amino acids, were synthesized by Zhang and coworkers.[134,135,136,137,138] Among these molecules, the surfactant-like peptides possessed a common motif with their polar region constructed of one or two charged amino acids and the non-polar of four or more consecutive hydrophobic amino acids.[137,138] The first surfactant peptide they reported on, the V_6D, was composed of a sequence of a negatively charged aspartic acid residue followed by six hydrophobic valine residues. The peptide self-assembles into aqueous solutions to afford tubular structures with diameters varying from 30 to 50 nm (as revealed by dynamic light scattering and transmission electron microscopy) and lengths of several microns (Figure 7.16). The observation of nanosized vesicles within these samples suggested a tunable dynamic behaviour. An interesting demonstration of the effect of the chirality at the molecular level in the superstructures was provided by studies on the eight-residue peptide KFE_8 (of sequence FKFEFKFE).[136] Left-handed ribbons with a regular pitch at the nanometer scale were formed by the right-handed peptide backbone in β-strand conformation. The precise mechanism ruling the nucleation and the growth of the fibers is not clear, though there is intense research in this area due to the relevance for protein conformational physiological and pathological processes. In this model system, three theoretical models have been proposed (a molecular model,[139] a semi-continuum model,[140] and third, a fully continuum, field-theoretic model[141]) which have helped in gaining some understanding of the assembling process. Experimental evidence has revealed that the fiber formation is a multi-stage process with distinct intermediate structures[142,143,144,145] and therefore efforts are now focusing on the

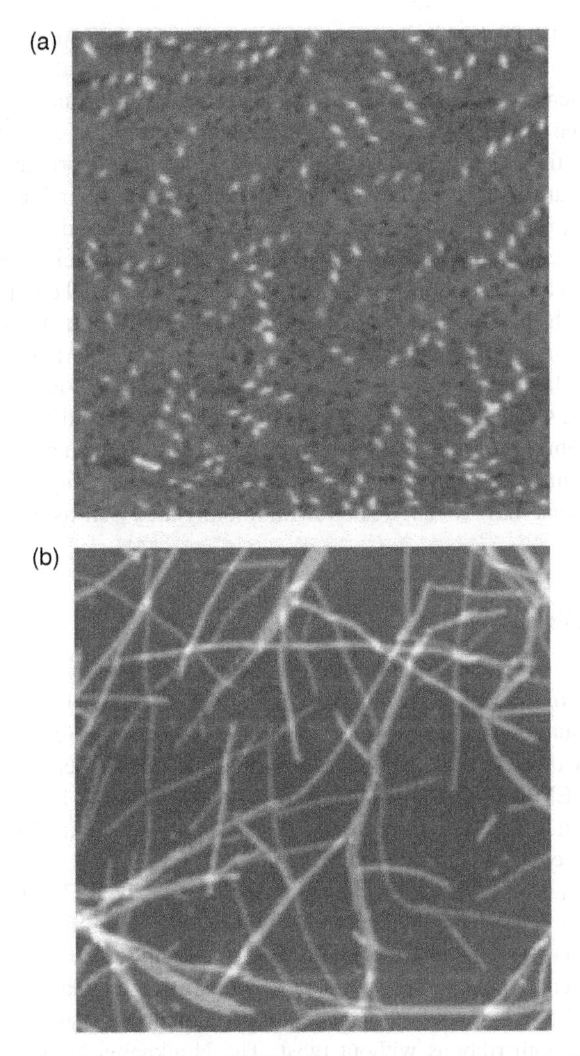

FIGURE 7.16. The peptide KFE_8 (of sequence FKFEFKFE) self-assembles in aqueous solution into left-handed helical ribbons. (a) Atomic force microscopy image (500 nm × 500 nm) of a peptide solution deposited over mica 8 min after preparation. (b) Same sample, 4 days after preparation (1 μm × 1 μm).

characterization of these intermediates for the understanding of the amyloid fibril formation. There are several diseases, including bovine spongi- form encephalopathy, Alzheimer's disease, type II diabetes and Creutzfeld–Jakob disease, connected with the formation of such amyloid fibrils. These nanofibers are very well ordered and possess remarkable regularity and, in some cases, helical periodicity. The mechanism whereby they undergo self-assembly follows on from many of the conclusions observed for molecular amphiphiles and is the focus of much attention.[146,147,148,149]

Various geometrical shapes have been observed upon addition of monovalent alkaline cations to the peptides.[135,150] SEM and AFM studies revealed interwoven nanofibers (with diameters of 10–20 nm) building up matrices with 50–200 nm pores.[135,136,137] The replacement of alanine with more hydrophobic residues (such as Val, Leu, Ile, Phe or Tyr) enhanced the tendency of the peptides to form matrices. These nanofibers are similar in scale to extracellular matrices that are crucial in allowing a variety of cells to adhere together to form functional tissues. Studies revealed that they indeed support various types of cell attachments, but that they also enable their proliferation and differentiation.[135,136]

The nanotubes made from the spontaneous assembly of peptides have numerous potential applications. Because of their natural architectures and their specific assembling properties, they can be used as biomaterials in medicine or as model compounds to investigate the origin of fibril-caused diseases. In addition they have application as bio-scaffolds in other areas of nanotechnology. One may envisage for example that they can serve as templates for metallization to form conducting wires immobilized onto a surface upon removing the organic scaffold.[134,151] In such an approach, Matsui and coworkers have functionalized the nanotubes formed by peptide bolaamphiphiles[152] with a metalloporphyrin[153] or the protein avidin.[154] The avidin coated tubes showed an ability to specifically bind to gold surfaces that have been treated with biotinylated self-assembled monolayers (SAMs).[154]

7.2.6. Gluconamides

Glucoamides are readily formed from the coupling of alkyl amines with simple sugars. Aqueous gels of N,n-alkyl aldonamides such as the N,n-octyl-D-gluconamide D-**28** (D-Glu-**8**) were found to readily form chiral superstructures when dissolved at high temperatures and cooled below 80 °C.[155] The morphology of these superstructures as revealed by freeze fracture EM and negative staining involves right-handed helically twisted ropes (diameter 125 Å, pitch 180 Å, gradient angle 35°) and stacks of "coins" (diameter 160–180 Å, thickness of the coin 70 Å).

Extensive studies on different gluconamide analogues (Figure 7.17) reveal that the resulting superstructures vary significantly. N-methylated gluconamides **29** do not form gels highlighting the importance of amide function for self-assembly. Gels consisting of rope-like structures are formed at lower temperatures by shorter chain gluconamides (hexyl, heptyl), whilst N-Alkanoyl-N′-gluconoyl-ethylenediamines, containing 2 non-methylated amide groups **30** gave smooth ribbons without twist. The N-alkanoyl-N-methyl-N′-gluconoyl-ethylenediamines **31** investigated (C6, C7, C8, C10) only the decanoyl derivative gave a gel, which, like that of the single non-methylated octyl gluconamide **28**, was found to contain right-handed helical ropes (diameter 100 Å). It is obvious that not only the van der Waals attractions between the alkyl chains but also the presence and nature of the intermolecular hydrogen bonds of the amide network(s) significantly alter the aggregation profiles. *The presence of multiple amide functions in the gluconamide monomer can not only induce stabilization but might also induce a negative misalignment which can destroy the formation of well-defined superstructures.* In the case of anhydrous N,n-octyl-D-gluconamide (D-Glu-**8**, **28a**) single X-ray diffraction studies revealed an enantiopolar crystal structure[156] formed by a head-to-tail crystal packing.[157] Investigations of the aggregates revealed the presence of 'bulgy double helix' assemblies, which consist of bilayer-thick single strands.[158] Detailed investigations of these structures (by electron microscopy using 1% phosphotungstate staining and image processing)[159] showed almost crystalline

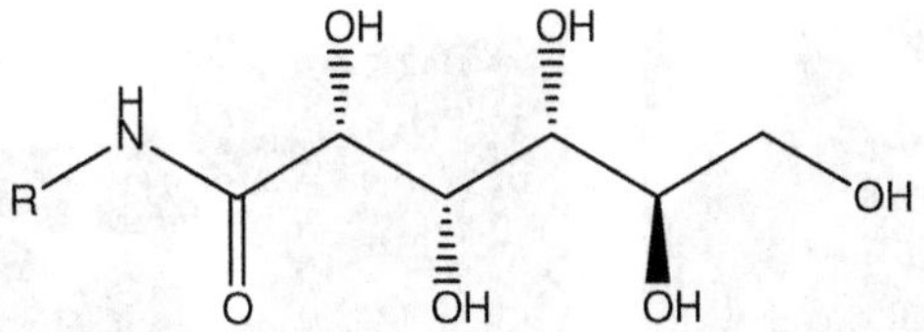

a, R=*n*-octyl
b, R=*n*-hexyl
c, R=*n*-heptyl
d, R=*n*-dodecyl
e, R=*n*-octadecyl

D-**28**, *D*-Gluconamide

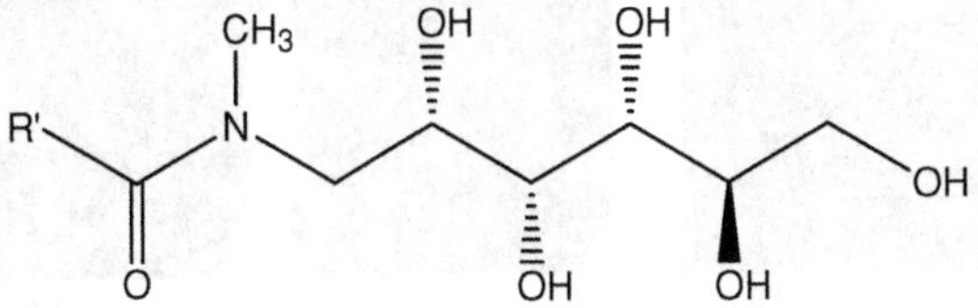

a, R'=*n*-pentyl
b, R'=*n*-hexyl
c, R'=*n*-heptyl
d, R'=*n*-nonyl
e, R'=*n*-undecyl

D-**29**, *D*-*N*-alkanoyl-*N*-methyl-glucamide

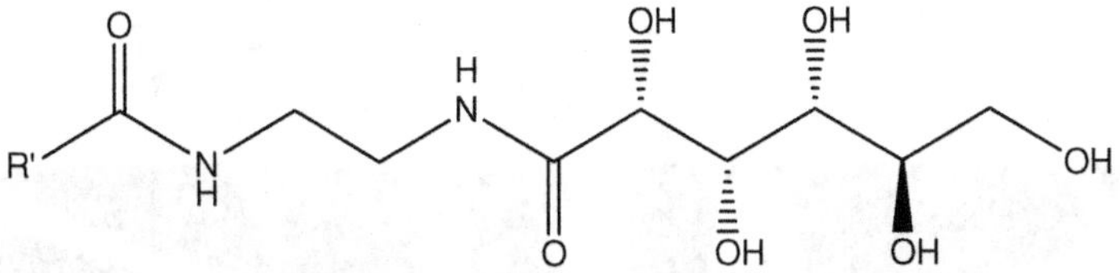

D-**30**, *D*-*N*-alkanoyl-*N'*-gluconoyl-ethylenediamine

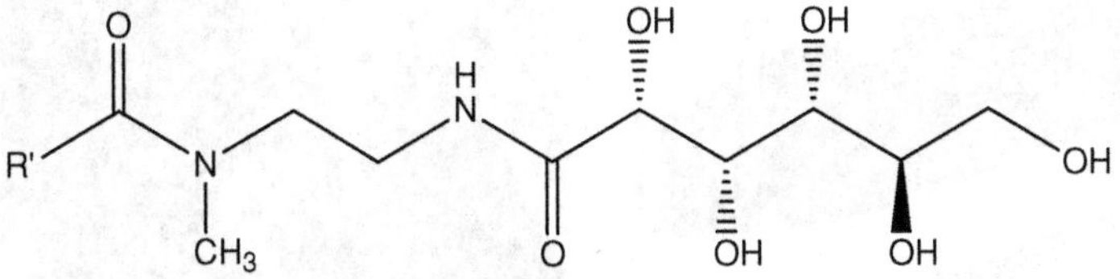

D-**31**, *D*-*N*-alkanoyl-*N*-methyl-*N'*-gluconoyl-ethylenediamine

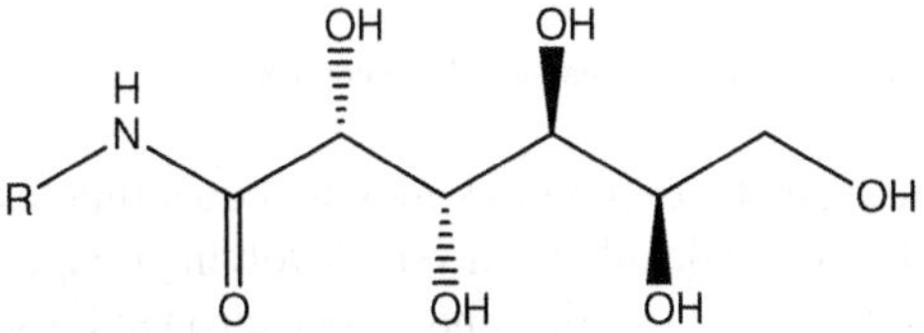

D-**32**, *D*-Galactonamide
a R = n-dodecyl
b R= n-dodeca-5,7-diynyl

FIGURE 7.17. (a)–(b). AFM height images of the rod-like structures of *N*-*n*-octyl-D-gluconamide (**90a**) on graphite. The cross-section profiles determined horizontally along the middle of (b) are shown below the image, where the vertical distances between the pairs of adjacent arrows are 9.6, 7.8, and 7.7 nm on going from left to right. (c) Zoomed-in part of the image in (a). The cross-section profile determined along the B-B line in (c) is shown above this image, where the vertical distance between the arrows is 8 nm and the small corrugations along this profile has a period of 9 nm. (d) Zoomed-in part of the image in (b) The contrast covers height variations in the 0–40 nm range in (a) and in the 0–50 nm range in (b). Reproduced with permission of the Royal Chemical Society.

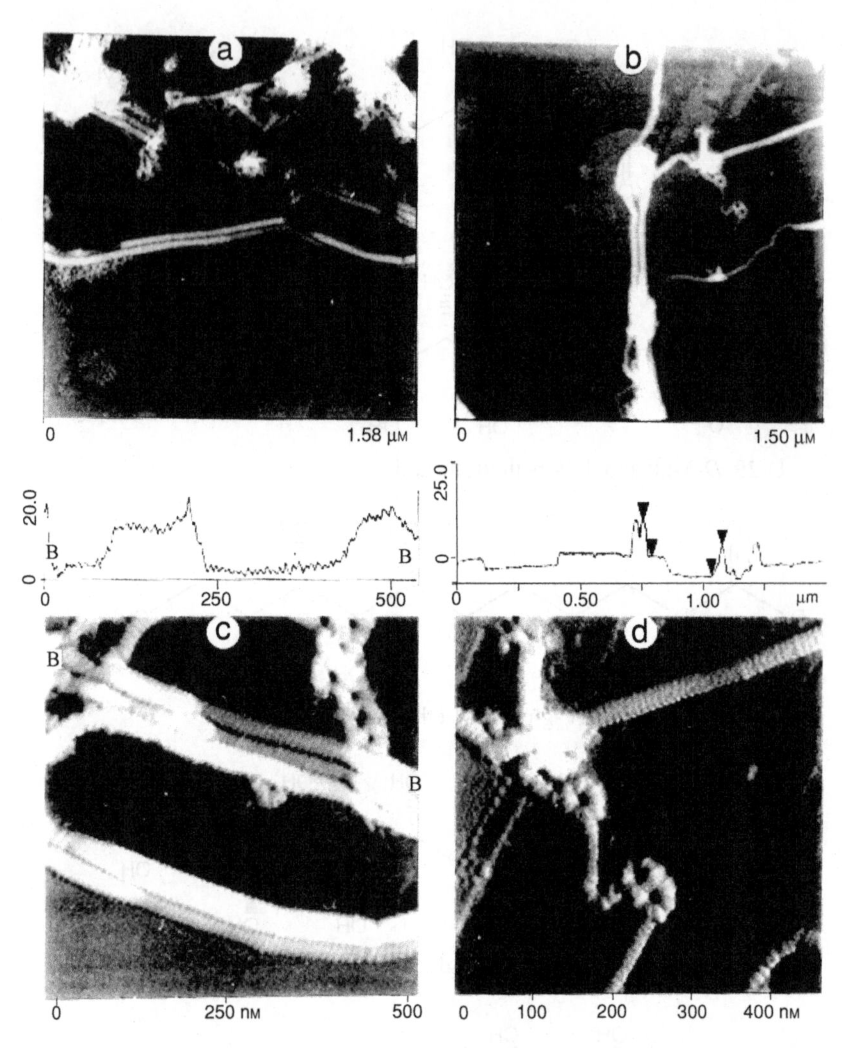

FIGURE 7.18. Some examples of gluconamide analogues.

2-dimensional arrays of fibers suggesting that the structures consist of quadrupole helices of threads. This model was later refined by more detailed analysis (involving images obtained by phosphotungstate staining followed by rapid freezing and cryoelectron microscopy) to a structure consisting of quadrupole helices of threads.[5,160] Interestingly, during this detailed analysis the handedness of the structures of D-Glu-8 obtained by phosphotungstate staining and cryomicroscopy was found to be opposite to that obtained during the measurements of Pt-shadowed dried gels. When D-Glu-**8** adsorbates on mica and graphite by AFM[161] were studied, left-handed fibers were also observed (Figure 7.18). A slightly tilted striation on the fibers, was considered to be reminiscent of 'rolls of coins' which preceded the fiber formation.

Enantiomorphic superstructures were also observed for the N,n-octyl-L-gluconamide, while the racemic mixture was found to lead only to non-fibrous, non-twisted platelets.[158]

The observed structures were explained by a "chiral bilayer effect" mechanism proposing that only the enantiomerically pure compounds can lead to the formation of helical fibers which in turn slowly rearrange to enantiopolar crystal layers (Scheme 7.1). Within the micellar fibers, the polar head groups are oriented toward the aqueous environment and must therefore go through an energetically unfavourable -slow- dehydration followed by a 180 ° to form the enantiopolar crystals.

A comparative study of all the diastereomers of N,n-octyl-gluconamide which aimed at resolving the effect of configuration of the stereocenters present in the carbohydrate head group led to interesting conclusions. In this study, all diasteromers possessing a D-(**28a**, **32a–38a**) or L-configuration, as well as the racemates of galactonamide (L-**32a**), mannonamide (L-**33a**), and gluconamide (L-**28a**) were examined in respect to their aggregation profiles.[162] The gels formed by both the galactonamide and mannonamide were significantly more stable than those studied in the aforementioned gluconamides. The D-enantiomer of the galactonamide **32a**, formed left-handed enantiomorphic 'whisker' type aggregates. Cochleate cylinders were formed from the mannonamide **33a** (in both water and 1,2-xylene). The talonamide **34a** crystallized rapidly in water forming occasionally whiskers or helical fibers upon slow cooling,[162] while the gulonamide **35a** also crystallized and could only give rolled-up sheets in 1,2-xylene. The allon-(**36a**), altron-(**37a**), and idon-(**38a**) amides displayed an excellent solubility in water while for the glucon-(**28a**), mannon-(**33a**) and galacton-(**32a**) amides the solubility was reduced. In fact for the mannon- and glucon-(**28a**), mannon-(**33a**) and galactonamide (**32a**) derivatives the precipitation of small platelets was in accordance with the proposed "chiral bilayer—effect". *From all the above comparative study revealed that the different aggregation behavior of the diastereomers is related to the degree of all-trans conformation that a certain isomer can acquire.*

The **32a** and **33a** derivatives for example are lacking the 1,3-*syn*-hydroxyl groups and therefore prefer a non-distorted all-*anti*-conformation of the hydrophobic chains. This conformation is extended and results in a preference for the formation of flat bilayer aggregates. On the other hand the long uniform ribbons are probably whiskers with screw dislocations caused by interaction of hydroxyl groups that belong in different layers. The fact that the gluconamides **28a** and **34a** form only micellar cylinders reflects probably a bend in the head group caused by 2,4-*syn*-interactions between hydroxyl groups.[163] This bend explains the formation of superstructures with a large curvature.

The presence of *syn*-postioned hydroxyl groups on the C3 and C5 of the N-octyl-D-gulon-(**35a**), altron-(**37a**), allon-(**36a**), and idon-(**38a**) amides, makes these compounds water-soluble and therefore does not allow the formation of aggregates. And induced a bent. This bent does not allow the formation of any regular chain amide hydrogen bonds due to the excessive hydration. The crystal structures of D-Gul-8 **35a**[164] and D-Tal-8 **34a**[165] have been reported and shown to contain tail-to-tail bilayers.

Various gluconamides possessing different chain lengths and head group stereochemistry were also compared regarding their aggregation properties during gelation with water. The observed structures and the size/stereochemistry/aggregation morphology relationships observed though are worth reviewing. A series of octyl chain derivatives (a) possessing carbohydrate head groups with differing stereochemistry, were compared to the corresponding dodecyl (d) derivatives. Both the racemic Glu-8 (**28a**) and the Glu-12 (**28d**) self-assembled into platelets. A mixture containing D-Glu-8 (**28a**) and D-Glu-12 (**28d**) was found to initially form clear non-turbid gels containing spherical aggregates, which rearranged into

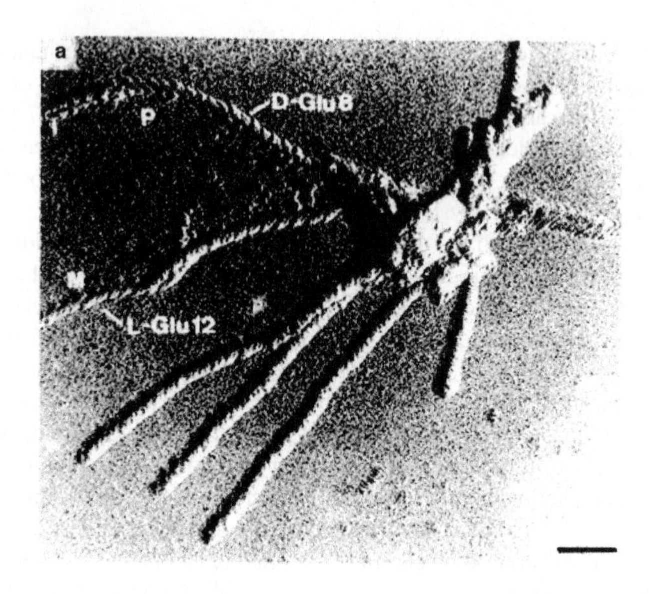

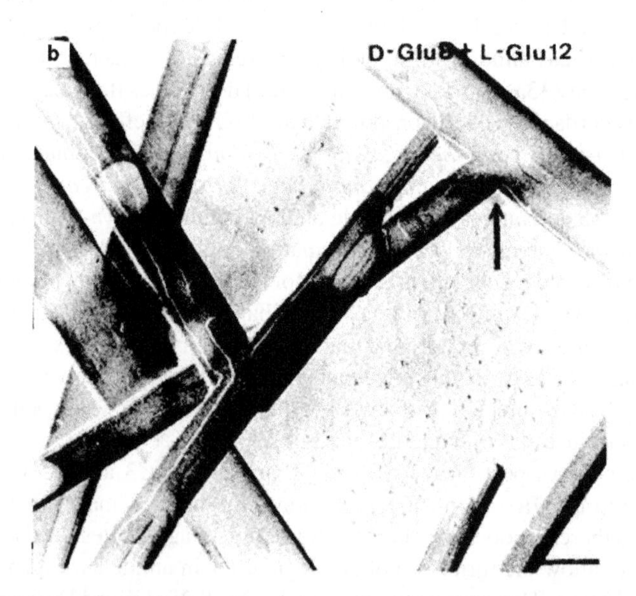

FIGURE 7.19. (*P*)-helices (D-Glu-8, **28a**) and (*M*)-helices (L-Glu-12, L-**28d**) first separate (a; bar, 100 nm) and then (b; bar 300 nm) unite to form elongated 'racemic' platelets. Reproduced with permission of the American Chemical Society.

helical fibers and knot-like structures and were also observed for the pure amphiphiles. Nevertheless no separation into Glu-8 and Glu-12 fibers was observed. The formation of P- and M-helices was observed though in a mixture of the D-Glu-8 (**28a**) and L-Glu-12. The P- and M-helices were studied on the basis of their handedness and the diameter and were unambiguously assigned[166] to pure D-Glu-8 and L-Glu-12 fibers respectively (Figure 7.19). It was therefore possible with these isomers to observe a "chain-length-induced racemate resolution". It should be noted though, that the "enantiomeric" fibers were only transiently observed, as thermodynamically favoured multilayered thin crystals ultimately dominated. TEM studies of shadowed crystal samples revealed an estimated layer thickness of about 5 ± 0.5 nm which is consistent with a D-Glu-8/L-Glu-12 bilayer. This means that the surfactants are first demixed into the homogeneous fibers and then combine to the non-chiral bilayers, exhibiting a "retarded chiral bilayer effect". *The highlighted comparative studies on alkylaldolamides clearly illustrate the importance of stereochemistry in the morphology of the resulting superstructures.* A set of rules governing the interactions between N-alkylaldonamides and predicting their aggregation motives was built up from these experimental results. Amphiphiles, which differ only in the length of the hydrophobic tail are not expected to separate but to result in mixed structures in some cases through separated short-lived intermediates. Racemic analogues are expected to crystallize together into platelets or tubes,[162] while diastereomers possessing opposed configurations at C3 and C5 are expected to separate into different morphologies formed exclusively by the individual components of the mixture.

The comparative studies on gluconamides included modifications that affect both the structure and the rigidity of the surfactants.[167,168] The rigid diacetylene unit, which has been extensively studied in the case of the phosphatidyl choline analogues was also incorporated in a number of hexonamides.[167,168] Some of the intrinsic characteristics that made its use attractive were its previously studied importance in the formation of well-defined aggregates (*e.g.* in the case phosphatidyl cholines[65]), the possibility of its topotactic polymerization mechanism leading to a red-purple conjugated enyne,[169,170] the prospect of stabilization of its aggregates by cross-linking and the ease of both following its reactions (UV spectroscopy) and observing its aggregates (using TEM without the need of staining).

A series of diacetylenic (N-dodeca-5,7-diynyl) aldonamides were compared with the corresponding non-acetylenic dodecyl compounds in terms of aggregation morphologies.[171] The enantiopure diynoic galactonamide **32b** assembled into enantiomorphic helical ribbons (c.f. Figure 7.31, right-handed helix for the D-enantiomer), as well as into closed hollow tubules (diameter approx. 1 μm). The corresponding racemate on the other hand, gave planar assemblies.[171] The head-to-tail enantiopolar packing observed in the crystal structure of the enantiopure N-dodeca-6,8-diynyl-D-gluconamide **28i**[168] suggests that the "*chiral bilayer effect*"[158] might account for this packing profile.

One of the most important features of these analogues is their ability to be further cross-linked. The feasibility of the post-polymerization was demonstrated by the application of UV light induced polymerisation of the diynoic galactonamide **32b**, which resulted in polymers retaining the superstructure of the surfactant aggregates.[167] Similar observations were made for the dodecyl galactonamides **32a**, which open up a route to the construction of pre-defined chiral nano-objects, which can be then stablized after assembly.

7.2.7. Redox-Active and Light-Sensitive Amphiphiles

The need to create functional nanosized architectures has led to the design and synthesis of surfactants that can be reversibly switched between two molecular states by incorporating redox-active or light sensitive functionalities.[172,173,174,175,176,177,178] In these systems the ability to tune the properties of the surfactant by either controlling the potential applied to

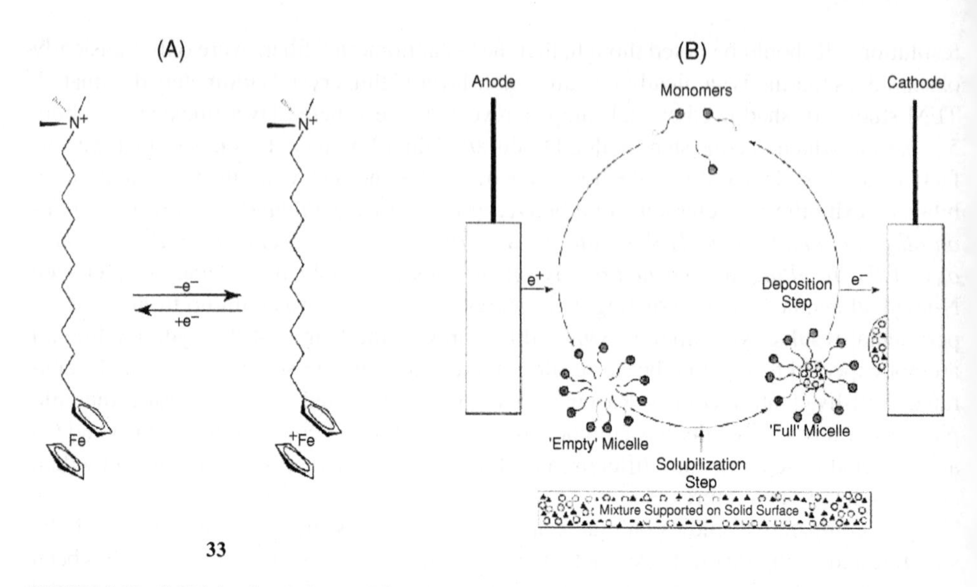

FIGURE 7.20. (A) Molecular structure of (11-ferrocenylundecyl) trimethylammonium bromide in reduced (left) and oxidized (right) states. (B) Schematic illustration of a separations scheme based on the electrochemical assembly and disassembly of micelles.

an electrode, or controlling the illumination of the solutions offers great opportunity for the creation of applied systems.[179,180]

Recent studies demonstrated that (11-ferrocenylundecyl) trimethylammonium bromide **33** undergoes a reversible change in micellization upon oxidation or reduction.[181] Upon aggregation, the reduced surfactant was shown to assemble into micelles capable of selectively solubilizing a mixture of hydrophobic compounds (Figure 7.20). Upon electrochemical disruption of the micelles, their contents were selectively deposited onto the electrodes. In order to determine the selectivity of the solubilization and the deposition, mixtures of two model drug-like compounds were utilized. It has been demonstrated that after 5 cycles of repeatedly solubilizing and depositing a mixture of initially equimolar fractions of Orange OT (*o*-tolueneazo-β-napthol) and Yellow AB (1-phenylazo-2-naphthylamine), the product contained >97% of Orange OT. Since the state of aggregation of the surfactant was found to be easily controlled in such a system, this represents a potentially useful system in which control over the aggregation of a surfactant leads to separation or analysis of compounds in micro-scale chemical process systems.

Kunitake *et al.*,[182] Whitten *et al.*[183,184] and Whitesides *et al.*[185] reported that the incorporation of an aromatic group into the hydrocarbon chain of an amphiphile induces strong non-covalent attraction between the aromatic chromophores leading to enhanced self-assembly in both aqueous solutions and at air–water interface. On the basis of the design principle of molecules that spontaneously form stable ordered structure on solid substrate, Liu *et al.* synthesised the chiral azobenzene amphiphile, N-[4-(4-dodecyloxyphenylazo)benzoyl]-*L*-glutamic acid (C_{12}-Azo-*L*-Glu, **34**, as shown in Figure 7.21),[186] which contained a chiral center and two carboxyl groups. The investigation of the self-organizing properties of C_{12}-Azo-*L*-Glu on solid substrates indicated multiple interactions existed between the molecules present in the resulting superstructures. While C_{12}-Azo-*L*-Glu underwent a reversible trans–cis photoisomerization in dilute solution, this

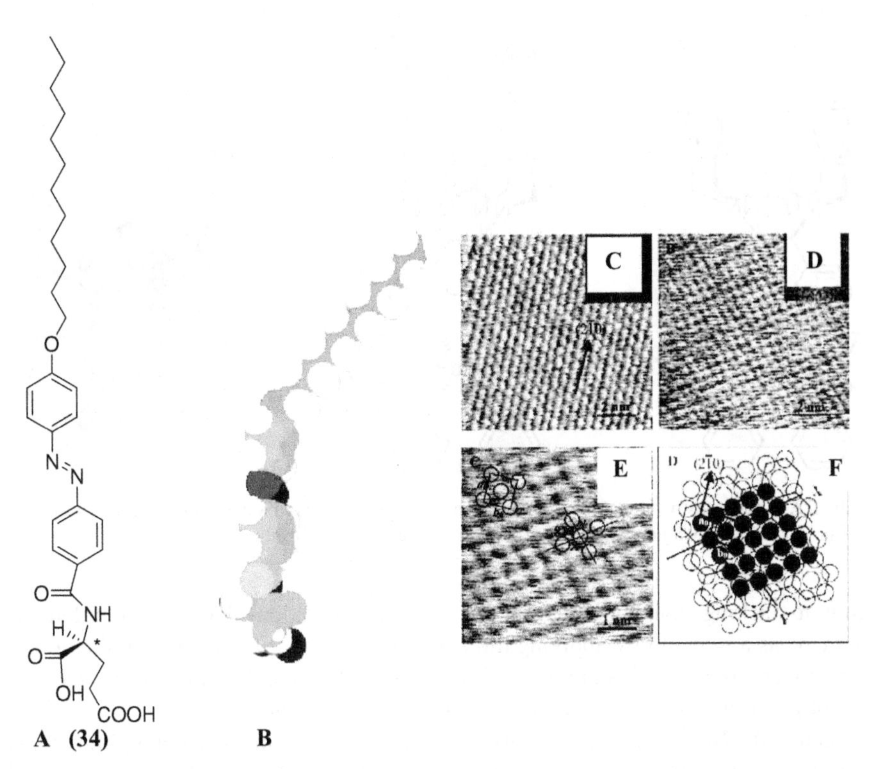

FIGURE 7.21. (A) Chemical structure of C_{12}-Azo-*L*-Glu. (B) The energy minimized configuration of C_{12}-Azo-L-Glu (the length of this molecule is about 30.5Å), (C) AFM image at the atomic scale of mica in the air obtained before dropping the solution on it (the inset shows the Fourier pattern that indicates a hexagonal crystal lattice). (D) High-resolution AFM image obtained from the surface of this flat-layered structure (the inset shows the Fourier pattern that indicates an oblique crystal lattice). (E) Zoomed-in image of (D), the crystal lattice can be observed more clearly. (F) Proposed molecular packing model of the layered structure.

photoisomerization was suppressed on solid substrate because of the H-aggregation, indicating the formation of a compact film. AFM studies revealed that upon casting the C_{12}-Azo-L-Glu from ethanol solution onto the hydrophilic surface of mica, a stable flat-layered structure formed spontaneously in large scale. High-resolution images allowed the identification of the relative orientation of molecular rows in the ordered thin film and the crystal lattice of mica and a model for the molecular packing of the layered structure was proposed. Furthermore it was found that mica induced a template effect upon the self-organization process. The results of the above mentioned studies suggested a cooperative/competitive effect between hydrogen bonding, π-π interactions and the presence of a chiral center in the molecule, all played a critcal role in the self-organization process and led to the formation of highly organized structures.

7.2.8. Finite Amphiphilic Assemblies

In Nature, self-assembly to form finite assemblies often involves the non-covalent organization of molecules containing not only amphiphilic character, but also specific information needed for additional intermolecular recognition processes to occur, *e.g.*, hydrogen

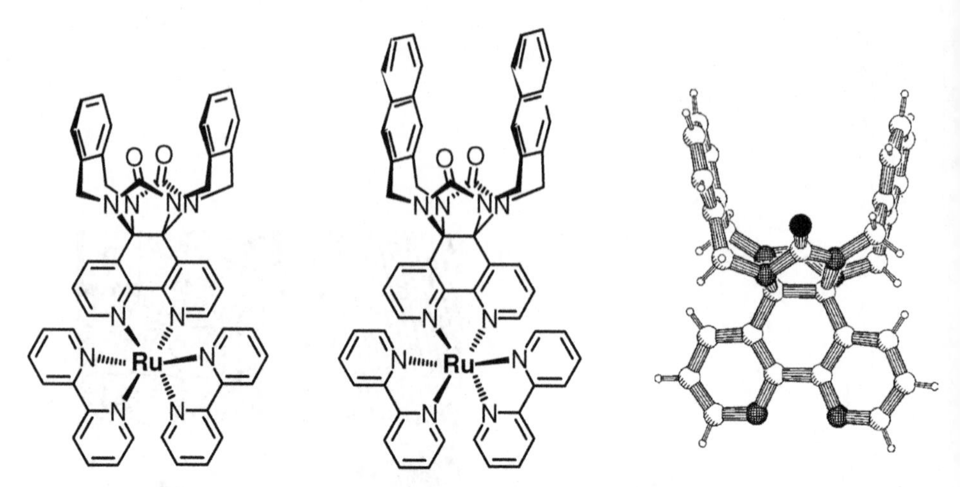

FIGURE 7.22. Chemical structure of the benzene and napthalene side-walled water-soluble metallo-hosts and the X-ray structure of the ruthenium free precusor (front) revealing the 'U-shaped cavity'.

bonding, π-stacking, van der Waals and electrostatic interactions. By using complementarity of shape as a tool and by applying directed intermolecular forces in combination with entropy-driven processes, Nature is capable of assembling building blocks into nanosized objects of precisely determined shape, structure and importantly function.

Recently the Nolte group has constructed a series of non-classical amphiphiles consisting of ruthenium-bipyridine centers functionalized with a receptor cavity viz., water-soluble metallo-hosts (Figure 7.22).[15]

In the case of the water-soluble metallo-host possessing a benzene side-walled, the cavities were observed to self-associate in water to form well-defined dimers which further self-assemble, although solutions in water remained clear up to relatively high concentrations (>30 mM). Samples of these solutions were studied with TEM and revealed rather ill defined, scroll-like mesoscopic assemblies with lengths up to 10 μm, and a typical width of approximately 100 nm. Enlargement of the receptor side-walls with naphthalene moieties, increases both the hydrophobic character of the hosts and the π-π interactions between the molecules. When the concentration of this naphthalene compound in water was increased to approximately 2 mM, the solution transformed into a turbid dispersion, which remained for days without any precipitation.

Proton NMR studies revealed that these molecules initially dimerize in a 'head-to-head' geometry ($K_{self} = 21000\ M^{-1}$), forming 'a seed' from which the aggregates grow in a 'head-to-tail' geometry.[187] When samples of this turbid solution (0.5%, w/v) were studied by TEM, discrete rectangular aggregates were observed (Figure 7.23), which were very monodisperse in both shape and size (typical dimensions 350 × 150 nm). Platinum shadowing revealed that the height of the rectangles was also relatively monodisperse, *viz.* 75 ± 10 nm. Somewhat surprisingly, in addition to the rectangular structures 'cigar-like' aggregates were observed (Figure 7.23), which were an order of magnitude larger (typical dimensions 4000 × 350 nm) than the rectangular ones. Analysis of these 'cigars' revealed that they were also highly monodisperse in both shape and size (aspect ratio length:width = 11 ± 2), and that they were built up from smaller subunits. These subunits appeared to have dimensions which

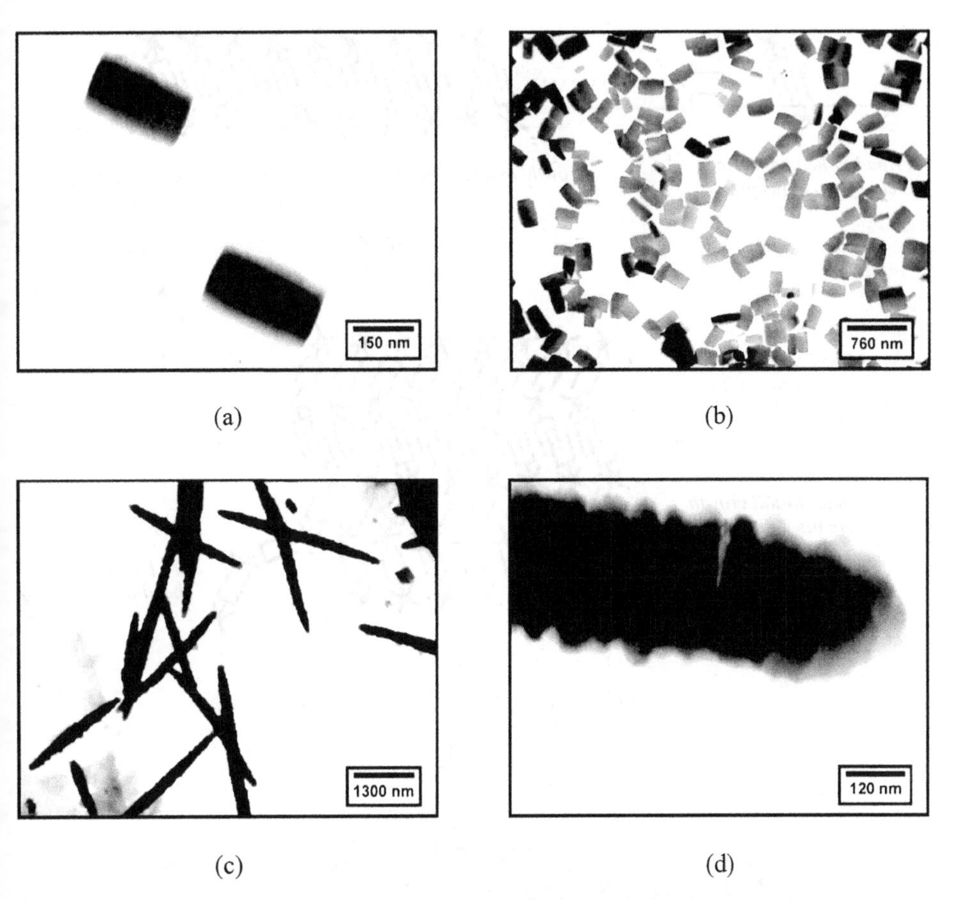

FIGURE 7.23. TEM images of the rectangular (a, b) and 'cigar-like' aggregates (c, d) formed by the naphthalene receptor amphiphile. (d) Magnifications of 'cigars', which show that they are built up from smaller subunits. Samples are not shadowed or stained.

were remarkably similar to the length and height of the rectangular aggregates (300–400 × 75 nm). It was proposed that the 'cigar-like aggregate' is a higher order assembly, which is constructed from a limited number (40–60) of rectangular aggregates. The organization of the molecules into rectangular and higher 'cigar-like' aggregates is a unique example of a hierarchical growth self-assembly process.

The electron microscopic studies also revealed that at high temperature the size of the rectangular aggregates becomes smaller and more monodisperse. The first effect was expected and can be explained by the fact that the strength of the interactions between the molecules, and consequently the aggregate size, decreases at higher temperature. The unexpected higher monodispersity is proposed to be the result of a self-repair process analogous to the repair processes observed in natural systems.[188]

A full analysis of the assembly process resulted in a growth model, in which a dimer acts as a nucleation point for further growth in two dimensions: as a result of the strong, hydrophobic interactions between the large naphthalene surfaces forming a so-called 'dimeric

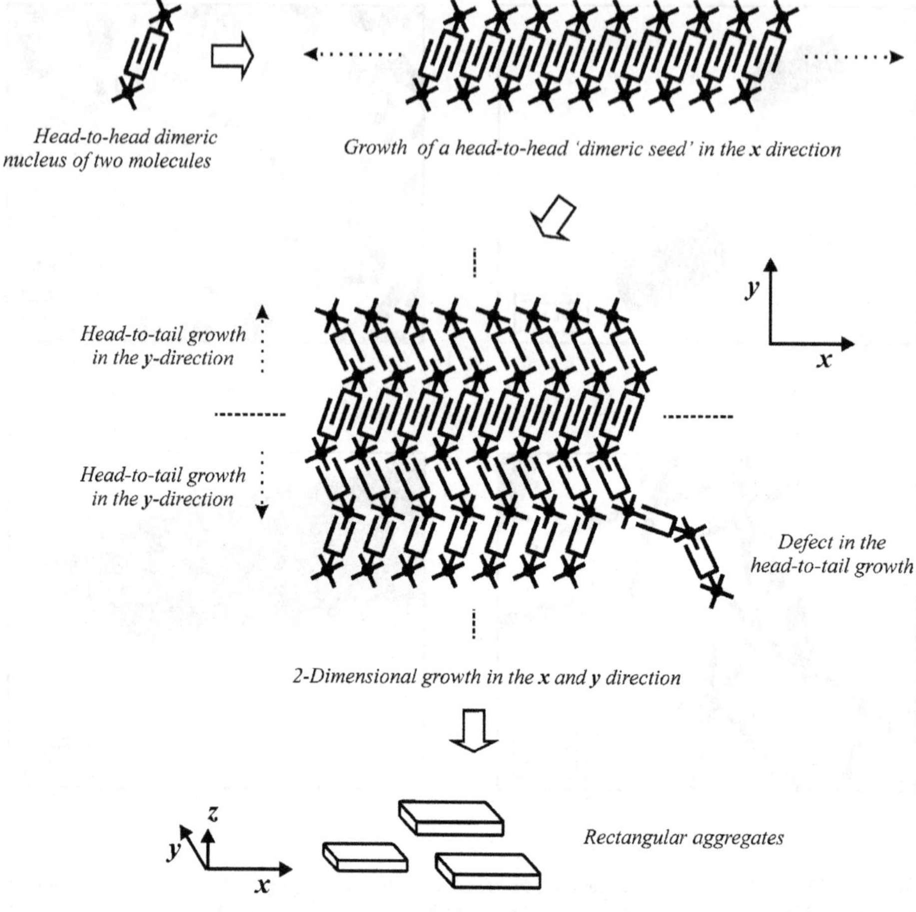

FIGURE 7.24. Proposed aggregation model for the self-assembly of water-soluble metallo-host amphiphiles.

layer', which has all the hydrophilic ruthenium-bipyridine units directed toward the aqueous phase. Simultaneously, further monomers attach themselves in a head-to-tail fashion to the bipyridine units of the above mentioned bilayer array of dimers, eventually resulting in the formation of a two-dimensional (2D) sheets. Then approximately 75 of these sheets stack. The finite nature of the aggregates is thought to be the result of the coherence between the molecules decreasing as the aggregates grow until at a certain moment it will be energetically no longer favorable to attach new monomers to the increasingly incoherent and relatively water-soluble exterior of the rectangular aggregates. At this point, the assembly stops growing. The final shape of the architecture is governed by a subtle balance between the enthalpy (strength of intermolecular interactions) and the entropy of the self-assembly process.

In the proposed model, Figure 7.24, the edges of the aromatic side-walls of the receptor amphiphile form the floor and the roof of the rectangles. A consequence of this architecture is that the rectangles in some cases organize themselves further, tilted on their edges, into a 'cigar-like' superstructure, in which all the relatively hydrophobic rectangular faces minimize their exposure to water. Similar hierarchical growth processes have been observed in natural as well as in artificial systems.[189,190]

7.2.9. *Fullerene-Based Amphiphiles*

Fullerenes possess excellent electronic properties since most of their derivatives have proven to be outstanding electron acceptors[191,192,193] and are also able to self-assemble into nanorods and vesicles depending on the structure of the monomer unit.[194,195] In fact, fullerenes have a strong tendency to form clusters of different sizes, especially in polar solvents.[196] Aggregation of C_{60} units may cause a significant change in their photochemical and photophysical properties, as compared with isolated molecules in solution.[192] Of course, this change can have a profound influence on fullerene based optical and electronic materials.[197,198] For instance, aggregation of fullerene spheroids was shown to play a crucial role in the preparation of photovoltaic cells.[199]

Also, biological tests of fullerenes, which are usually carried out in aqueous solutions, are heavily affected by aggregation.[200] Basically, dissolution of unmodified C_{60} or mono-functionalized organofullerenes is always accompanied by a high degree of clustering. On the other hand the extremely hydrophobic C_{60} fullerene can be made soluble in water by connecting it with functional chargeable groups such as carboxylic acids[201] or amines.[202] Hydrophilic behavior can also be introduced by an elegant and less obvious approach, in which polarizable phenyl groups are added to C_{60} to stabilize its anion.[203] These properties make fullerenes ideal candidates for the construction of novel functional amphiphiles where the chemical functionality drives a self-assembling process, which conserves or possibly enhances the properties of the monomer.

Studies of the association behavior of the potassium salt of pentaphenyl fullerene ($Ph_5C_{60}K$, 41) in water, performed by the group of Nakamura by laser light scattering (LLS), revealed that the hydrocarbonanions $Ph_5C_{60}^-$ associate into bilayers, forming stable spherical vesicles (Figure 7.25). By using a combination of static and dynamic light scattering measurements, the size of $Ph_5C_{60}K$ (in terms of both the radius of gyration and the hydrodynamic radius), the size distribution, and the shape of the associated particles were determined. The spherical vesicles were found to have an average hydrodynamic radius and a radius of gyration of about 17 nanometers at a very low critical aggregation concentration of less than 10^{-7} moles per liter.[204] The average aggregation number of associated particles in these large spherical vesicles is about 1.2×10^4.

Prato *et al.* synthesized four ionic fullerene derivatives (Figure 7.26), which are relatively soluble in polar solvents and showed that they organize into morphologically different nanoscale structures as verified via transmission electron microscopy.[205] Spheres, nanorods, and nanotubules form in water depending on the side chain appendage of the fullerene spheroid. Also, computer simulations were used for investigating the relative spatial arrangements. More specifically they demonstrated that the combination of a hydrophobic fullerene core, with hydrophilic ammonium groups and also with other self organizing groups, assembled into three fundamentally different low-dimensional shapes and, in particular, it was demonstrated that the linkage with a porphyrin macrocycle introduces a substantial morphological refinement at the mesoscopic level.

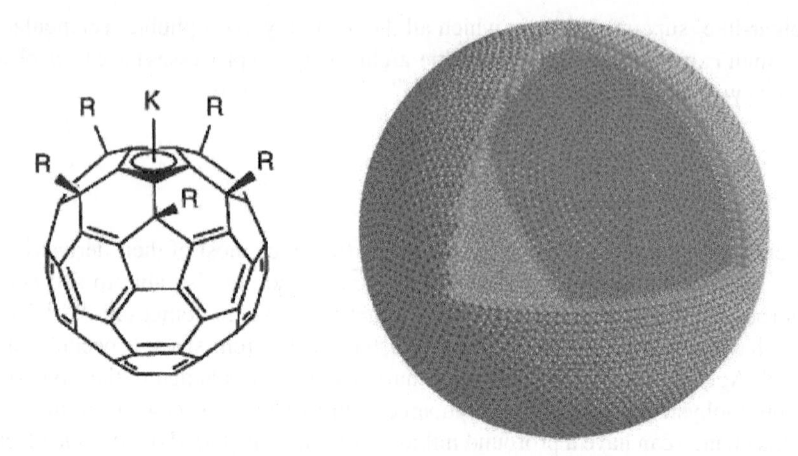

FIGURE 7.25. Chemical structure of the penta-substituted fullerene potassium salt. In the present work, only the compound with R = C_6H_5 has been studied. A bilayer vesicle model, consisting of $N_{outer} = 6693$ molecules in an outer shell of radius $R_{outer} = 517.6$ nm plus $N_{inner} = 5973$ molecules in an inner shell of radius $R_{inner} = 16.7$ nm. A sector has been cut out for clarity. The hydrophobic fullerene bodies are shown in green, the hydrophilic charged cyclopentadienide regions are in blue, and the five substituents are schematically represented as yellow sticks.

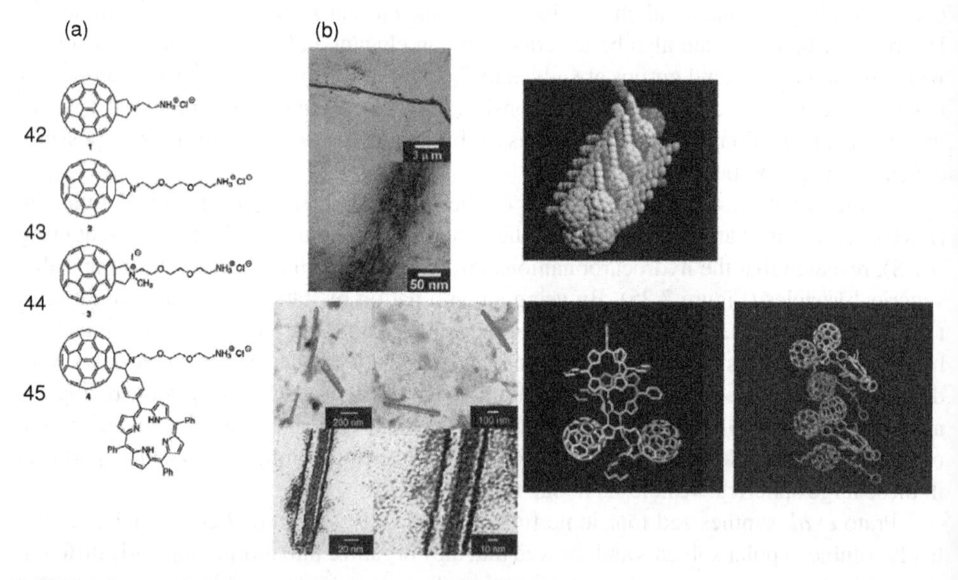

FIGURE 7.26. (a) Molecular structure of compounds 42–45 (b) TEM images of a bundle of nanorods formed by compound 43 (*Upper*), and closer view of the bundle (*Lower*). (c) Pictorial representation of the nanorods formed by self-assembly of salt 43 in water, (d) Nanotubules formed by compound 45 observed at different scales. The measures of the two focused tubules are 35 × 530 and 50 × 470 nm. (e) Most stable dimeric structures of 45. Note that *Upper* is 6 kcal/mol per molecule more stable than *Lower.* (*Upper*) Interlayered structure (see text). (*Lower*) Porphyrin stacked structure. Notice that the porphyrin stacked structure is 2 kcal/mol per molecule more stable than the interlayered structure.

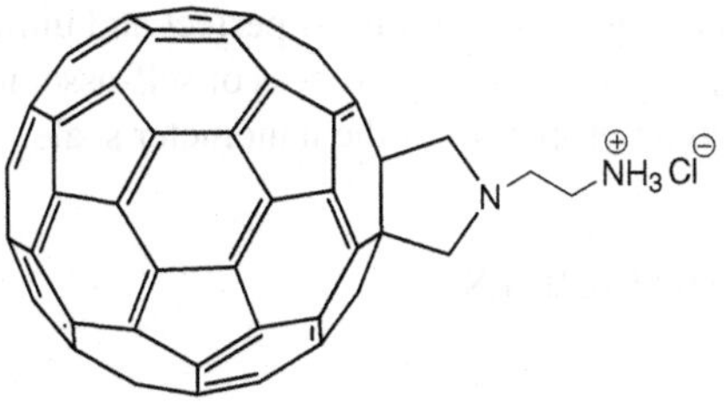

42

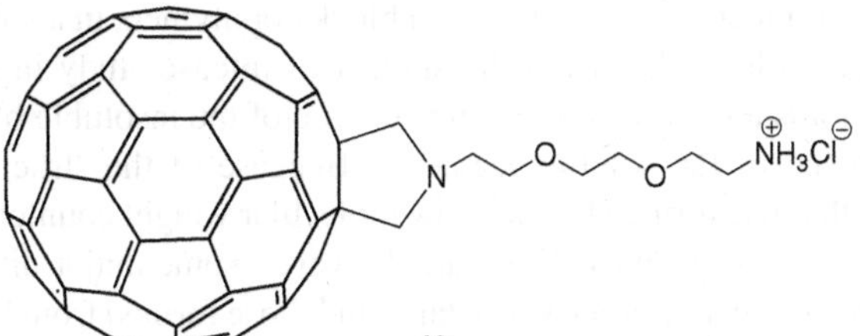

43

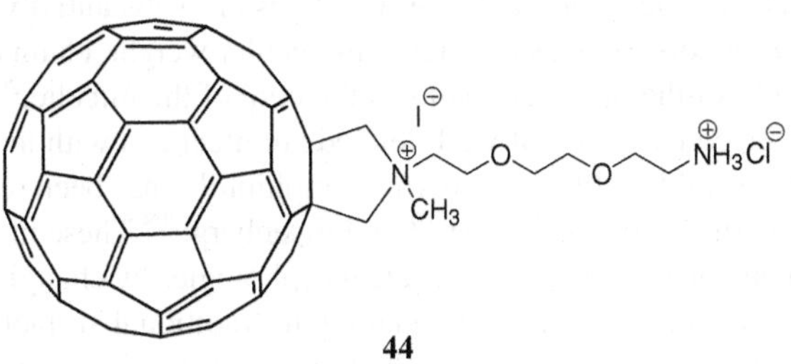

44

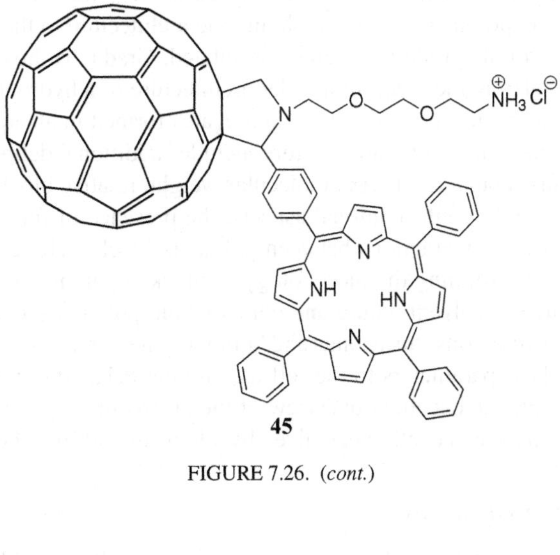

45

FIGURE 7.26. *(cont.)*

This efficient method for the fabrication of almost perfect and uniformly shaped nanotubular crystals, which order spontaneously by a process of self-assembly, opens a simple route to exploitation of the fullerene properties at the nanometer scale.

7.3. BLOCK COPOLYMER AMPHIPHILES

7.3.1. Introduction

The formation of micellar aggregates from block copolymers in a solvent selective for one of the constituent blocks has been the subject of intense study in the past decades. These spherical colloids are built from a compact core of the insoluble blocks, surrounded by a corona composed of the soluble segments. Because of the structural similarity of these aggregates with those formed by their low molecular weight counterparts (*vide infra*), they are also referred to as micelles. There are, however, some major differences between micelles formed from more traditional surfactants and those formed from block copolymers; the critical micellar concentration (cmc) of the latter is usually much lower, block copolymer micelles are more stable and their exchange dynamics are substantially slower. This is a result of a number of factors, such as the higher molecular weight, chain entanglement and the decreased mobility of the polymer chains in the core of the micelle.[206]

More recently the construction of well-defined architectures with nanometer-sized dimensions from self-assembling block copolymer amphiphiles has been reported in groundbreaking studies from the groups of Meijer[207] and Eisenberg.[208] These investigators proved that it is possible to obtain a variety of aggregate morphologies by changing the structure of the constituent block copolymers, again in analogy to traditional surfactants (*vide supra*). Whereas the design rules for the synthesis of low molecular weight surfactants in relation to the aggregate morphologies are better established as described above, in the case of amphiphilic diblock copolymers the rules relating the architecture to the block copolymer composition are still in the process of being formulated. Gradually, however a consensus is being developed relating the macromolecular architecture of a hydrophobic/hydrophilic diblock copolymer to the morphology of the aggregates formed in an aqueous dispersion. In this context the macromolecular architecture includes additional definitions when compared to the molecular counterpart, viz.; molecular weight, relative block length, structure (*e.g.* linear vs. branched), conformational aspects, the presence of functional groups and the possibility of specific interactions between polymers blocks. These are, however, not the only parameters determining the morphology of block copolymer aggregates, also solution conditions such as solvent nature and composition, polymer concentration and the presence of additives (*e.g.* ions, surfactants and homopolymers) play an important role. The contribution of all these parameters to the self-assembling behaviour is beyond the scope of this chapter, however an excellent overview of the effects on aggregate morphology of the solution conditions, has recently been given by Choucair and Eisenberg.[209]

7.3.2. Theoretical Considerations

It is general considered that the driving force for the self-assembly of amphiphilic molecules is a solvophobic effect, more specific in an aqueous environment, this is referred to as the hydrophobic effect. The type of aggregate morphology formed can be predicted

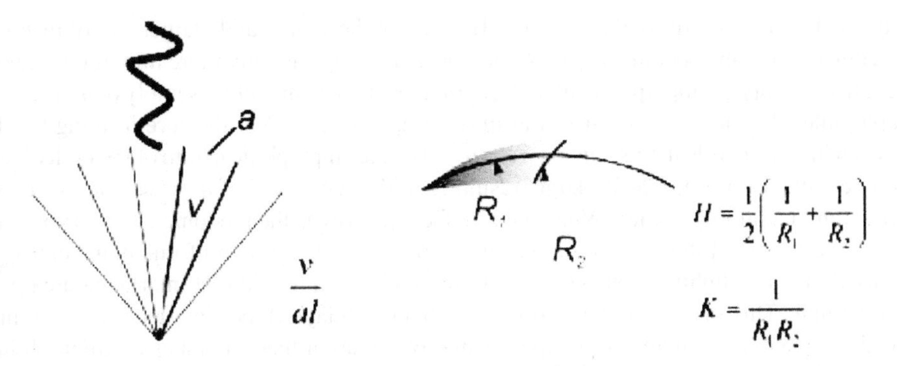

FIGURE 7.27. Definition of the interfacial mean curvature (H) and Gaussian curvature (K) and the packing parameter (P). Reproduced with permission from WILEY-VCH Verlag GmbH & Co.).

using Israelachvilli's theory in the case of the self-assembly of traditional surfactants. This theory is mainly based on the geometric relation between the hydrophilic and hydrophobic parts of the molecule (*vide infra*). In the case of block copolymer amphiphiles, however, the situation is more complex. Purely based on similar geometric considerations a comparable packing parameter can be defined, but because of the bigger dimensions of the molecules also the curvature of the hydrophobic-hydrophilic interface becomes an important parameter.[210] The packing parameter is consequently defined as:[211]

$$P = v/al = 1 + Hl + Kl^2/3$$

Here v is the hydrophobic volume of the amphiphile, a is the interfacial energy and l is the chain length. The parameters describing the hydrophobic-hydrophilic interface are the mean curvature H and the Gaussian curvature K (Figure 7.27).

The most straightforward morphologies (*i.e.* spheres, cylinders and bilayers) are obtained by the combination of parameters given in Table 7.3. From this table it is evident that an increasing hydrophobic/hydrophilic ratio results in a change in aggregate morphology from spherical to rod-like micelles to vesicles.

If not only geometric, but also thermodynamic parameters are taken into consideration, the difference between the self-assembly of polymeric amphiphiles compared to low molecular weight surfactants is even more pronounced. The two major contributions to the free energy of the system are; 1) the loss of entropy when flexible parts of the amphiphile are enforced in the restricted environment of the aggregates, and 2) the interfacial energy

TABLE 7.3. Packing parameter, mean curvature (H) and Gaussian curvature (K) for different aggregate morphologies.

Morphology	v/al	H	K
Spherical	1/3	1/R	$1/R^2$
Cylindrical	½	1/(2R)	0
Bilayers	1	0	0

of the hydrophobic-hydrophilic interface. Because of the significantly larger size of macromolecular amphiphiles both these contributions play a substantially more important role in the self-assembly of amphiphilic diblock copolymers than in the self-assembly of molecular amphiphiles. In particular the glass transition temperature (Tg) of the core forming block plays an important role in this context. As a result of the amphiphilic nature of the molecules the interfacial energy is usually large, resulting in the separation of these segments in the self-assembled morphologies. When the interfacial energy is the predominant contribution to the free energy of the system, *i.e.* when polymers having a low conformational entropy are used, the amphiphiles organize into morphologies with the lowest interfacial area per volume unit. This gives a preference to form planar (bilayer type) interfaces over cylindrical or spherical domains. The experimental observation that a number of amphiphilic macromolecules containing a rigid or conformationally restricted segment form bilayer type architectures is in line with this prediction. In the former category rod-coil type block copolymers have been also investigated (Section 3.4.), while in the latter case examples in which one of the constituent blocks has a specific interaction (*e.g.* hydrogen bonding, ionic interactions, etc) are given in Section 3.6.[210] Polymeric amphiphiles that are highly flexible, but have defined branching points, *i.e.* dendritic-linear diblock copolymers (Section 3.5) can also be considered conformationally restricted.

A situation where specific interactions are present between the solvophobic segments building up the amphiphilic macromolecules often results in highly ordered domains in the formed aggregates. This is the case when for example mesogenic groups are introduced in the macromolecules or when secondary interactions between these non-soluble parts are present. To a lesser extent it is also these basic principles that explain the change in aggregate morphology, when the ion type or concentration is changed in the case of an amphiphilic block copolymer containing a polyelectrolyte segment (Section 3.6).

7.3.3. Linear Diblock Copolymers: Coil-Coil's

The most studied class of amphiphilic diblock copolymers consists of macromolecules having two flexible segments. The first studies on the morphological aspects of self-assembling coil-coil diblock copolymers were reported by Zhang and Eisenberg.[208] Using solvent conditions in which (near) thermodynamic equilibrium was assumed, a series of colloidal dispersions were prepared from a family of systematically varied polystyrene-*block*-poly(acrylic acid) (PS-*b*-PAA) copolymers. As a consequence of the relatively high T*g* of polystyrene, the core forming block, the aggregates were prepared in a water/dimethylformamide (DMF) solvent mixture. Subsequently the DMF was removed by extensive dialysis, freezing in the equilibrium morphology of the formed aggregates. It was found that with a decreasing degree of polymerization of the corona-forming PAA block the aggregate morphology changed from spherical micelles, to cylindrical micelles, to vesicles and eventually to large spherical objects (Figure 7.28). It was postulated that the latter large spheres were built from clustered reversed micelles (*i.e.* having a hydrophobic corona) with an overall hydrophilic surface, referred to as large compound micelles (LCMs).[212] These observations nicely follow the predicated structural changes based on the packing parameter (Section 3.2).

So far these trends are analogous to the aggregation of traditional low molecular weight surfactants. It was, however, also shown by Eisenberg and coworkers that the hydrophobic

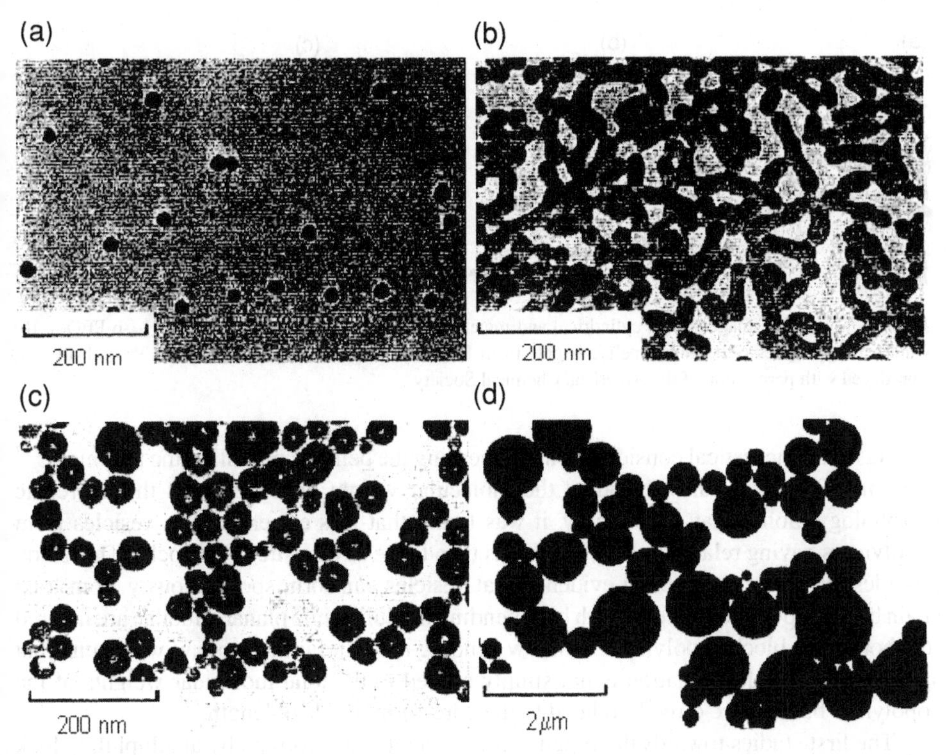

FIGURE 7.28. Transmission Electron Micrograph (TEM) images of different morphologies formed by PS-b-PAA diblock copolymers: (a) PS_{200}-b-PAA_{21} spherical micelles; (b) PS_{200}-b-PAA_{15} rod-like micelles; (c) PS_{200}-b-PAA_8 vesicles and (d) PS_{200}-b-PAA_4 large compound micelles. Reproduced with permission from American Association for the Advancement of Science (1995).

volume (or the degree of stretching of the core forming chains) is an important parameter in the self-assembly of diblock copolymers amphiphiles. When 10% (w/w) PS was added to vesicle forming PS_{410}-*b*-PAA_{16} micelles were predominantly observed.[212] In these cases the surface area increases from 6.6 nm^2 per corona chain for the vesicles to 12 nm^2 per corona chain for the micellar architectures.

The same trend in morphological transition from spherical to bilayer aggregates with decreasing hydrophilic length has also been observed in aqueous solution of very different diblock copolymer family, *i.e.* polystyrene-*block*-poly(ethylene oxide) (PS-*b*-PEO).[213] A co-solvent was in this case also required to obtain an equilibrium state, because of the high T*g* of the PS segment (Figure 7.29). More recent work on block copolymer systems with a lower glass transition temperature, which do not require the use of an organic co-solvent, showed that multiple morphologies are integrated parts of the phase diagrams.[210] For example vesicular architectures were formed from poly(ethylene glycol)-*block*-poly(propylene glycol)-*block*-poly(ethylene glycol) (PEO_5-*b*-PPO_{68}-*b*-PEO_5)[214] and polybutadiene-*block*-poly(ethylene glycol) (PB-*b*-PEO)[215] block copolymers. The extensive studies by Eisenberg and coworkers on block copolymer systems consisting of PS-*b*-PAA,[208] PS-*b*-PEO[213] and PS-*b*-poly(vinyl pyridine) displaying a wide range of possible morphologies (Figure 7.30).[216]

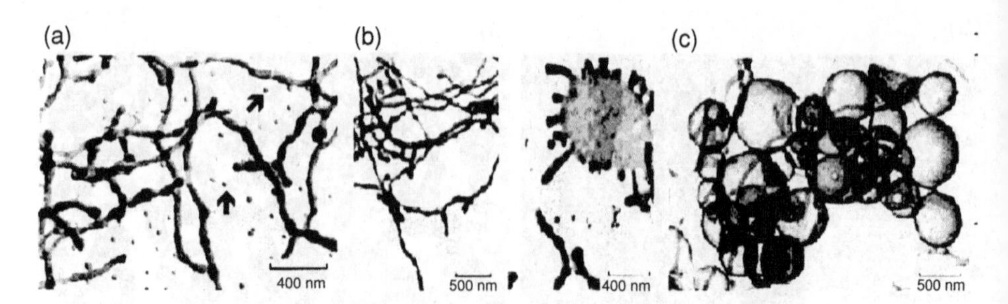

FIGURE 7.29. TEM images of (a) cylindrical and micellar (black arrow) aggregates from PS_{240}-b-PEO_{80}; (b) cylindrical aggregates (left) and lamellae (right) from PS_{240}-b-PEO_{45} and (c) vesicles from PS_{240}-b-PEO_{15}. Reproduced with permission of the American Chemical Society.

Based on theoretical considerations concerning the bending moduli of the polymers,[217] Shen and Eisenberg further studied the molecular weight dependence of the aggregate morphology. Somewhat surprisingly, it was found that it is easier to form vesicles from copolymers having relatively long PS blocks than those having short PS blocks. This study provides the first experimental evidence that vesicles can form spontaneously from long chain block copolymer systems with high bending moduli, while planar lamellae are favored for short chain block copolymers with low bending moduli. Furthermore, it was found that spontaneous vesicle formation is not simply related to the total molecular weights of the copolymer but is more directly related to the core-forming block length.[218]

The first studies towards the functionality of aggregates formed by amphiphilic block copolymers were performed by Discher *et al.*[219,220] and by Meier and coworkers.[221] The

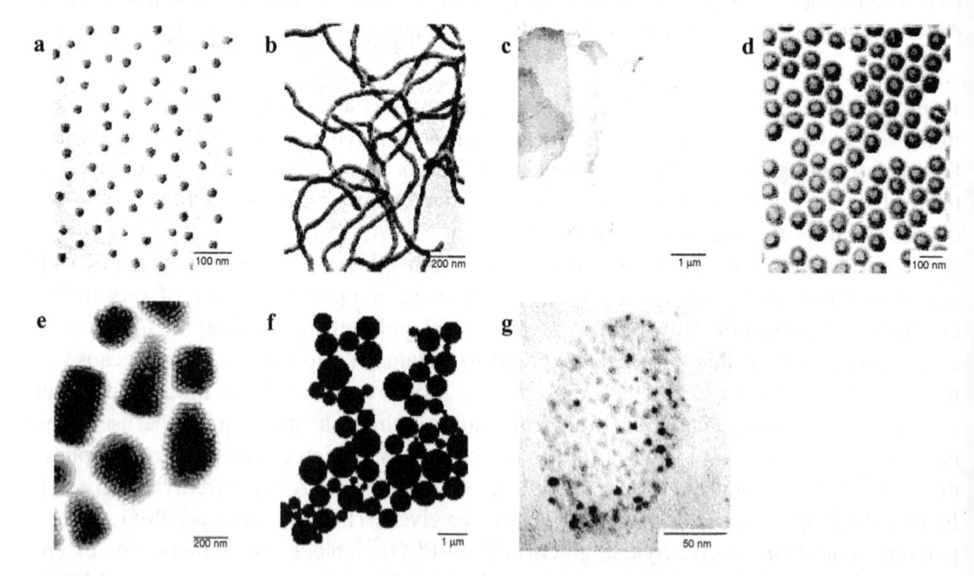

FIGURE 7.30. Different morphologies found upon self-assembly of PS-*b*-PAA copolymers. (a) micellar spheres; (b) micellar rods; (c) lamellae; (d) vesicles; (e) hexagonally packed hollow hoop (HHH) structures; (f) LCMs and (g) the internal structure of a LCM. Reproduced with permission from the Canadian Chemical Society.

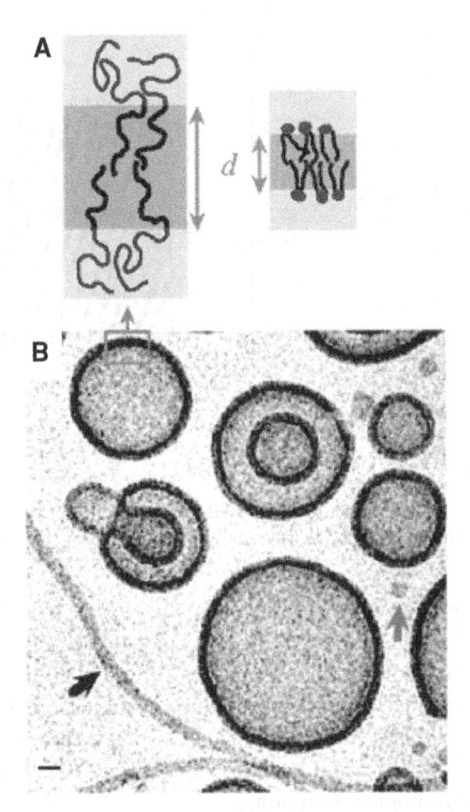

FIGURE 7.31. (A) Schematic representation of the bilayer membrane formed by PEE_{37}-b-PEO_{40} and for comparison a similar representation of a phopholipid bilayer. (B) Dispersion in water of PEE_{37}-b-PEO_{40} showing next to vesicles, rod-like micelles (grey arrow) and spherical micelles (black arrow). Reproduced with permission of the American Association for the Advancement of Science.

Discher group performed detailed investigations on the formation and stability of vesicular architectures formed by polyethylethene-*block*-poly(ethylene glycol) (PEE-*b*-PEO) copolymers. Upon dispersion in aqueous solution by hydration and subsequent preparative vitrification, worm-like and spherical micelles were found coexisting with small ($D < 200$ nm) vesicles (Figure 7.31). The Discher group also demonstrated that giant vesicles with diameters of 20–50 μm could be prepared by electroformation which allowed a detailed characterization of the materials properties and also the mechanical properties using micropipette aspiration. From the observed microdeformation of these vesicles it was concluded that these block copolymer 'polymersomes' are highly deformable, but also possess an enhanced toughness as compared to conventional lipid membranes. The permeability was determined by osmotic swelling/shrinkage and when compared to classical phospholipids vesicles, the permeability of the polymer membrane for water molecules was found to be at least 10 times less (*i.e.* 2.5 μm/s for the polymersomes and 25 to 100 μm/s for the liposomes).

The membrane formed by the self-assemby of an amphiphilic ABA triblock copolymers was used by Meier and coworkers to reconstitute membrane proteins in artificial

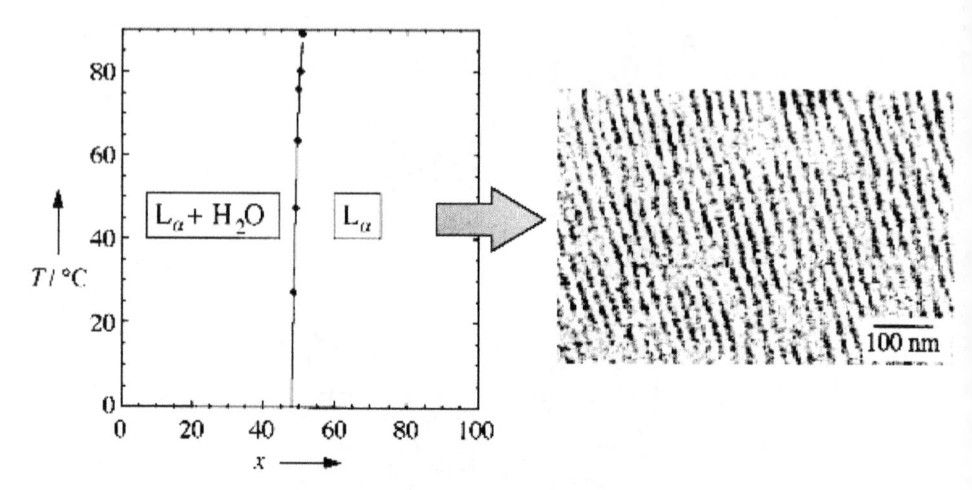

FIGURE 7.32. Phase diagram of PMOXA-b-PDMS-b-PMOXA in water and Cryo-TEM image of the lamellar phase formed at x = 50. x = polymer fraction in water in % w/w; L_α = lamellar liquid crystalline phase.

ultra-thin films.[221] Based on earlier studies on the vesicle forming block copolymer poly(2-methyloxazoline)-*block*-poly(dimethylsiloxane)-*block*-poly(2-methyloxazoline)[222] (PMOXA-*b*-PDMS-*b*-PMOXA) stable freestanding films were prepared. In Figure 7.32 the phase behavior of the block copolymer in aqueous solution is displayed between 0 °C and 90 °C, together with a micrograph of the lamellar phase formed. Both bacterial porins OmpF and maltoporin were reconstituted in the nanosized block copolymer membrane and found to retain their biological function.

7.3.4. Linear Diblock Copolymers: Rod-Coil's

The second class of amphiphilic macromolecules, which forms well-defined colloidal dispersions, is that of the rod-coil block copolymers. Jenekhe and Chen[223,224] described the preparation of giant vesicles in organic solvents by the self-assembly of polystyrene-*block*-poly(phenylquinoline) (PS-*b*-PPQ) copolymers. Long-range interactions between individual aggregates were clearly seen, whilst the use of selective solvents for the rigid PPQ rod led to the formation of large micrometer-sized aggregates with a wide variety of morphologies. The observed spherical and cylindrical structures contained a large hollow cavity as a result of the close packing of the rigid rod blocks. The micelle-like aggregates were demonstrated to be able to encapsulate large amounts of C_{60} fullerenes within the inner cavity and the PS core. The use of a selective solvent for the flexible PS coil instead of for the PPQ yielded exclusively hollow spherical architectures, several micrometers in size (Figure 7.33). Further long-range, close-packed self-ordering of the micelles produced periodic microporous materials of which the microstructure and optical properties could be tuned by the addition of small amounts of fullerenes, which were incorporated in the PS corona, opening the way toward the construction of unique optically active nano-objects.

Investigations on diblock copolypeptide amphiphiles by Deming and coworkers further elegantly proved the dependence of the aggregation behavior on the chain conformation of the polymer blocks.[225] Well-defined block copolymers were synthesized containing

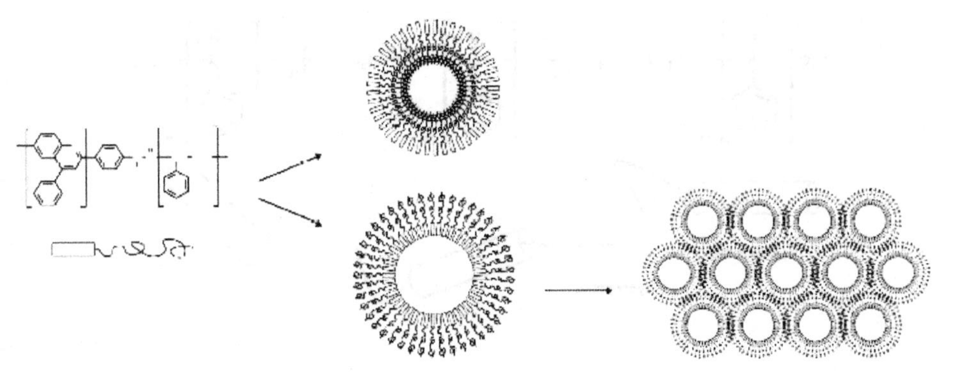

FIGURE 7.33. Schematic representation of solvent dependent aggregation of PPQ-PS block copolymers.

different polypeptide segments combining a random coil with a α-helical, β-strand or random coil segment. It was found that the macromolecular conformation had a strong effect on the hydrogel formation upon dispersion of these amphiphilic polymers in water. Klok, Lecommandoux and coworkers have also investigated oligomers built from a polystyrene segment and an α-helical polypeptide in the bulk material.[226] The same synthetic concept was used to prepare polybutadiene-*block*-poly(L-glutamatic acid), which was found to also form vesicular aggregates in water.[227] In an independent study by Kukula *et al.* the same assembly properties for these hybrid block copolymers were also found. In the latter study it was additionally shown that no serious change of the vesicle morphology was observed during the pH induced helix-coil transistion of the poly(L-glutamate) block, indicating that in this case other factors than chain conformation play a dictating role in the aggregate formation.[228]

Charged peptide analogous rod-coil diblock copolymers (Figure 7.34) were prepared by polymerizing peptide-derived isocyanides using a polystyrene macro-initiator. Under optimised conditions with respect to pH, the macromolecules of negatively charged polystyrene-*block*-poly(isocyanoalanyl alanine) (PS_{40}-*b*-$PIAA_{20}$) self-assembled in aqueous solution to yield micellar rods having lengths up to several micrometers. Zwitter ionic block copolymer polystyrene-*block*-poly(isocyanoalanyl histidine) (PS_{40}-*b*-$PIAH_{20}$) displayed a similar behavior. Interestingly, in this case it was possible to change the stiffness of the supramolecular rods by varying the type of counter ion.

The ratio between the hydrophobic and hydrophilic segments in the macromolecules also in this case strongly influences the aggregation behavior of the obtained 'super-amphiphiles.' Block copolymers PS_{40}-*b*-$PIAA_{30}$ and PS_{40}-*b*-$PIAH_{30}$ did not form any clear morphology upon dispersion in an aqueous buffer. Compounds PS_{40}-*b*-$PIAA_{10}$ and PS_{40}-*b*-$PIAH_{15}$ formed in addition to bilayer-type assemblies like plates and vesicles, also helical assemblies (*e.g.* superhelices, Figure 7.34). The observed differences in the supramolecular structures generated by the negatively charged and zwitterionic copolymers suggest that the helices are formed by a complex assembly mechanism, involving the hierarchical transfer of chirality from the monomeric building blocks via the secondary helical structure of the polyisocyanide to the final chiral superstructures.[229] These studies exemplify the importance of the block copolymer structure (*i.e.* the packing parameter) in determining the aggregate

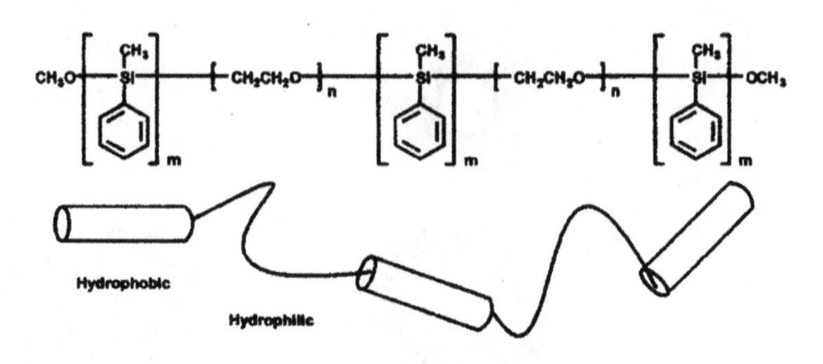

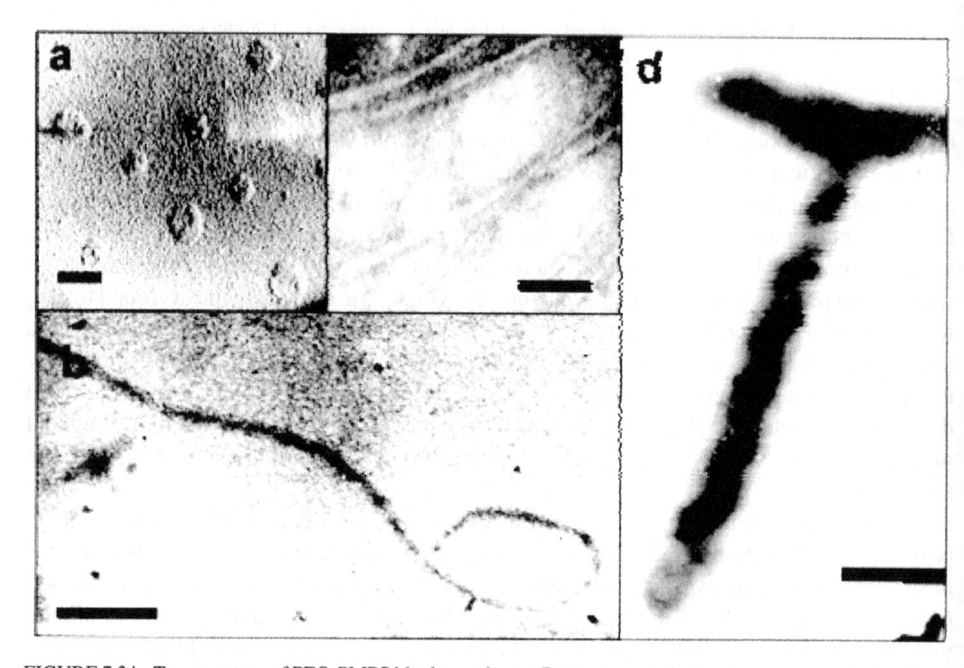

FIGURE 7.34. Top: structure of PEO-PMPS block copolymer. Bottom: transmission electron micrographs showing the formation of (a) vesicles, (b, c), micellar fibers, and (d) superhelices from this block copolymer. Reproduced with permission from the American Chemical Society.

morphology, but also show that fine tuning the interactions between copolymer segments can also play an important role (*vide supra*).

A slightly modified block copolymer, containing a rigid isocyanide segment, *i.e.* polystyrene-*block*-poly(isocyanoalanyl aminoethyl thiophene) (PS-*b*-PIAT) (Figure 7.34), was also synthesized with the aim of creating aggregates which could be stabilized after self-assembly by cross-linking of the thiophene moieties, resulting in a (conducting) polythiophene outer surface. Upon dispersion of PS_{40}-*b*-$PIAT_{50}$ in both organic (*e.g.* chloroform, toulene) and aqueous solution vesicular aggregates were formed. As was proven by fluorescence experiments, the thiophene groups which reside in the corona of these vesicles in water could be polymerised using a ruthenium(II) type catalyst. When vesicular assemblies were prepared in the presence of the lipase enzymes, their statistic inclusion led to the formation of a new type of functional polymersome nano-reactor. Fluorscence studies using a pro-fluorescent substrate revealed that the included enzyme remains active for many months and also that the polymer membrane is still permeable for substrate molecules, which were converted by the enzymes present in the interior of the vesicle.[230] This approach to functional and stable nano and micro-reactors holds much promise in the field of lab on a chip.

7.3.5. Dendritic-Linear Diblock Copolymers

The first report of a block copolymer possessing the structural distinct linear and dendritic segments was by Fréchet and coworkers.[231] The linear block in these macromolecules has a random coil conformation, which contrasts with the restricted conformation of the dendritic part. Aggregation studies in (aqueous) solution on dendritic-linear diblock copolymers were performed by Meijer *et al.*[207] using poly(propylene imine)-polystyrene copolymers. In these studies it was shown that with higher generations of the dendritic segment the increasing size of the water soluble part led to apparent changes in the morphologies of the aggregates formed. Upon going from the first generation up to the fourth generation, the morphology changed from inverted micelles to vesicles, then to micellar rods, and finally micelles. This structural acrobatics can be explained again by using the packing parameter (*vide infra*) or the more simplified Isrealachvilli's theory. Whilst in the superamphiphilic macromolecules described by Meijer the linear segment was hydrophobic and the dendritic segment hydrophilic, the opposite is the case in the linear poly(ethylene glycol) and dendritic poly(benzyl ether) block copolymers studied by the Fréchet group.[232] As a result of the large apolar segment these amphiphilic copolymers behave as unimolecular micelles below the critical micellar concentration (*cmc*), with the PEG chains forming a hydrophilic shell around the dendron. At concentration above the cmc, multimolecular micelles are formed. Similar dendritic-linear copolymers were synthesized by connecting a hydrophobic carbosilane dendritic wedge to a PEG tail.[233] The cmc of these block copolymers decreased with increasing generation of the dendrimer part, leading to complete insolubility at generation three. Additionally, it was observed that this increase in generation led to the formation of larger aggregates. Similar carbosilane dendritic wedges were combined with a rigid polyisocyanopeptide segment by Nolte *et al.*[234] to yield a block copolymer of which the third generation derivative self-assembled into micelles. These micelles transformed into micellar cylinders upon the addition of silver ions, presumable due to stronger interactions between the polyisocyanopeptide segments between which the silver binds.

7.3.6. Specific Interactions between Block Copolymer Segments

As briefly discussed in the above examples of polyisocyanopeptide-containing block copolymers, the aggregation behavior of block copolymers can be altered by changing the interactions between the different segments. A clear example of architectural control using ionic interactions was demonstrated by Harada and Kataoka. They used poly(ethylene glycol)-*block*-poly(α,β-aspartic acid) and poly(ethylene glycol)-*block*-poly(L-lysine) as a pair of polymers which are oppositely charged. Pairs with a comparable length of the charged segment formed neutral polyion complexes upon mixing. It was assumed first bimolecular complexes were formed, which in a second step grow into larger PIC micelles. In a complex mixture of matching and non-matching pairs, these PIC micelles were only observed for the matched pairs. Block copolymers containing a charged segment with non-matching length remained in solution as individual entities. The driving force for this self-selection assembly process is assumed to be the increased stability of the micelles over the individual PICs.[235,236]

Ionic interactions can also be employed to prepare vesicles from block copolymers as demonstrated by Schlaad *et al*[237] Based on the relative ratio between the two component blocks and the hydrophilic character of these blocks, no vesicles were expected to form. It was found, however, that because of the polyelectrolyte complex formation between the negatively charged methacrylate block in polybutadiene-*block*-poly(cesium methacrylate) (PB_{216}-*b*-PMA_{29}) and the positively charge pyridinium block in polystyrene-*block*-poly(1-methyl-4-vinylpyridinium iodide) ($PS_{211} - b\text{-}PVP_{33}$) asymmetric vesicles formed in THF with a microphase separated membrane (Figure 7.35). Changing the interaction between these charged blocks led to an array of different aggregate morphologies. The addition of low molecular weight acids (*i.e.* malonic acid), which specifically binds to one of the constituent block, disrupted the vesicular architecture while changing the solvent conditions resulted in an inversion of the original structure.

These examples emphasize that not only the macromolecular architecture plays an important role in the determination of the aggregate morphology, but that also interactions between block copolymer segments can strongly influence the final structure of the assemblies formed by these types of amphiphiles as seen for the smaller molecular amphiphiles.

7.4. BIOHYBRID ASSEMBLIES

In the previous paragraphs we reviewed a wide variety of natural and synthetic amphiphiles focusing on the factors that govern their self-assembling properties and their use in the creation of novel functional nanostructures. It is evident that the amphiphilic character of a vast variety of molecules, varying from simple molecular amphiphiles up to block copolymers, has already been extensively studied, and that by varying the chemical structure of the head group and/or the hydrophobic tail(s), assemblies possessing different morphologies can be readily created. Furthermore, it has been demonstrated that depending on the individual structure of the monomer and on the conditions employed, different properties of the basic building block can be expressed at the supramolecular level. As the need for the creation of functional nanoassemblies increases, researchers are focusing not only on the creation of novel architectures but also on the incorporation of functionality into the monomeric structure and the study of its expression in terms of material properties of the superstructure. These efforts have lead to the creation of a wide range of interesting functional nanostructures.

In the rapidly expanding area of nanochemistry, chemists often use Nature as inspiration to create their systems and in trying to mimic Nature, both in the efficiency and the perfection of the structures a number of "*natural*" building blocks have already been incorporated into synthetic materials. We have already reviewed how structurally simple natural building blocks (such as lipids, amino acids, peptides, carbohydrates and natural polymers)

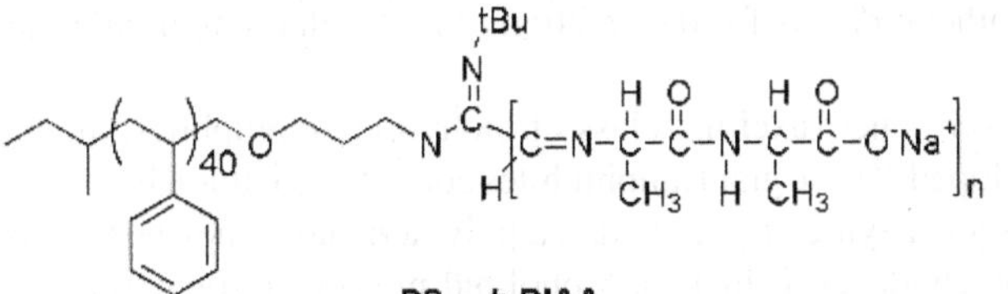

PS$_{40}$-*b*-PIAA$_n$

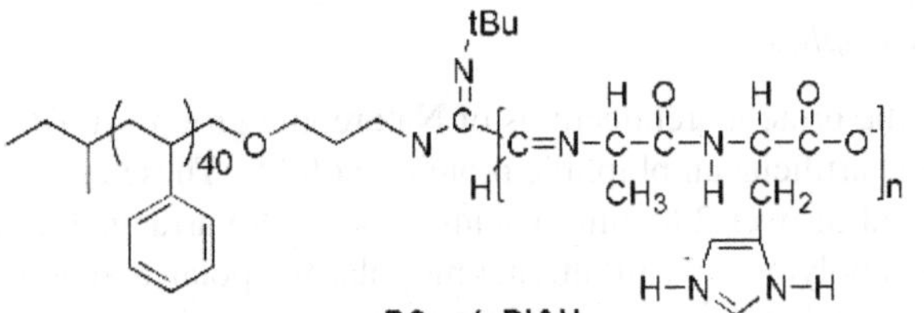

PS$_{40}$-*b*-PIAH$_n$

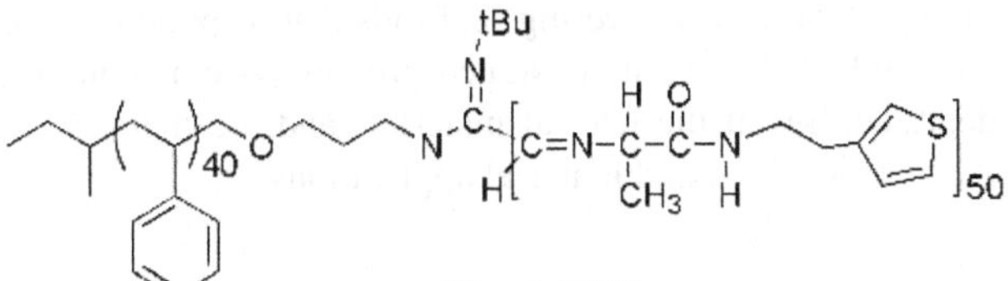

PS$_{40}$-*b*-PIAT$_{50}$

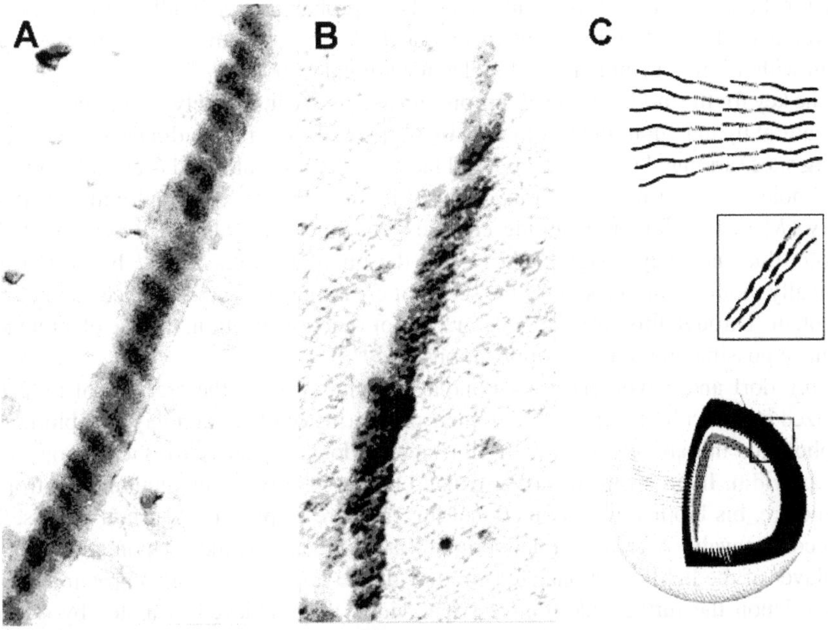

FIGURE 7.35. Top: structure of polystyrene-block-polyisocyanopeptides. Bottom: transmission electron micrographs of superhelices formed by (A) PS$_{40}$-b-PIAA$_{10}$ and (B) PS$_{40}$-b-PIAH$_{15}$ and (C) schematic representation of vesicles formed by PS$_{40}$-b-PIAT$_{50}$.

were combined with synthetic blocks for the creation of molecular amphiphiles and block copolymers.

Very recently however, a new exciting class of biohybrid amphiphiles, the *giant amphiphiles*, has been developed. These giant amphiphiles consist of a natural biomacromolecular head group, such as an enzyme or protein and a polymeric tail. They possess the same hydrophilic/hydrophobic character as their phospholipid molecular counterparts but have dimensions many times larger (Section 4.3).

7.4.1. Artificial Protein Assemblies

In an effort to mimic the protein architectures of Nature, the groups of Tirrell[238] and Deming[239] have synthesized artificial amphiphilic protein scaffolds. These genetically prepared polypeptides consisted of over 200 amino acids which were arranged in a manner similar to that of diblock copolymers (with alternating polar-nonpolar regions) and their structure was designed to resemble those found in structural proteins that support internal cell shapes. The designed proteins self-assemble in solution and form hydrogels with considerable mechanical strength and fast-recovering scaffolds that respond to pH changes[238] or are heat resistant up to 90 °C.[239,240] The designed protein systems combine both the functionality and biodegradability of the natural materials and are therefore destined to prove useful in a variety of biomedical and material applications.

7.4.2. Streptavidin-Conjugates–Supramolecular Clicking

The most commonly used protein, often ultilized as a versatile supramolecular building block for the synthesis of biohybrid materials, is probably streptavidin. Streptavidin was discovered in the bacterium *Streptomyces avidinii*,[241,242,243] and is an homo-tetrameric protein with a 2-fold symmetry, and molecular weight of 60 kDa.[244,245,246]

The key for the modification of this protein and its use in a variety of applications[247,248] is its natural ligand, vitamin H (d-biotin, **46** in Figure 7.2). Streptavidin possess four biotin binding sites (arranged in pairs at opposed faces of the molecule) and is capable of binding biotin molecules in a non-cooperative way (Figure 7.36)[244,249,250] and with a very high affinity ($K_a \sim 10^{15}$ for the resulting complex).[251] Since the valeric acid carboxyl group of biotin does not play a significant role in the binding process,[251,247] biotin is usually chemically modified at this site for the scope of creating new ligands, which can bind with the protein. Through this very versatile method of functionalization, the use of streptavidin was made possible in numerous applications.[247,248]

Ringsdorf and coworkers used monolayer techniques for the creation of lipid functionalized streptavidin layers.[252,253] In their pioneering work, a monolayer of biotinylated phospholipids initially created a two-dimensional docking matrix for the immobilization of streptavidin. In an effort to further expand this system while maintaining control over its structure, bis-biotinylated bifunctional linkers were employed. Addition of these linkers to concanavalin A created a new protein-based bidentate linker. Upon addition to the monolayer of the lipid-functionalized streptavidin a second biotilynated protein layer was created. Upon the further addition of streptavidin, they achieved in a step-by-step construction process well-defined alternating protein triple layers of streptavidin and concanavalin A (Figure 7.37b). A similar approach has more recently been used to crosslink

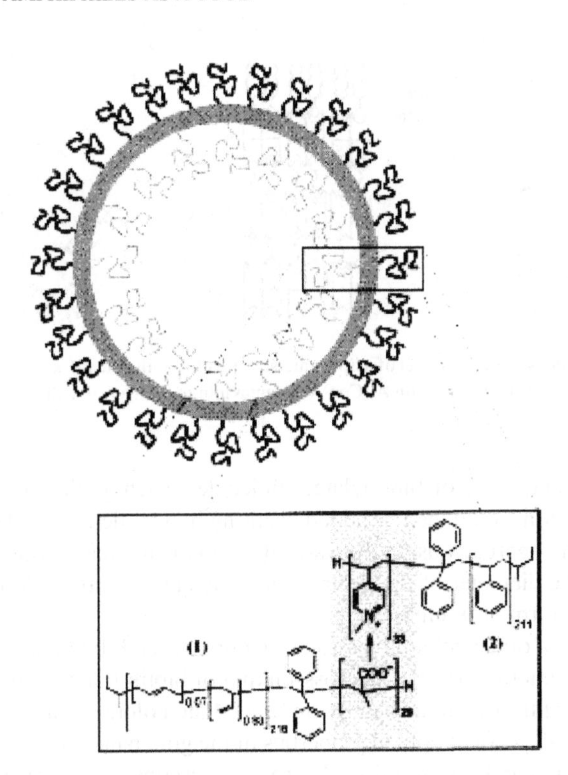

FIGURE 7.36. Schematic model of a vesicle with a corona of segregated polymer chains, as formed in a mixture of oppositely charge block copolymers. Reproduced with permission from the American Chemical Society (2003).

vesicles derived from mixtures of non-functionalized and biotin-functionalized lipids (Figure 7.37c).[253,254,255,256] The great advantage of using this site-specific binding between the vesicles is that their spherical shape and membrane morphology were preserved while with more conventional methods (liposomes that have been aggregated under osmotic stress[257] or by addition of calcium ions)[258] vesicles often deform and become fragile and leaky.

A central backbone of straight tubular structures formed by the self-assembly of biotinylated lipids was used as a scaffold in order to construct functionalized three-dimensional, highly ordered helical arrays of streptavidin (Figure 7.38).[259,260,261] Using these structures

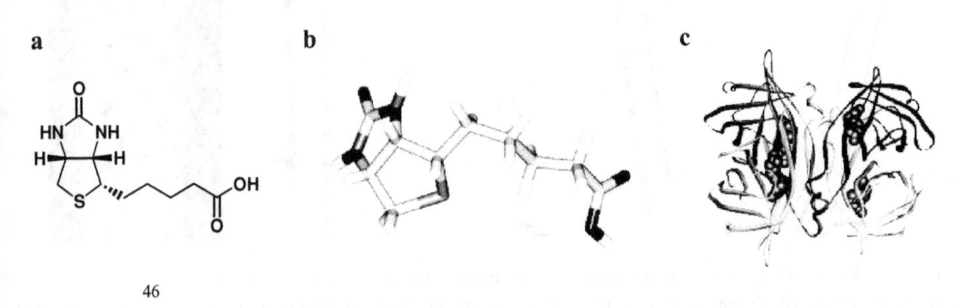

FIGURE 7.37. Chemical structure of biotin (**a**), X-ray structures of biotin (**b**) and biotin saturated streptavidin (**c**).

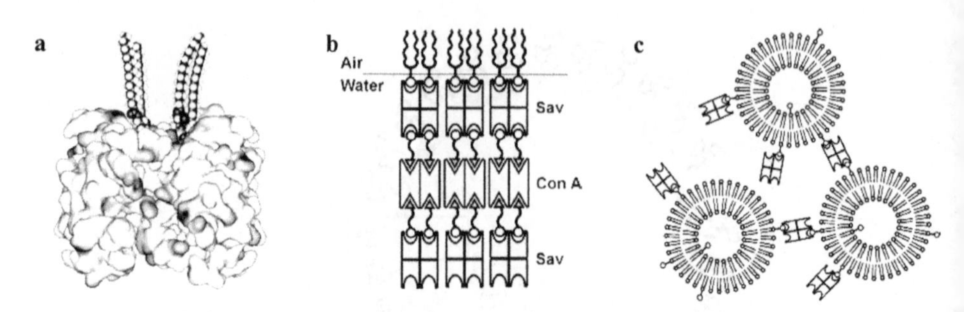

FIGURE 7.38. Computer generated model of a liposome-streptavidin conjugate (a), a schematic representation of a liposome-streptavidin (Sav)-concanavalin A (Con A)-streptavidin multilayer (b) and of streptavidin crosslinked vesicles (c).

as a matrix, a second layer of biotinylated molecules, such as the biotin-appended iron storage protein ferritin, could then be added resulting in well-defined and stable nanowires. Following the same approach using derivatized carbon nanotubes acting as a backbone, it was proven that the diameter of the central core plays an essential role in the subsequent architecture of the formed assemblies.[262]

In an effort to organize gold colloids, Connolly and Fitzmaurice also used the streptavidin-biotin interaction.[263] The cross-linking of biotin functionalized gold upon the addition of streptavidin was indicated by an immediate colour change (from red to blue) due to the distance-dependent optical properties of the gold nanoparticles,[264,265] whilst dynamic light scattering and transmission electron microscopy studies revealed the presence of cross-linked networks, containing an average of 20 interconnected particles separated by about 5 nm, which correlates well with the diameter of streptavidin (Figure 7.39).

More recently the application of the biotin-streptavidin approach has been extensively used by Niemeyer and coworkers in order to construct and manipulate supramolecular assemblies of DNA.[266] In the formation of three-dimensional networks of DNA, streptavidin acted predominantly as a bivalent linker defining the spatial arrangement of the DNA fragments (Figure 7.40).[267] Interestingly, thermal treatment led to well defined nanocircles within which one streptavidin molecule connecting both ends of the DNA.[268,269] Although

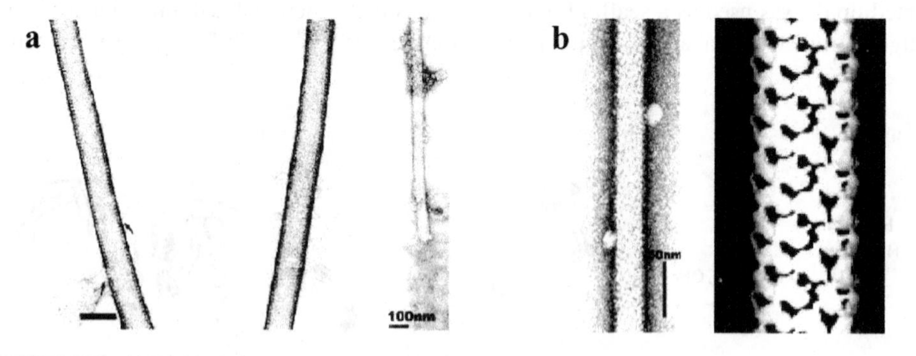

FIGURE 7.39. (*a*) Lipid tubular structures covered with streptavidin (left, bar = 50 nm) and with streptavidin and ferritin (right). (**b**) Streptavidin organized on a carbon nanotube (left) and the corresponding three dimension model (right).

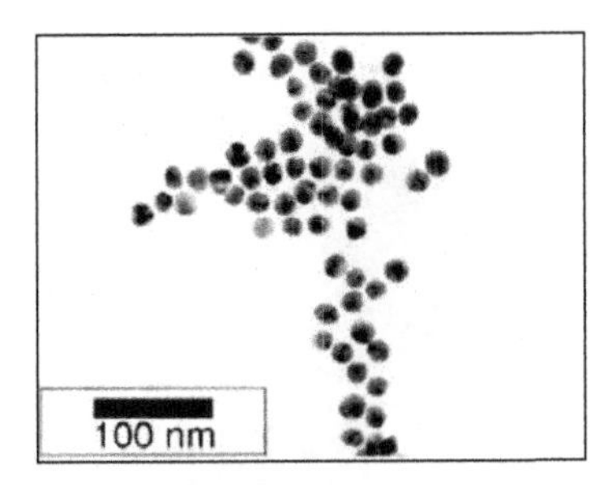

FIGURE 7.40. (a) TEM image of biotin-modified gold particles after streptavidin-induced aggregation.

ot exactly amphiphilic in character, the concept of polymer-protein hybrids has led to the ecent development of the *giant amphiphiles*, using synthetic hydrophobic polymers as tails.

.4.3. Giant Amphiphiles

The *giant amphiphiles* represent a new class of surfactants combining interesting elf-assembling properties and expression of catalytic activity of the head group in the *ano*assemblies. Giant amphiphiles were designed to consist of an enzyme or protein acting s a head group and a synthetic hydrophobic polymer as the tail. These biohybrid polymers iffer from other protein-polymer conjugates in the sense that the protein to polymer-ratio ; predefined and the position of the conjugation site is precisely known (Figure 7.41). *iant amphiphiles* are in fact block copolymers, which have by design significantly higher nolecular weights and volumes than their completely synthetic counterparts. Furthermore, aking into account that Nature synthesizes its biopolymers with high efficiency, the giant mphiphiles have the intrinsic structural advantage over the synthetic block copolymers of ossessing a monodisperse block (the protein). Four distinct approaches have been devel-ped, as discussed in the next sections.

he Biotin-Streptavidin Approach The synthesis of giant amphiphiles was first reported by ne group of Nolte, using the well-established biotin-streptavidin approach (Figure 7.38).[270]

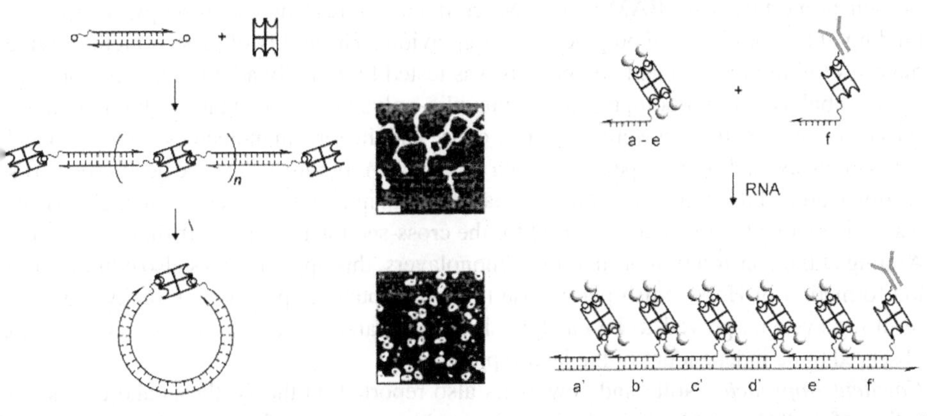

IGURE 7.41. Formation of DNA-streptavidin networks (left, bars = 100 nm) and the spatial arrangement of old (a–e) and antibody (f) bounded streptavidin molecules by DNA-RNA interactions (right).

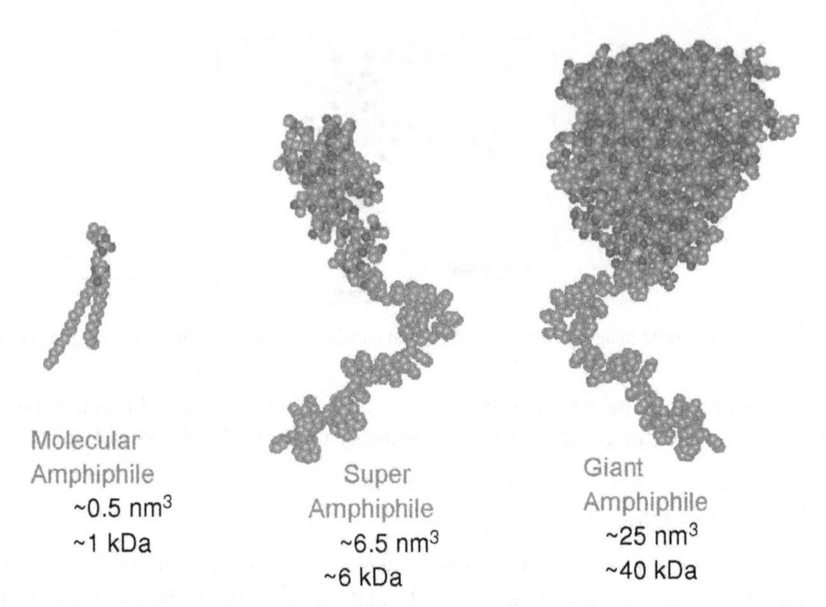

FIGURE 7.42. Computer generated models of the molecular, super and giant amphiphiles and their respective dimensions.

The hybrid amphiphiles were prepared by the association of two molecules of monobiotinylated polystyrene (polystyrene was coupled on the valeric acid carboxyl group of biotin) with streptavidin. As already mentioned, the affinity between streptavidin and biotin is so high (Figure 7.37, Ka $= 10^{15}M^{-1}$; 21 kcal mol^{-1}) that the complex formation can be regarded as irreversible. The protein-polymer hybrids were constructed using monolayer techniques, which allowed the monitoring of their formation through the recorded compression isotherms. More specifically, the biotinylated polymer was spread at the air/water interface of a Langmuir trough and the addition of streptavidin to the sub-phase followed. Compression of the resulting monolayers in the presence and absence of streptavidin and characterization of their properties (with BAM and AFM, confocal fluorescent spectroscopy) verified the binding of two biotinylated polymers per streptavidin. The ability of these protein-polymer matrices to bind biotinylated compounds was tested by initially adding ferritin molecules to the subphase of non-compressed streptavidin-polystyrene conjugates. The monolayers formed upon compression were uniformly dense and the ferritin molecules could be visualized as closely packed in respect with each other (TEM, Figure 7.42). The head group cross sectional area of the conjugates was calculated as 126 nm^2 (63 nm^2 per polymer chain), and was in agreement with values reported for the cross-sectional area of ferritin (133 nm^2).[271] Aiming at the construction of functional monolayers, this approach was also employed for the formation of Horseradish peroxidase (HRP)/streptavidin/polystyrene biohybrids. The resulting systems proved to be catalytically active, and the activity of bound HRP was observed to be independent of the lateral pressure.

Covalent Approach Nolte and coworkers also reported on the synthesis and characterization of well-defined biohybrids through covalent coupling of a polymer directly to the enzyme.[272] In order to achieve the covalent coupling in a predefined position, a disulfide

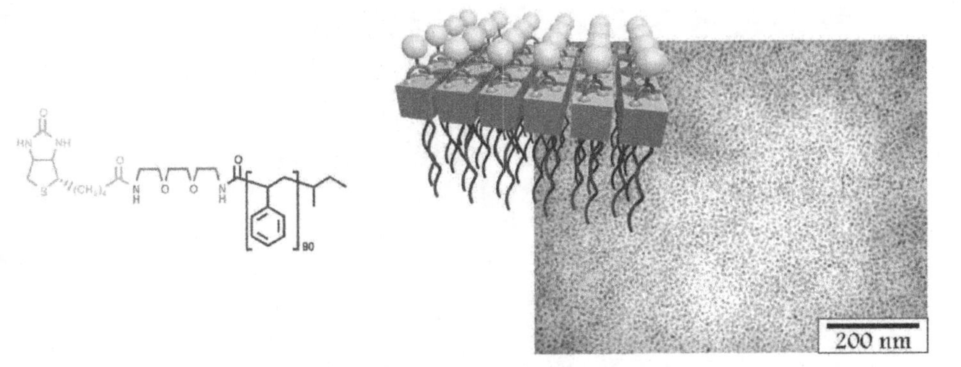

FIGURE 7.43. Addition of streptavidin to the biotinylated polystyrene amphiphile (left) results in a biohybrid monolayer, which can be further funtionalized by the addition of biotinylated proteins/enzymes (schematic representation in the middle). TEM image of a ferritin-streptavidin-polystyrene monolayer. Each black spot represents a single ferritin.

bridge positioned at the outer shell of the lipase B from *Candida antarctica* (CALB) was specifically reduced to provide two readily functionalizable free thiol groups. A single maleimide –capped polystyrene of 40 repeat units, was added to the enzyme using both monolayer techniques and reaction in solution to afford the lipase-polystyrene giant amphiphiles. In the first case, the increase of the cross sectional area upon compression of the monolayer indicated the presence of CALB as head group in the layer. In the latter case, a THF/water mixture was used as solvent in order to achieve the coupling between monodispersed species. TEM studies of the resulting giant amphiphiles revealed the formation of μm long fibers consisting of bundles of micellar rods (Figure 7.43). The individual rods possessed a diameter between 25 and 30 nm, closely corresponding to that predicted for a micellar architecture. It was therefore the first example in which a giant amphiphile was shown to exhibit self-assembling properties defined by the structure of the individual hybrids and similar to that seen for the molecular amphiphiles and block-copolymers.

The micellar bioassemblies however only exhibited 6–7% of the initial activity of the head group CALB. The authors ascribe this loss of activity partially to destabilization of the protein but predominantly to the bundling of individual rods, which makes most of the head groups inaccessible to the substrate (see Figure 7.43).

Cofactor Reconstitution Approach An alternative approach is to use the self-assembly of Nature by constructing a giant amphiphile through the direct coupling of the polymer to the cofactor of an enzyme and a subsequent reconstitution of the apoenzyme around the functionalized cofactor. This method was employed by Nolte and coworkers in order to create polystyrene Horseradish peroxidase biohybrids.[273] The desired biohybrid amphiphiles were prepared by adding a THF solution of the heme-appended polymer to an aqueous solution of the apoenzyme. The formation of the HRP-polystyrene surfactants was indicated by UV/Vis spectroscopy and electrophoresis. Electron microscopy (TEM, Cryo-TEM, Figure 7.44) revealed the formation of vesicular aggregates with diameters of 80–400 nm. In most cases these aggregates enclosed spherical objects, often located away from the center of the aggregates. To explain these structures, the authors assume that the heme-functionalized polymer first forms aggregates on to which the apo-HRP can subsequently

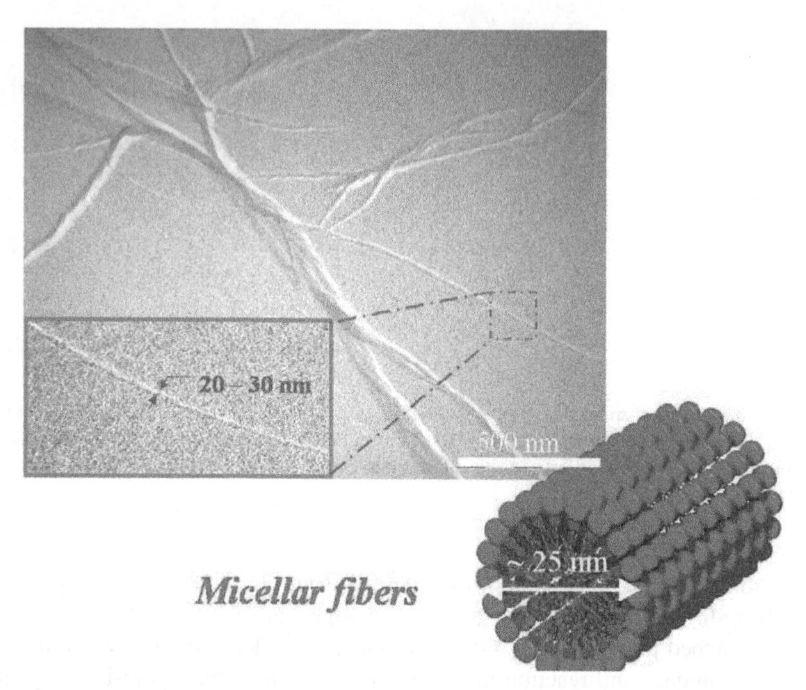

FIGURE 7.44. Transmission electron micrographs of the micellar assemblies formed by the aggregation of a lipase-polystyrene giant amphiphile in water. Expansion reveals a single micellar fiber with a diameter of 20–30 nm. Schematic representation of the micellar rod which possesses a polystyrene core.

be reconstituted in the form of the biohybrid. More specifically, these structures arise from vesicles growing from the polystyrene aggregates in such a way that they enclose the initial aggregate. Concerning the activity of these bio-assemblies, no activity could be observed when HRP was reconstituted at 4 °C. However, when the reconstitution of apo-HRP was carried out at 22 °C the enzyme-polymer hybrid surprisingly regained much of its activity.
Supramolecular Coupling Approach In order to readily vary the bio-architectures formed and in turn their properties, a modular approach to the construction of biohybrid amphiphiles was recently developed. This new strategy for the synthesis of precisely defined, monodispersed enzyme-polymer hybrids uses the supramolecular approach of metal-to-ligand coordination.[274]

The metal-to-ligand coordination coupling chosen for the construction of a family of giant amphiphiles was that of a metal-terpyridine complex, which is known to form stable bis-complexes with a variety of transition-metal ions.[275] The versatility of this metal-to-ligand system has been recently demonstrated in the construction of a series of coordination polymers and block copolymers.[276]

The components of the modular biohybrid amphipiles were constructed by the specific covalent coupling of a 2,2′:6′,2″-terpyridine-maleimide linker to a variety of enzymes, (the lipase B from *Candida antarctica* (CALB) and a protein, the bovine serum albumin (BSA)), using the same methodology developed for the construction of covalent biohybrids.[272] Once functionalized, a series of and homo-enzyme-enzyme and hetero enzyme-enzyme and enzyme-protein assemblies are in principal readily accessible via the

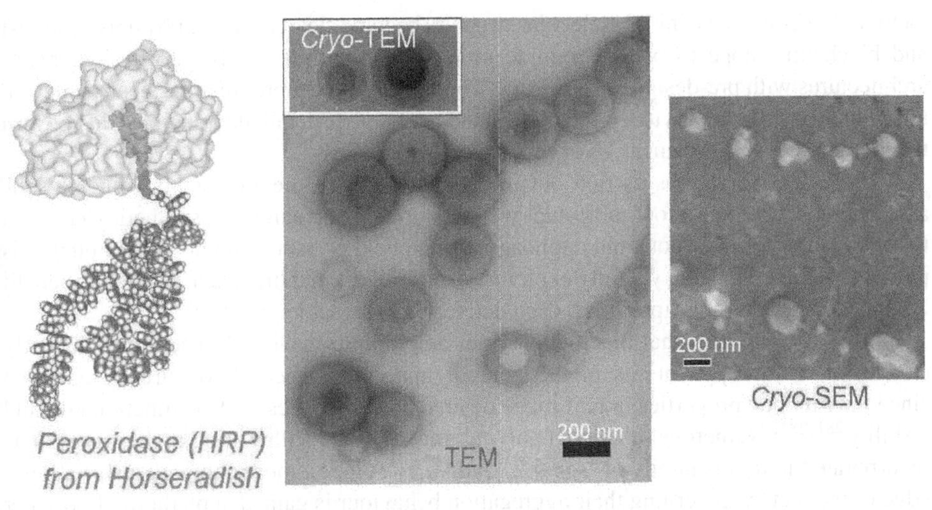

FIGURE 7.45. Schematic representation of the reconstituted polystyrene-horseradish peroxidase giant amphiphile (left) and the vesicular architectures it forms in solution (middle TEM image, right CryoSEM image).

reaction with Ru(III)Cl_3 and with $FeCl_2$ in different solvent conditions. The synthesis of the biohybrids was carried out using two approaches. In the first approach, different metals (Ru or Fe) were added to the terpyridine-enzyme hybrids to afford the mono-coordinated species which was in turn coupled with a 2,2′:6′,2″-terpyridine-capped polystyrene to afford the giant amphiphiles (Figure 7.4). In an alternative synthetic approach, the 2,2′:6′,2″-terpyridine-enzyme adducts were directly coupled to metallo-mono-terpyridine-polystyrene units (Figure 7.45). These metallo-biohybrids showed very interesting aggregation morphologies (micelles and vesicles were observed with TEM), which depended not only on the head group employed but also on the specific synthetic methodology that was followed. Very importantly, such biohybrids were also found to retain part of their catalytic activity.[277] The versatility of this supramolecular approach has yet to be fully realized but the combination of redoxactive (Cu^{2+}) and photoactive (Ru^{3+}, Ir^{3+}) linker units with self-assembled biological active proteins and enzymes offer many intriguing possibilities.

It is clear even from the few systems studied that the field of biohybrid giant amphiphiles has enormous potential. Nature provides a wide variety of proteins and enzymes possessing a continuum of material and functional properties. With the rapid development of molecular biology and biotechnology (protein expression, site directed mutagenesis) this variety of building blocks becomes increasing more readily available and more easily modified. The mutation of a desired protein for example with a single activated functional group can effortlessly lead to a library of biohybrids with unique properties.

7.5. CONCLUSION

The study of the self-assembly behavior of amphiphilic molecules, macromolecules and biohybrid systems, has led to the discovery of numerous spectacular and stunning nanometer and micrometer sized architectures. Only when the basic rules relating the resultant nano-architecture to the initial molecular structure become clearer, can the physicist, chemist and biochemist hope to pre-programme molecules to self-assembly into desired nano-architectures with pre-designed properties. Though significant progress has been achieved, the challenge still remains to try and understand in complete detail the principles that govern the hierarchical self-organization of molecules.

The extensive studies carried out in the latter half of the 21st century on the self-assembling behaviour of low-molecular weight amphiphiles (natural or synthetic) have laid the foundation for their potential application in the field of nanotechnology. Although this potential is far from being yet fully explored, nanosciences and life-sciences already benefit by their utilization (*e.g.* liposomes, drug delivery, nanoreactors, etc.).[278,279,280]

In spite of the fact that block-copolymer amphiphiles have only been more recently developed, many applications in the field of material sciences have already emerged since their unique properties impart many desired characteristics such as functionality and stability.[281,282,] Numerous functional polymer materials are still waiting in the wings to be incorporated in to this family of '*super-amphiphiles*'. Once a more comprehensive knowledge of the factors governing their aggregation behaviour is gained, a plethora of polymer based architectures with valuable nanoscience applications can be anticipated.

The successful synthesis of the biohybrid members of the family of amphiphiles, the *giant amphiphiles*, has also attracted the spotlight of fundamental sciences. The incorporation of biological molecules (proteins and enzymes) in to the amphiphilic structure mimics the self-assembly seen in Nature by amphiphilic proteins. Though enzymes and proteins have been extensively funtionalized in the past, the realization that they can express assembling properties similar to that of their molecular and polymeric counterparts offers new methods for the construction of functional biomimetic assemblages. The future for amphiphiles appears to be unlimited.

REFERENCES

1. B. Alberts, D. Bray, J. Lewis, M. Raff, K. Roberts, J. D. Watson, *Molecular Biology of the Cell* (Garland, New York, 1983).
2. P. Kostyuk, *Plasticity in Nerve Cell Function* (Oxford University Press, Oxford, UK, 1998).
3. J.-M. Lehn, *Science* **260**, 1762 (1993).
4. "Synkinetic" means involuntary movement, the performing of an unintended movement when making a voluntary one.
5. J.-H. Fuhrhop, J. Köning, *Membranes and Molecular Assemblies: The Synkinetic Approach* (The Royal Society of Chemistry: Cambridge, 1994).
6. D. Phili, J. F. Stoddart, *Angew. Chem. Int. Ed. Eng.* **35**, 1154 (1996).
7. A. E. Rowan and R. J. M. Nolte, *Angew. Chem. Int. Ed. Engl.* **37**, 63 (1998).
8. C. Tanford, *The Hydrophobic Effect, Formation of Micelles and Biological Membranes* (Wiley Interscience, New York 1973).
9. C. Tanford, *The Hydrophobic Effect*, 2nd edition (Wiley, NewYork, 1980).
10. W. Kauzmann, *Adv. Protein Chem.* **14**, 1 (1959).
11. W. Blokzijl and J. B. F. N. Engberts, *Angew. Chem. Int. Ed. Engl.* **32**, 1545 (1993).
12. D. F.; Evans and H. Wennerström, *The Colloidal Domain, where Physics, Chemistry, Biology, and Technology meet.* (VCH, New York/Weinheim, Cambridge, 1994).
13. T. Kunitake, *Angew. Chem. Int. Ed. Engl.* **31**, 709 (1992).
14. F. M. Menger, *Angew. Chem. Int. Ed. Engl.* **30**, 1086 (1991).
15. J. A. A. W. Elemans, R. de Gelder, A. E. Rowan, R. J. M. Nolte, *Chem. Commun.* 1553 (1998).

16. N. A. J. M. Sommerdijk, P. J. J. A. Buynsters, A. M. A. Pistorius, M. Wang, M. C. Feiters, R. J. M. Nolte, and B. Zwanenburg, *J. Chem. Soc. Chem. Commun.* 1941 (1994).
17. J. H. Hafkamp, M. C. Feiters, and R. J. M. Nolte, *Angew. Chem.* **106**, 1055–1055 (1994).
18. J. M. Schnur, *Science* **262**, 1669–1676 (1993).
19. N. Nakashima, S. Asakuma, and T. Kunitake, *J. Am. Chem. Soc.* **107**(2), 509–510 (1985).
20. D. A. Frankel and D. F. O'Brian, *J. Am. Chem. Soc.* **116**(22) 10057–10069 (1994).
21. J.-H. Fuhrhop and W. Helfrich, *Chem. Rev.* 1565–1582 (1993).
22. H. Yanagawa, Y. Ogawa, H. Furuta, and K. Tsuno, *J. Am. Chem. Soc.* **111**, 4567–4570 (1989).
23. J. N. Israelachvili, D. J. Mitchel, and B. W. Ninham, *J. Chem. Soc. Faraday Trans. 2*, **72**, 1525 (1976).
24. When the packing considerations are taken into account, this model practically predicts that amphiphiles with a single alkyl chain are will form micelles or bilayers, those with two alkyl chains bilayers, and those with three alkyl chains inverted hexagonal phases.[0]
25. T. Kunitake, Y. Okahata, M. Shimomura, S. Yasunami, and K. Takarabe, *J. Am. Chem. Soc.* **103**, 5401 (1981).
26. J. C. M. Van Hest, M. W. P. L. Baars, D. A. P. Delnoye, M. H. P. van Genderen, and E. W. Meijer, *Science* **268**, 1592 (1995).
27. D. A. Frankel and D. F. O'Brian, *J. Am. Chem. Soc.* **107**, 509–510 (1994).
28. A. Sein and J. B. F. N. Engberts, *Langmuir*, **11**, 455 (1995).
29. T. Kunitake and Y. Okahata, *J. Am. Chem. Soc.* **102**, 549 (1980).
30. Y. Okahata and T. Kunitake, *Ber. Bunsenges. Phys. Chem.* **84**, 550 (1980).
31. H. Engelkamp, S. Middelbeek, and R. J. M. Nolte, *Science* **284**, 785–788 (1999).
32. T. Goto and T. Kondo, *Angew. Chem. Int. Ed. Engl.* **30**, 17 (1991).
33. J.-M. Lehn, *Supramolecular Chemistry* (VCH Press, New York, 1995).
34. G. M. Whitesides, J. P. Mathias, and C. T. Seto, *Science* **254**, 1312–1319 (1991).
35. S. I. Stupp, V. LeBonheur, K. Walker, L. S. Li, K. E. Huggins, M. Keser, and A. Amstutz, *Science* **276**, 384–389 (1997).
36. E. R. Zubarev, M. U. Pralle, L. M. Li, and S. I. Stupp, *Science* **283**, 523–526 (1999).
37. E. R. Zubarev, M. U. Pralle, E. D. Sone, and S. I. Stupp, *J. Am. Chem. Soc* **123**, 4105–4106 (2001).
38. Y.-Y. Won, H. T. Davis, and F. S. Bates, *Science* **283**, 960–963 (1999).
39. S. A. Johnson, P. J. Ollivier, and T. E. Mallouk, *Science* **283**, 963–965 (1999).
40. R. D. Piner, J. Zhu, F. Xu, S. Hong, and C. A. Mirkin, *Science* **283**, 661–663 (1999).
41. T. Yokoyama, S. Yokoyama, T. Kamikado, Y. Okuno, and S. Mashiko, *Nature* (London) **413**, 619–621 (2001).
42. K. Holmberg, *Current Opinion in Colloid & Interface Science*, **6**, 148–159 (2001).
43. S. Budavari, (Ed.) *The Merck Index*, 11 edition (Merck&Co. Rahway, 1989), p. 1328).
44. S. B. Mahato, B. C. Pal, and A. K. Nandy, *Tetrahedron* **48**, 6717 (1992).
45. H. Brockerhoff, in *Bio-organic Chemistry*, edited by E. E. van Tamelen, (Academic Press, New York, 1977), Vol. 3, p. 1.
46. A. D. Bangham and R. W. Horne, *J. Mol. Biol.* **8**, 660 (1964).
47. T. Kunitake and Y. Okahata, *J. Am. Chem. Soc.* **99**, 3860 (1977).
48. T. Kunitake, Y. Okahata, M. Shimomura, S. Yasunami, and K. Takarabe, *J. Am. Chem. Soc.* **103**, 5401 (1981).
49. T. Kunitake, Templating, self-assembly, and self-organization, in *Comprehensive Supramolecular Chemistry*, edited by J. L. Atwood, J. E. D. Davies, D. D. Macnicol, F. Vögtle, J.-M. Lehn (series editor), J.-P. Sauvage and M. W. Hosseini (volume editors) (Pergamon, Elsevier Science Ltd. 1996), Vol. IX, p. 351.
50. J. H. Fendler, *Membrane Mimetic Chemistry*, (Wiley-Interscience, New York, 1982).
51. H. Ringsdorf, B. Schlarb, and J. Venzmer, *Angew. Chem. Int. Ed. Engl.* **27**, 113 (1988).
52. T. Kunitake and Y. Okahata, *Chem. Lett.* 1337 (1977).
53. T. Kunitake and Y. Okahata, *Bull. Chem. Soc. Jpn.* **51**, 1877 (1978).
54. T. Kunitake, N. Nakashima, S. Hayashida, and K. Yonemori, *Chem. Lett.* 1413 (1979).
55. E. J. R. Sudholter, J. B. F. N. Engberts, and D. Hoekistra, *J. Am. Chem. Soc.* **102**, 2467 (1980).
56. J.-H. Fuhrhop, H. Bartsch, and D. Fritzsch, *Angew. Chem. Int. Ed. Engl.* **20**, 804 (1981).
57. T. Kunitake, Y. Okahata, and S. Yasunami, *J. Am. Chem. Soc.* **104**, 5547 (1982).
58. F. Giulieri, M.-P. Krafft, and J. G. Riess, *Angew. Chem. Int. Ed. Engl.* **33**, 1514 (1994).

59. F. Giulieri, F. Guillod, J. Greiner, M.-P. Krafft, and J. G. Riess, *Chem. Eur. J.* **2**, 1335 (1996).
60. Y. Ishikawa, H. Kuwahara, and T. Kunitake, *J. Am. Chem. Soc.* **111**, 8530 (1989).
61. Y. Ishikawa, H. Kuwahara, and T. Kunitake, *Chem. Lett.* 1737 (1989).
62. Y. Ishikawa, H. Kuwahara, and T. Kunitake, *J. Am. Chem. Soc.* **116**, 5579 (1994).
63. J.-H. Fuhrhop, T. Bedurke, A. Hahn, S. Grund, J. Gatzmann, and M. Riederer, *Angew. Chem. Int. Ed. Engl.* **33**, 350 (1994).
64. J.-H. Fuhrhop, P. Schnieder, J. Rosenberg, and E. Boekema, *J. Am. Chem. Soc.* **109**, 3387–3390 (1987).
65. P. Yager and P. E. Schoen, *Mol. Cryst. Liq. Cryst.* **106**, 371 (1984).
66. P. Yager, P. E. Schoen, C. Davies, R. Price, and A. Singh, *Biophys. J.* **48**, 899 (1985).
67. Singh A. and J. M. Schnur, *Polym. Prepr. Am. Chem. Soc. Div. Polym. Chem.* **26**, 184 (1985).
68. A. Singh, J. Schnur, and M. Synth, *Commun.* **16**, 847 (1986).
69. J. M. Schnur, *Science* **262**, 1669 (1993).
70. M. A. Markowitz and J. M. Schnur, A. Singh, *Chem. Phys. Lipids* **62**, 193 (1992).
71. M. A. Markowitz, S. Baral, S. Brandow, and A. Singh, *Thin Solid Films*, **224**, 242 (1993).
72. J. H. Georger, A. Singh, R. R. Price, J. M. Schnur, P. Yager, and P. E. Schoen, *J. Am. Chem. Soc.* **109**, 6169 (1987).
73. B. N. Thomas, C. R. Safinya, R. J. Plano, and N. A. Clark, *Science* **267**, 1635 (1995).
74. A. Singh, T. G. Burke, J. M. Calvert, J. H. Georger, B. Herendeen, R. R. Price, P. E. Schoen, and P. Yager, *Chem. Phys. Lipids* **47**, 135 (1988).
75. E. M. Arnett and J. M. Gold, *J. Am. Chem. Soc.* **104**, 636 (1982).
76. A. I. Boyanov, B. G. Tenchov, R. D. Koyanova, and K. S. Koumanov, *Biochim. Biophys. Acta*, **732**, 711 (1983).
77. M. S. Spector, J. V. Selinger, A. Singh, J. M. Rodriguez, R. R. Price, and J. M. Schnur, *Langmuir* **14**, 3493 (1998).
78. F. Giulieri, M.-P. Krafft, and J. G. Riess, *Angew. Chem. Int. Ed. Engl*, **33**, 1514 (1994).
79. H. Ringsdorf, B. Schlarb, and J. Venzmer, *Angew. Chem. Int. Ed. Engl.* **27**, 113 (1988).
80. M. S. Spector, J. V. Selinger, and J. M. Schnur, Chiral and molecular melf-mssembly, in *Topics in Stereochemistry*: *Materials-Chirality*, edited by M. M. Green, R. J. M. Nolte, and E. W. Meijer, S. E. Denmark, J. Siegel (Wiley Interscience, Hoboken, New Jersey, 2003), Vol. 24, p. 281–371.
81. P. Yager, R. R. Price, J. M. Schnur, P. E. Schoen, A. Singh, and D. G. Rhodes, *Chem. Phys. Lipids* **46**, 171 (1988).
82. W. J. Helfrich, *Chem. Phys.* **85**, 1085 (1986).
83. W. Helfrich and J. Prost, *Phys. Rev. A.* **38**, 3065 (1988).
84. J.-H. Fuhrhop and W. Helfrich, *Chem. Rev.* **93**, 1565 (1993).
85. P. Nelson and T. Powers, *Phys. Rev. Lett.* **69**, 3409 (1992).
86. P. Nelson and T. Powers, *J. Phys. II (Fr.)* **3**, 1535 (1993).
87. J. V. Selinger and J. M. Schnur, *Phys. Rev. Lett.* **71**, 4091 (1993).
88. J. V. Selinger, F. C. MacKintosh, and J. M. Schnur, *Phys. Rev. E* **53**, 3804–3818 (1996).
89. A. S. Rudolph, B. R. Singh, A. Singh, and T. G. Burke, *Biochim. Biophys. Acta* **943**, 454 (1988).
90. B. N. Thomas, C. M. Lindemann, and N. A. Clark, *Phys. Rev. E. Stat. Phys. Plasmas Fluids Relat. Interdiscip. Top.* **59**, 3040–3047 (1999).
91. B. N. Thomas, R. C. Corcoran, C. L. Cotant, C. M. Lindemann, J. E. Kirsch, and P. J. Persichini, *J. Am. Chem. Soc.* **120**, 12178–12186 (1998).
92. B. N. Thomas, R. C. Corcoran, C. L. Cotant, C. M. Lindemann, J. E. Kirsch, and P. J. Persichini, *J. Am. Chem. Soc.* **124**, 1227–1233 (2002).
93. A. Singh, T. G. Burke, G. M. Calvert, J. H. Georger, B. Herendeen, R. R. Price, P. E. Schoen, and P. Yager, *Chem. Phys. Lipids* **47**, 135–147 (1988).
94. M. S. Spector, J. V. Selinger, A. Singh, J. M. Rondriguez, R. Price, and J. M. Schnur, *Langmuir* **14**, 3493–3500 (1998).
95. J. V. Selinger, M. S. Spector, and J. M. Schnur, *J. Phys. Chem. B.* **105**, 7157–7169 (2001).
96. U. Seifert, J. Shillcock, and P. Nelson, *Phys. Rev. Lett.* **77**, 5237–5240 (1996).
97. S. Pakhomov, R. P. Hammer, B. K. Mishra, and B. N. Thomas, *PNAS* **100**, 3040–3042 (2003).
98. M. M. Green, N. C. Peterson, T. Sato, A. Teramoto, R. Cook, and S. Lifson, *Science* **268**, 1860–1866 (1995).

99. S. Lifson, C. Andreola, N. C. Peterson, and M. M. Green, *J. Am. Chem. Soc.* **111**, 8850–8858 (1989).
100. D. R. Link, G. Natale, R. Shao, J. E. Mackennan, N. A. Clark, E. Körblova, and D. M. Walba, *Science* **278**, 1924–1927 (1997).
101. J. V. Selinger and R. L. B. Selinger, *Phys. Rev. Lett.* **76**, 58–61 (1996).
102. N. A. J. M. Sommerdijk, M. C. Feiters, R. J. M. Nolte, and B. Zwanenburg, *Recl. Trav. Chim. Pays-Bas* **113**, 194 (1994).
103. N. A. J. M. Sommerdijk, T. H. L. Hoeks, M. Synak, M. C. Feiters, R. J. M. Nolte, and B. Zwanenburg, *J. Am. Chem. Soc.* **117**, 4338 (1997).
104. N. A. J. M. Sommerdijk, M. H. L. Lambermon, M. C. Feiters, R. J. M. Nolte, and B. Zwanenburg, *Chem. Commun.* 1423 (1997).
105. N. A. J. M. Sommerdijk, M. H. L. Lambermon, M. C. Feiters, R. J. M. Nolte, and B. Zwanenburg, *Chem. Commun.* 455 (1997).
106. N. A. J. M. Sommerdijk, P. J. J. A. Buynsters, H. Akdemir, D. G. Geurts, R. J. M. Nolte, and B. Zwanenburg, *J. Org. Chem.* **62**, 4955, and correction 9388 (1997).
107. N. A. J. M. Sommerdijk, P. J. A. A. Buynsters, A. M. A. Pistorius, M. Wang, M. C. Feiters, R. J. M. Nolte, and B. Zwanenburg, *J. Chem. Soc. Chem. Commun.* 1941(1994), and addition 2736 (1994B).
108. N. A. J. M. Sommerdijk, P. J. J. A. Buynsters, H. Akdemir, D. G. Geurts, M. C. Feiters, R. J. M. Nolte, and B. Zwanenburg, *Chem. Eur. J.* **4**, 127 (1998).
109. Z. Reich, L. Zaidman, S. B. Gutman, T. Arad, and A. Minski, *Biochemistry* **33**, 14177 (1994).
110. P. J. J. A. Buijnsters and N. A. J. M. Sommerdijk (unpublished results).
111. N. Nakashima, S. Asakuma, and T. Kunitake, *Chem. Lett.* 1709 (1984).
112. K. Yamada, H. Ihara, T. Ide, T. Fukumoto, and C. Hirayama, *Chem. Lett.* 1713 (1984).
113. T. Shimizu and M. Hato, *Thin Solid Films* **180**, 179 (1989).
114. T. Shimizu, M. Mori, H. Minamikawa, and M. Hato, *Chem. Lett.* 1341 (1989).
115. T. Shimizu, M. Mori, H. Minamikawa, and M. Hato, *J. Chem. Soc. Chem. Commun.* 183 (1990).
116. V. Madison, C. M. Deber, and E. R. Blout, *J. Am. Chem. Soc.* **99**, 4788 (1977).
117. T. Shimizu and M. Hato, *Biochim. Biophys. Acta* **1147**, 50 (1993).
118. N. Nakashima, H. Fukushima, and T. Kunitake, *Chem. Lett.* 1207 (1981).
119. N. Nakashima, S. Asakuma, and T. Kunitake, *J. Am. Chem. Soc.* **107**, 509 (1985).
120. Y. Ishikawa, T. Hishimi, and T. Kunitake, *Chem. Lett.* 25 (1990).
121. H. Ihara, M. Takafuji, C. Hirayama, and D. F. O'Brien, *Langmuir* **8**, 1548 (1992).
122. T. Kuo and D. F. O'Brien, *Macromol.* **23**, 3225 (1990).
123. T. Kuo and D. F. O'Brien, *Langmuir* **7**, 584 (1991).
124. D. G. Rhodes, D. A. Frankel, T. Kuo, and D. F. O'Brien, *Langmuir* **10**, 267 (1994).
125. H. Zepik, E. Shavit, M. Tang, T. R. Jensen, K. Kjaer, G. Bolbach, L. Leiserowitz, I. Weissbuch and M. Lahav, *Science* **295**, 1266–1269 (2002).
126. J. Schneider, C. Messerschmidt, A. Schulz, M. Gnade, B. Schade, P. Luger, P. Bombicz, V. Hubert, and J. H. Fuhrhop, *Langmuir* **16**, 8575–8584 (2000).
127. J. Schneider, C. Messerschmidt, A. Schulz, M. Gnade, B. Schade, P. Luger, P. Bombicz, V. Hubert, and J.-H. Fuhrhop, *Langmuir* **16**, 8575–8584 (2000).
128. T. Gore, Y. Dori, Y. Talmon, M. Tirrell, and H. Bianco-Peled, *Langmuir* **17**, 5352–5360 (2001).
129. M. D. Shultz, M. J. Bowman, Y. W. Ham, X. M. Zhao, G. Tora, and J. Chmielewski, *Angew. Chem. Int. Edit.* **39**, 2710–2713 (2000).
130. R. Zutshi and J. Chmielewski, *Bioorg. Med. Chem. Lett.* **10**, 1901–1903 (2000).
131. F. Reichel, A. M. Roelofsen, H. P. M. Geurts, T. I. Hamalainen, M. C. Feiters, and G.-J. Boons, *J. Am. Chem. Soc.* **121**, 7989–7997 (1999).
132. J. D. Hartgerink, E. Beniash, and S. I. Stupp, *Science* **294**, 1684–1688 (2001).
133. J. D. Hartgerink, E. Beniash, and S. I. Stupp, *Proc. Natl. Acad. Sci. USA* **99**, 5133–5138 (2002).
134. S. Zhang, D. M. Marini, and W. S. Hwang Santoso *Opinion in Chemical Biology* **6**, 865–871 (2002).
135. T. C. Holmes, S. Delacalle, X. Su, A. Rich, and S. Zhang, *Proc. Natl. Acad. Sci. USA* **97**, 6728–6733 (2000).
136. D. M. Marini, Hwang, W. D. A. Lauffenburger, S. Zhang, and R. D. Kamm, *Nano Letters* **2**, 295–299 (2002).
137. M. R. Caplan, P. N. Moore, S. Zhang, R. D. Kamm, and D. A. Lauffenburger, *Biomacromolecules* **1**, 627–631 (2000).

138. S. Vauthey, S. Santoso, H. Gong, N. Watson, and S. Zhang, *Proc. Natl. Acad. Sci. USA* **99**, 5355–5360 (2002).
139. W. Hwang, D. M. Marini, R. D. Kamm, and S. Zhang, *J Phys Chem B* **106**, 2002 in press.
140. I. A. Nyrkova, A. N. Semenov, A. Aggeli, and N. Boden, *Eur Phys J B* **17**, 481–497 (2000).
141. J. V. Selinger, M. S, Spector, and J. M. Schnur, *J. Phys. Chem. B* **105**, 7157–7169 (2001).
142. H. A. Lashuel, D. Hartley, B. M. Petre, T. Walz, and P. T. Lansbury, Jr., *Nature* **418**, 291 (2002).
143. H. A. Lashuel, S. R. LaBrenz, L. Woo, L. C. Serpell, and J. W. Kelly, *J. Am. Chem. Soc.* **122**, 5262–5277 (2000).
144. T. Kowalewski and D. M. Holtzman, *Proc. Natl. Acad. Sci. USA* **96**, 3688–3693 (1999).
145. V. Munoz and L. Serrano, *Biochemistry* **34**, 15301–15306 (1995).
146. W. Hwang, D. Marini, R. Kamm, and S. Zhang, *J. Chem. Physics* **118**, 389–397 (2003).
147. M. G. Ryadnov and D. N. Woolfson, *Nat. Mater.* **2**, 329–332 (2003).
148. S. Lee and D. Eisenberg, *Nat. Struct. Biol.* **10**, 725–730 (2003).
149. F. Chiti, M. Stefani, N. Taddei, G. Ramponi, and C. M. Dobson, *Nature* **424**, 805–808 (2003).
150. J. Kisiday, M. Jin, B. Kurz, H. Hung, C. Semino, S. Zhang, and A. J. Grodzinsky, *Proc. Natl. Acad. Sci. USA* **99**, 9996–10001 (2002).
151. M. Antonietti, *Curr. Opin. Coll. Inter. Sci.* **6**, 244–248 (2001).
152. H. Matsui and B. Gologan, *J. Phys. Chem. B* **104**, 3383–3386 (2000).
153. H. Matsui and R. MacCuspie, *Nano Letters* **1**, 671–675 (2001).
154. H. Matsui, P. Porrata, and G. E. Douberly, *Nano Letters* **1**, 461–464 (2001).
155. B. Pfannemüller and W. Welte, *Chem. Phys. Lipids* **37**, 227 (1985).
156. L. Addadi, Z. Berkovitch-Yellin, I. Weissbuch, J. van Mil, L. J. W. Shimon, M. Lahav, and L. Leisorowitz, *Angew. Chem.* **24**, 466 (1985).
157. V. A. Zabel Müller-Fahrnow, R. Hilgenfeld, W. Saenger, B. Pfannemüller, V. Enkelmann, and W. Welte, *Chem. Phys. Lipids* **39**, 313 (1986).
158. J.-H. Furhhop, P. Schnieder, J. Rosenberg, and E. Boekema, *J. Am. Chem. Soc.* **109**, 3387 (1987).
159. J. Köning, C. Boettcher, H. Winkler, E. Zeitler, Y. Talmon, and J.-H. Furhrhop, *J. Am. Chem. Soc.* **115**, 693 (1993).
160. H. Stark, F. Zemlin, and C. Boettcher, *Ultramicroscopy* **63**, 75–79 (1996).
161. I. Tuzov, K. Crämer, B. Pfannemüller, S. N. Magonov, and M.-H. Whangbo, *New. J. Chem.* **20**, 37 (1996).
162. J.-H. Fuhrhop, P. Schnieder, E. Boekema, and W. Helfrich, *J. Am. Chem. Soc.* **110**, 2861 (1988).
163. Z. Horton, Z. Walaszek, and I. Ekiel, *Carbohydr. Res.* **119**, 263 (1983).
164. C. André, P. Luger, S. Svenson, and J.-H. Fuhrhop, *Carbohydr. Res.* **230**, 31 (1992).
165. C. André, P. Luger, S. Svenson, and J.-H. Fuhrhop, *Carbohydr. Res.* **240**, 47 (1994).
166. It is worth noting that the diameter of the M-helix (L-Glu-12) was approximately 1.5 times that of the P-helix (D-Glu-8).
167. D. A. Frankel and D. F. O'Brien, *J. Am. Chem. Soc.* **113**, 7436 (1991).
168. J.-H. Fuhrhop, P. Blumtritt, C. Lehmann, and P. Luger, *J. Am. Chem. Soc.* **113**, 7437 (1991).
169. G. Wegner, *Makromol. Chem.* **154**, 35 (1972).
170. R. H. Baughman and R. R. Chance, *Ann. N. Y. Acad. Sci.* **313**, 705 (1978).
171. D. A. Frankel and D. F. O'Brien, *J. Am. Chem. Soc.* **116**, 10057 (1994).
172. J. Y. Shin and N. L. Abbott, *Langmuir* **15**, 4404–4410 (1999).
173. L. I. Jong and N. L. Abbott, *Langmuir* **14**, 2235–2237 (1998).
174. Y. Orihara, A. Matsumura, Y. Saito, N. Ogawa, T. Saji, A. Yamaguchi, H. Sakai, and M. Abe, *Langmuir* **17**, 6072–6076 (2001).
175. Y. Kakizawa, H. Sakai A. Yamaguchi, Y. Kondo, N. Yoshino, and M. Abe, *Langmuir* **17**, 8044–8048 (2001).
176. N. Aydogan, C. A. Rosslee, and N. L. Abbott *Colloid Surf A* **201**, 101–109 (2002).
177. B. S. Gallardo, V. K. Gupta, F. D. Eagerton, L. I. Jong, V. S. Craig, R. R. Shah, and N. L. Abbott, *Science* **283**, 57–60 (1999).
178. M. A. Susan, M. Begum, Y. Takeoka, and M. Watanabe *Langmuir* **16**, 3509–3516 (2000).
179. C. A. Rosslee and N. L. Abbott. *Curr. Opin. Colloid Interf. Sci.* **5**, 81–87 (2000).
180. Y. J. Shin, L. I. Jong, N. Aydogan, N. L. Abbott, In *Reaction and Synthesis of Surfactant System*, 1st edition, edited by J. Texter, (Marcel Dekker Inc., New York 2001), 155–173.

181. C. A. Rosslee and N. L. Abbott, *Anal. Chem.* **73**, 4808–4814 (2001).
182. T. Kunitake, J.-M. Kim, and Y. J. Ishikawa, *J. Amer. Soc. Perkin Trans.* **2**, 885 (1991).
183. D. G. Whitten, L. Chen, H. C. Geiger, J. Perlstein, and X. Song, *J. Phys. Chem. B* **102**, 10098 (1998).
184. D. G. Whitten, *Acc. Chem. Res.* **26**, 502 (1993).
185. G. M. Whitesides, E. E. Simanek, J. P. Mathlas, C. T. Seto, D. N. Chin, M. Mammen, and D. M. Gordon, *Acc. Chem. Res.* **28**, 37 (1995).
186. Y. Zhang, C. Tan, Q. Liu, R. Lu, Y. Song, L. Jiang, Y. Zhao, T. J. Li, and Y. Liu, *Applied Surface Science* **220**, 224–230 (2003).
187. J. N. H. Reek, A. Kros, and R. J. M. Nolte, *Chem. Commun.* 245 (1996).
188. P. Madrich, *Ann. Rev. Biochem.* **56**, 435 (1987).
189. S. A. Jenekhe and X. L. Chen, *Science* **283**, 372 (1999).
190. P. Madrich, *Ann. Rev. Biochem.* **56**, 435 (1987).
191. H. Imahori and Y. Sakata, *Eur. J. Org. Chem.* 2445–2457 (1999).
192. D. M. Guldi and M. Prato, *Acc. Chem. Res.* **33**, 695–703 (2000).
193. N. Martín, L. Sànchez, B. Illescas, and I. Pérez, *Chem. Rev.* **98**, 2527–2547 (1998).
194. K. Prassides, M. Keshavarz, E, C. B. Beer, R. Gonzalez, Y. Murata, F. Wudl, A. K. Cheetham, and J. P. Zhang, *Chem. Mater.* **8**, 2405–2408 (1996).
195. S. A. Jenekhe and X. L. Chen, *Science* **279**, 1903–1907 (1998).
196. Y. P. Sun and C. Bunker, *Nature* (*London*) **365**, 398–401 (1993).
197. M. Prato, *J. Mater. Chem.* **7**, 1097–1109 (1997).
198. M. Prato, *Top. Curr. Chem.* **199**, 173–188 (1999).
199. C. Brabec, N. Sariciftci, and J. Hummelen, *Adv. Funct. Mater.* **11**, 15–26 (2001).
200. T. Da Ros, M. Prato, *Chem. Commun.* 663–669 (1999).
201. H. Tokuyama, S. Yamago, E. Nakamura, T. Shiraki, and Y. Sugiura, *J. Am. Chem. Soc.* **115**, 7918 (1993).
202. S. Takenaka, K. Yamashita, M. Takagi, T. Hatta, A. Tanaka, and O. Tsuge, *Chem. Lett.* 319–320 (1999).
203. M. Sawamura, H. Iikura, and E. Nakamura, *J. Am. Chem. Soc.* **118**, 12850 (1996).
204. S. Zhou, C. Burger, B. Chu, M. Sawamura, N. Nagahama, M. Toganoh, U. E. Hackler, H. Isobe, and E. Nakamura, *Science* **291**, 1944–1947 (2001).
205. V. Georgakilas, F. Pellarini, M. Prato, D. M. Guldi, M. Melle-Franco, and F. Zerbetto, *Proc. Natl. Acad. Sci.* **99**, 5075–5080 (2002).
206. For an overview on the micellization of block copolymers see: M. Moffit, K. Khougaz, and A. Eisenberg, *Acc. Chem. Res.* **29**, 95–102 (1996).
207. J. C. M. Van Hest, D. A. P. Delnoye, M. W. L. P. Baars, M. H. P. van Genderen, and E. W. Meijer, *Science* **268**, 1592 (1995).
208. L. Zhang and A. Eisenberg, *Science* **268**, 1728 (1995).
209. A. Choucair and A. Eisenberg, *Eur. Phys. J. E* **10**, 37–44 (2003).
210. M. Antonietti and S. Förster, *Adv. Mater.* **15**, 1323–1333 (2003).
211. S. T. Hyde, *J. Phys. (Paris)* **51**, C7209 (1990).
212. L. Zhang, A. Eisenberg, *J. Am. Chem. Soc.* **118**, 3168–3181 (1996).
213. K. Yu and A. Eisenberg, *Macromolecules* **29**, 6359–6361 (1996).
214. K. Schillen, K. Bryskhe, and Y. S. Mel'nikova, *Macromolecules* **32**, 6885–6888 (1999).
215. M. Maskos and J. R. Harris, *Macromol. Rapid Commun.* **22**, 271 (2001).
216. N. S. Cameron, M. K. Corbierre, and A. Eisenberg, *Can. J. Chem.* **77**, 1311–1326 (1999).
217. H. Shen and A. Eisenberg, *J. Phys. Chem. B.* **103**, 9488–9497 (1999).
218. H. Shen and A. Eisenberg, *Macromolecules* **33**, 2561–2572 (2000).
219. B. M. Discher, Y.-Y. Won, D. S. Ege, J. C.-M. Lee, F. S. Bates, D. E. Discher, and D. A. Hammer, *Science* **284**, 1143–1146 (1999).
220. B. M. Discher, H. Bermudez, D. A. Hammer, D. E. Discher, Y.-Y. Won, and F. S. Bates, *J. Phys. Chem. B.* **106**, 2848 (2002).
221. W. Meier, C. Nardin, and M. Winterhalter, *Angew. Chem. Int. Ed.* **39**, 4599 (2000).
222. C. Nardin, T. Hirt, J. Leukel, and W. Meier, *Langmuir* **16**, 1035 (2000).
223. S. A. Jenekhe and X. L. Chen, *Science* **279**, 1903–1907 (1998).
224. S. A. Jenekhe and X. L. Chen, *Science* **283**, 372–375 (1999).

225. A. P. Nowak, V. Breedveld, L. Pakstis, B. Ozbas, D. J. Pine, D. Pochan, and T. J. Deming, *Nature* **417**, 424 (2002).
226. S. Lecommandoux, M. F. Archer, J. F. Langenwalter, and H.-A. Klok, *Macromolecules* **34**, 9100 (2001).
227. F. Chécot, S. Lecommandoux, Y. Gnanou, and H.-A. Klok, *Angew. Chem. Int. Ed.* **41**, 1340 (2002).
228. H. Kukula, H. Schlaad, M. Antonietti, and S. Förster, *J. Am. Chem. Soc.* **124**, 1658 (2002).
229. J. J. L. M. Cornelissen, M. Fischer, N. A. J. M. Sommerdijk, and R. J. M. Nolte, *Science* **280**, 1427 (1998).
230. D. M. Vriezema, J. Hoogboom, K. Velonia, K. Takazawa, P. C. M. Christianen, J. C. Maan, A. E. Rowan, and R. J. M. Nolte, *Angew. Chem. Int. Ed.* **42**, 772 (2003).
231. I. Gitsov, K. L. Wooley, and J. M. J. Fréchet, *Angew. Chem.* **104**, 1282 (1992).
232. J. M. J. Fréchet, I. Gitsov, T. Monteil, S. Rochat, J.-F. Sassi, C. Vergelati, and D. Yu, *Chem. Mater.* **11**, 1267 (1999).
233. Y. Chang, Y. C. Kwon, S. C. Lee, and C. Kim, *Macromolecules* **33**, 4496 (2000).
234. J. J. L. M. Cornelissen, R. van Heerbeek, P. C. J. Kamer, J. N. H. Reek, N. A. J. M. Sommerdijk, and R. J. M. Nolte, *Adv. Mater.* **14**, 489 (2002).
235. A. Harada and K. Kataoka, *Macromolecules* **28**, 5294 (1995).
236. A. Harada and K. Kataoka, *Science* **283**, 65 (1999).
237. S. Schrage, R. Sigel, and H. Schlaad, *Macromolecules* **36**, 1417 (2003).
238. W. A. Petka, J. L. Harden, K. P. McGrath, D. Wirtz, and D. A. Tirrell, *Science* **281**, 389–392 (1998).
239. A. P. Nowak, V. Breedveld, L. Pakstis, B. Ozbas, D. J. Pine, D. Pochan, and T. J. Deming, *Nature* **417**, 424–428 (2002).
240. E. R. Welsh and D. A. Tirrell, *Biomacromolecules* **1**, 23–30 (2000).
241. L. Chaiet and F. J. Wolf, *Arch. Biochem. Biophys.* **106**, 1 (1964).
242. L. Chaiet, T. W. Miller, F. Tausig, and F. J. Wolf, *Antimicrob. Agents Chemother.* **161**, 28 (1963).
243. E. O. Stapley, J. M. Mata, I. M. Miller, T. C. Demny, and H. B. Woodruff, *Antimicrob. Agents Chemother.* **161**, 20 (1963).
244. P. C. Weber, D. H. Ohlendorf, J. J. Wendoloski, and F. R. Salemme, *Science* **243**, 85 (1989).
245. F. C. Bernstein, T. F. Koetzle, G. J. Williams, E. E. Meyer, Jr. M. D. Brice, J. R. Rodgers, O. Kennard, T. Shimanouchi, and M. Tasumi, *J. Mol. Biol.* **112**, 535 (1977).
246. C. Rosano, P. Arosio, and M. Bolognesi, *Biomol. Eng.* **16**, 5 (1999).
247. M. Wilchek and E. A. Bayer, *Anal. Biochem.* **171**, 1 (1988).
248. M. Wilchek and E. A. Bayer, *Methods Enzymol.* **184**, 14 (1990).
249. M. L. Jones and G. P. Kurzban, *Biochemistry* **34**, 11750 (1995).
250. T. Sano and C. R. Cantor, *Proc. Natl. Acad. Sci. U. S. A.* **92**, 3180 (1995).
251. N. M. Green, *Adv. Protein Chem.* **29**, 85 (1975).
252. W. Müller, H. Ringsdorf, E. Rump, G. Wildburg, X. Zhang, L. Angermaier, W. Knoll, M. Liley, and J. Spinke, *Science* **262**, 1706 (1993).
253. H. Ringsdorf and J. Simon, *Nature* **371**, 284 (1994).
254. S. Chiruvolu, S. Walker, J. Isrealachvili, F.-J. Schmitt, D. Leckband, and J. A. Zasadzinski, *Science* **264**, 1753 (1994).
255. S. A. Walker, M. T. Kennedy, and J. A. Zasadzinski, *Nature* **387**, 61 (1997).
256. E. T. Kisak, M. T. Kennedy, D. Trommeshauser, and J. A. Zasadzinski, *Langmuir* **16**, 2825 (2000).
257. C. A. Helm, J. N. Israelachvili, and P. M. McGuiggan, *Biochemistry* **31**, 1794 (1992).
258. S. W. Hui, S. Nir, T. P. Stewart, L. T. Boni, and S. K. Huang, *Biochim. Biophys. Acta* **941**, 130 (1988).
259. P. Ringler, W. Muller, H. Ringsdorf, and A. Brisson, *Chem. Eur. J.* **3**, 6 (20 (1997).
260. P. Huetz, S. vanNeuren, P. Ringler, F. Kremer, J. F. L. vanBreemen, A. Wagenaar, J. Engberts, J. Fraaije, and A. Brisson, *Chem. Phys. Lipids* **89**, 15 (1997).
261. I. Reviakine and A. Brisson, *Langmuir* **17**, 8293 (2001).
262. F. Balavoine, P. Schultz, C. Richard, V. Mallouh, T. W. Ebbesen, and C. Mioskowski, *Angew. Chem. Int. Ed.* **38**, 1912 (1999).
263. S. Connolly and D. Fitzmaurice, *Adv. Mater.* **11**, 1202 (1999).
264. C. A. Mirkin, R. L. Letsinger, R. C. Mucic, and J. J. Storhoff, *Nature* **382**, 607 (1996).
265. R. Elghanian, J. J. Storhoff, R. C. Mucic, R. L. Letsinger, and C. A. Mirkin, *Science* **277**, 1078 (1997).
266. C. M. Niemeyer, *Chemistry* **7**, 3188 (2001).

267. C. M. Niemeyer, M. Adler, B. Pignataro, S. Lenhert, S. Gao, L. Chi, H. Fuchs, and D. Blohm, *Nucleic Acids Res.* **27**, 4553 (1999).
268. C. M. Niemeyer, M. Adler, S. Gao, and L. Chi, *Angewandte Chemie, International Edition* **39**, 3055 (2000).
269. C. M. Niemeyer, R. Wacker, and M. Adler, *Angew. Chem. Int. Ed.* **40**, 3169 (2001).
270. J. M. Hannink, J. J. L. M. Cornelissen, J. A. Farrera, P. Foubert, F. C. De Schryver, A. J. Nico M. Sommerdijk, and R. J. M. Nolte, *Angew. Chem. Int. Ed.* **40**, 4732–4734 (2001).
271. H. Yoshimura, T. Scheybani, W. Baumeister, and K. Nagayama, *Langmuir* **10**, 3290–3295 (1994).
272. K. Velonia, A. E. Rowan, and R. J. M. Nolte, *J. Am. Chem. Soc.* **124**, 4224–4225 (2002).
273. M. J. Boerakker, J. M. Hannink, P. H. H. Bomans, P. M. Frederik, R. J. M. Nolte, E. M. Meijer, and N. A. J. M. Sommerdijk, *Angew. Chem. Int. Ed.* **41**, 4239–4241 (2002).
274. B. G. G. Lohmeijer and U.S. Schubert, *Angew. Chem. Int. Ed. Eng.* **41**, 3825–3829 (2002).
275. E. C. Constable, *Adv. Inorg. Chem. Radiochem.* **30**, 69–121 (1986).
276. B. G. G. Lohmeijer and U. S. Schubert, *Angew. Chem. Int. Edit.* **41** (20), 3825–3829 (2003).
277. K. Velonia, P. Thordarson, P. R. Andres, U. S. Schubert, A. E. Rowan, and R. J. M. Nolte, *Polymer Preprints* **44**, 648 (2003).
278. Y.-Y. Luk and N. L. Abbott, *Curr. Op. in Colloid & Interface Science* **7**, 267–275 (2002).
279. S. S. Santoso, S. Vauthey, and S. Zhang, *Curr. Op. in Colloid & Interface Sci.* **7**, 262–266 (2002).
280. R. S. Makkar and S. S. Cameotra, *App. Microbiol. and Biotech.* **58**, 428–434 (2002).
281. S. Foerster, *Top. Curr. Chem.* **226**, 1–28 (2003).
282. M. L. Adams, A. Lavasanifar, and G. S. Kwon, *J. Pharm, Sci.* **92**, 1343–1355 (2003).

8

Self-Assembly of Colloidal Building Blocks into Complex and Controllable Structures

Joe McLellan, Yu Lu, Xuchuan Jiang, and Younan Xia

8.1. INTRODUCTION

Colloids are small particles with at least one of their dimensions in the range of a few nanometers to one micrometer, where Brownian motion plays a critical role.[1–3] They are analogous to giant molecules in some respects, and behave in fair agreement with statistical mechanics.[4] Since the pioneering work by Faraday and Graham more than 140 years ago, colloids has become a subject of great importance to many fields that include chemistry, biology, materials science, condensed matter physics, applied optics, and fluid dynamics.[5–15] The enormous impact of colloids can also be appreciated by their extensive use in a variety of commercial products such as foods, drinks, inks, paints, toners, coatings, papers, cosmetics, photographic films, and magnetic recording media.

Fundamental studies on colloids usually require monodisperse samples, where the particles are uniform in size, shape, composition, and surface chemistry. Hence, most advances have been brought about by the improvement of or addition to the many synthetic strategies capable of generating monodisperse colloids in relatively copious quantities.[16] As confined by the interfacial energy, "sphere" represents perhaps the simplest symmetry that a colloidal particle can easily assume during nucleation and growth. In the past several decades, many methods have been developed for synthesizing spherical colloids. Many theoretical models concerning colloidal behaviors are also based on the spherical symmetry. For example, the Mie theory that describes the optical scattering properties of individual colloidal particles[17] and the Derjaguin-Landau-Vervey-Overbeek (DLVO) model[18–19] that deals with the interactions between particles are both derived from spherical colloids. Thanks to the continuous efforts of many research groups, a wealth of colloids can now be synthesized from various

Department of Chemistry, University of Washington, Seattle, WA 98195, USA.
Corresponding author. E-mail: xia@chem.washington.edu, Tel: 206-543-1767; Fax: 206-685-8665.

organic polymers and inorganic materials as truly monodispersed entities, in which the size and shape of the particles, as well as the charges chemically fixed on surfaces, are all identical to within 1–2%.[16,20–22] In comparison, only a limited number of methods have been demonstrated for generating nonspherical colloids,[23] with typical shapes including rods, spheroids, and plates. Only a few of these methods were able to produce truly monodispersed samples.

Spherical colloids are traditionally exploited as a model system to investigate the light scattering properties,[24] interparticle interactions,[25] and hydrodynamic properties associated with colloidal particles.[26,27] In recent years, they have also been actively explored as a class of simple and versatile building blocks for self-assembly to generate long-range ordered lattices such as colloidal crystals.[28–35] When spherical colloids are arranged into a periodic lattice, interesting phenomena are often observed. Opals are good examples of this. The attractive iridescent colors of opals are caused by their three-dimensionally periodic lattices of silica and zirconia colloids that are colorless by themselves.[36] A relatively new field of research has grown out of the study of this class of materials, which is often referred to as photonic crystals or photonic bandgap structures (PBG).[37,38] These materials are technologically important, as they may lead to advances in the fabrication of light sources, detectors, and waveguiding structures. In other demonstrations, spherical colloids have also been examined as a class of simple building blocks to generate highly complex structures via self-assembly. Typical examples include polygonal, polyhedral, and helical aggregates containing one or more than one types of spherical colloids. The complexity and functionality of these self-assembled structures can be further enhanced by incorporating core-shell and/or nonspherical colloids as the building blocks.

The objectives of this chapter are the following: *i*) to briefly discuss a number of methods that have been demonstrated for the facile synthesis of spherical and nonspherical colloids with well-controlled sizes, shapes, and properties; *ii*) to address experimental issues related to the self-assembly of spherical colloids into well-defined aggregates; *iii*) to demonstrate the potential of spherical colloids in producing three-dimensionally periodic lattices; and *iv*) to assess a number of intriguing applications associated with periodic arrays of spherical colloids.

8.2. COLLOIDAL BUILDING BLOCKS

This section briefly discusses several typical examples of colloidal building blocks that have been prepared as monodisperse entities, with the variation in size, shape, composition, and surface charges all controlled within 1–2%. The most commonly used colloidal building blocks have spherical symmetry. Many of them are readily available either commercially or through well-developed synthetic methodologies. Nonspherical building blocks have also become the focus of many research groups in recent years, and a number of methods have been demonstrated with different levels of success. As limited by space, here we only concentrate on new colloidal systems that our group demonstrated in the past several years.

8.2.1. Spherical Colloids

Figure 8.1A shows a TEM image of polystyrene beads, which are commercially available in a wide variety of sizes ranging from tens of micrometers down to less than

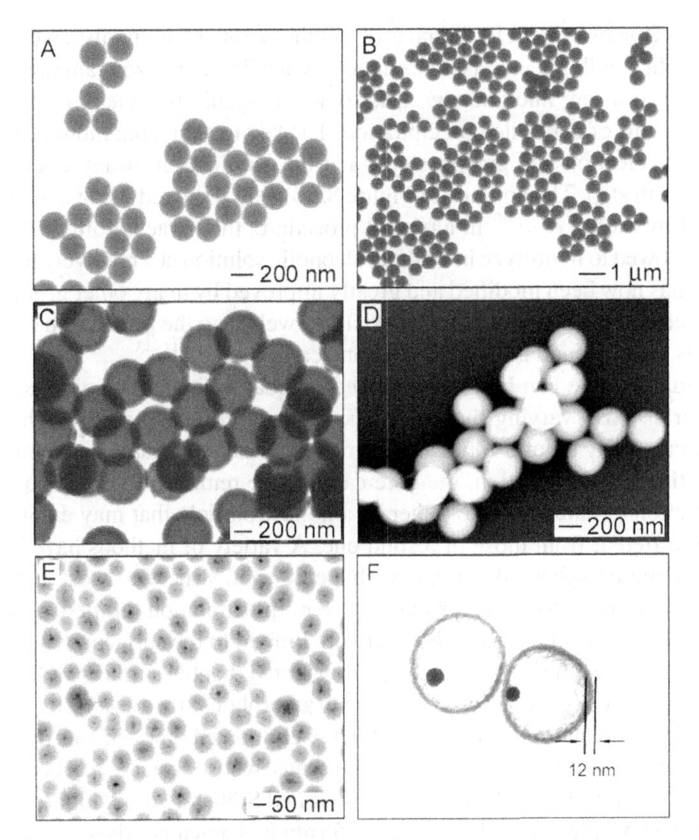

FIGURE 8.1. Electron microscopy images of several typical examples of spherical colloids: (A) a TEM image of polystyrene beads; (B) a TEM image of silica beads; (C) a TEM image of polystyrene@silca core-shell spheres; (D) a TEM image of gold@silca core-shell particles; (E) a TEM image of core-shell particles with magnetic iron oxide as the cores and silica as the shells; and (F) a backscattering SEM image of silica shells containing movable gold cores.

50 nm. These polymer latexes are usually prepared using a process known as emulsion polymerization.[21,22] At least four components are involved in a typical process: the monomer; the dispersion medium (usually water); the surfactant (or emulsifier); and the initiator. The majority of monomers exist as emulsion droplets stabilized by the emulsifier, and only a small portion is trapped in the micelles self-assembled from the surfactant molecules. The formation of polymer latexes begins with formation of a burst of primary free radicals through the decomposition of the water-soluble initiator. These radicals quickly initiate the formation of oligomeric nuclei by polymerizing the small amount of monomer dissolved in the aqueous phase. These tiny nuclei then diffuse into the micelles and eventually grow into larger particles until all monomer dissolved in each micelle has been polymerized. The diffusion of monomer molecules from emulsion droplets to micelles ensures the continuation of polymer growth, which will not be terminated until all monomers in the reaction medium have been depleted. In addition to emulsion polymerization, monodisperse polystyrene beads can now be more conveniently prepared using an emulsifier-free process, which is often called precipitation polymerization.[39]

Figure 8.1B shows the TEM image of another class of monodisperse colloids: silica spheres. Silica colloids are also commercially available in sizes ranging from tens of nanometers up to a few micrometers. Like other inorganic particles, they are generally produced by means of controlled precipitation. The key to achieving monodispersity is the separation of nucleation and growth steps, and the elimination of any nucleation events during the growth step. The colloids in Figure 8.1B were prepared using a method initially demonstrated by Stöber *et al.*[40] In a typical procedure, the tetraethylorthosilicate (TEOS) precursor is allowed to hydrolyze in a dilute, alcoholic solution at a relatively high pH value. This method has now been modified and greatly improved by many other groups, and it has also been extended to other inorganic materials, as well as to the production of cubes, rods, and ellipsoids from a range of metal oxides and carbonates.[17,41–43]

A colloidal particle can be functionalized by coating its surface with shell made of another material.[44] By varying the size, structure, and composition of such a core-shell particle, one can easily tailor its optical, electrical, thermal, mechanical, magnetic, and catalytic properties.[45–52] In addition, the core can also be removed in a subsequent step via solvent extraction or calcination to generate a hollow particle that may exhibit properties substantially different from those of a solid one. A variety of methods have been demonstrated for generating spherical colloids with a core-shell structure. Most of them involve the use of a controlled adsorption or reaction (*e.g.*, precipitation, grafted polymerization, or sol-gel condensation) that could be effectively limited to the surface vicinity of a spherical colloid.[53–55] For example, it has been demonstrated that the surfaces of latex, metal, or metal oxide colloids could be directly coated with amorphous silica as long as the concentration of colloidal templates was sufficiently high to eliminate homogeneous nucleation.[56–58] Figure 8.1C shows a TEM image of polystyrene beads, whose surfaces had been coated with uniform shells of amorphous silica through a modified Stöber method. Figure 8.1D shows a backscattering SEM image of gold colloids (~50 nm in diameter), whose surfaces had been coated with silica shells using a similar method. The incorporation of gold colloids renders these core-shell particles with the capability to strongly scatter light in the visible region due to surface plasmon resonance (SPR). Figure 8.1E shows the TEM image of another example of core-shell sample: superparamagnetic magnetite particles whose surfaces had been coated with uniform shells of silica. In this, case, the use of a magnetic-active material as the core allows for the manipulation of the core-shell particles with an external magnetic field. Figure 8.1F shows a TEM image of spherical, core-shell colloids with movable cores. Such complex particles were prepared by coating gold colloids with silica shells, followed by silanation with an initiator for atom transfer radical polymerization (ATRP), and the formation of polymer shells via ATRP of the monomer benzyl methacrylate (BzMA).[59–65] After the silica shell had been removed by wet-etching with an aqueous HF solution, the polymer shell containing a movable core would be formed. It is worth mentioning that all these samples could be prepared as stable suspensions (in alcohols or water) in relatively copious quantities.

Many methods have also been demonstrated to process materials other than polystyrene and silica into spherical colloids as monodisperse samples. For example, a new method was recently reported that allowed for the generation of titania spherical colloids with uniform diameters controllable in the range of 200 to 500 nm.[66] Titania is a semiconductor and has a much higher refractive index (2.6 for anatase and 2.9 for rutile) when

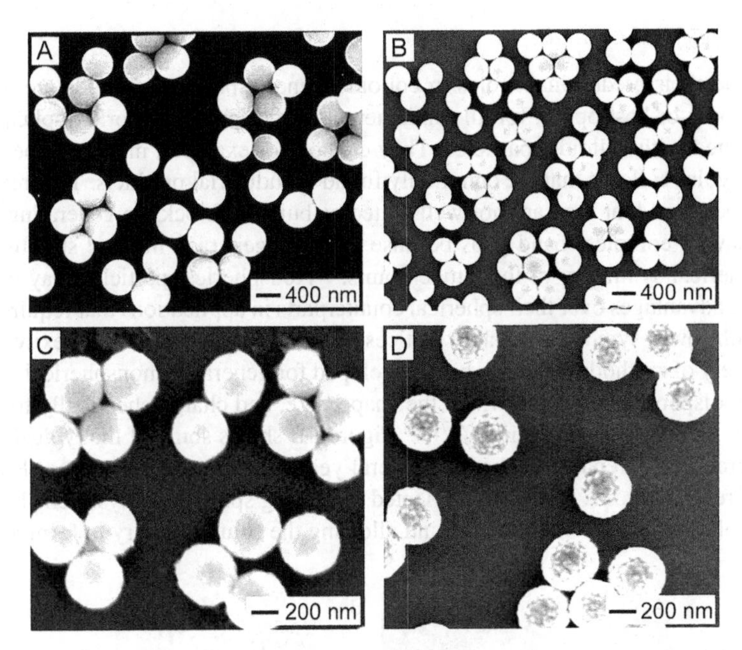

'IGURE 8.2. SEM images of titania-based spherical colloids that were synthesized using a polyol-mediated rocess: (A, B) spheres of titania glycolates with two different diameters; and (C, D) spheres in the anatase and utile phase of titania, respectively.

ompared to polystyrene (1.6) or silica (1.5). Colloids made of titania are of particu-ar interest for a range of applications: for example, as pigments or paper whiteners,[67] hotocatalysts,[68] and optical coatings.[69] In a typical synthesis, tetraalkoxyltitaniums, or Ti(OR)$_4$ (with R = -C_2H_5, iso-C_3H_7, and n-C_4H_9), were dissolved in ethylene glycol o form titanium glycolates and thus to greatly reduce the hydrolysis rate of an conven-ional alkoxide. When such a complex was poured into acetone, it underwent homoge-eous nucleation and growth to form monodisperse spherical colloids of the titanium glycolate. These colloids could be harvested and then redispersed in ethanol and wa-er without the assistance of any surfactant. They could also be readily assembled into hree-dimensionally crystalline lattices with strong optical diffraction. It was possible to control the sizes of these spherical colloids by adjusting the concentration of the glyco-ate precursor in acetone. Figures 8.2A and 8.2B show SEM images of spherical colloids of different diameters, and both of them were made of the titania glycolate. By anneal-ng at different temperatures, the titanium glycolate could be converted to the anatase or utile phase of titania without changing the spherical morphology of the Se colloidal par-icles. Figure 8.2C shows the SEM image of a sample that was calcined at 500 °C in air o obtain the anatase phase. Further annealing at 950 °C converted the titania glycolate colloids to the rutile phase of titania. The SEM image of a sample that had been trans-formed into the rutile phase is shown in Figure 8.2D. As a result of crystallization, the surfaces of these colloidal particles had become rougher as compared to the amorphous ones.

8.2.2. *Nonspherical Colloids*

Despite their predominant roles in colloid science, spherical colloids are not necessarily the only or best option for all fundamental studies and real-world applications that are associated with colloidal particles. They cannot, for example, model the behaviors of highly irregular colloids that are commonly found in industrial products. Theoretical studies have indicated that they are not well-suited as building blocks in generating photonic crystals having complete band gaps because of the degeneracy in band structure caused by the spherical symmetry of the lattice points.[23] Nonspherical particles may offer some immediate advantages over their spherical counterparts in applications that require building blocks with lower symmetries and/or lattices with higher levels of complexity.[70] To this end, a variety of methods have also been developed for generating nonspherical colloids as truly monodisperse samples, in which the shape, size, and charge chemically fixed on the surface are all identical to within 2%.[71–85] Figure 8.3 shows some of the typical examples that our group has prepared in the past several years. Figures 8.3A-C show SEM images of polystyrene spheroids that were fabricated by adding spherical polystyrene beads to an aqueous solution of polyvinyl alcohol and allowing the solution to dry to form a thin film.

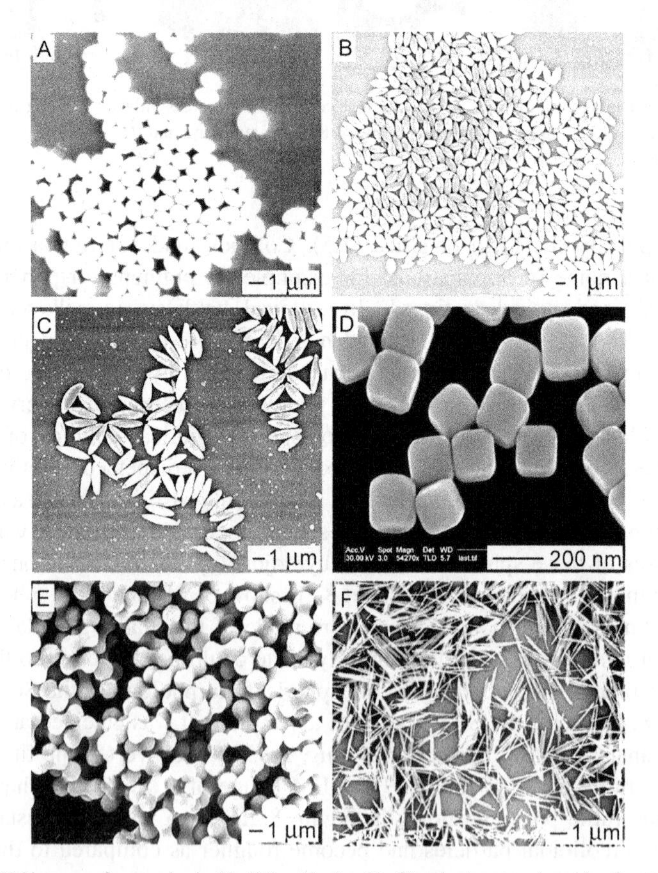

FIGURE 8.3. SEM images of non-spherical building blocks: (A–C) polystyrene spheroids of various aspect ratios; (D) silver nanocubes, (E) iron oxide (α-Fe_2O_3) peanuts; and (F) Te nanorods.

This film was then heated above the glass transition temperatures of both polymers, uniaxially stretched, and finally allowed to cool down to room temperature. The stretched film was finally dissolved in hot water to release the polystyrene spheroids. The aspect ratios of resultant spheroids could be conveniently tuned by adjusting the magnitude of elongation.

In addition to organic polymers, a range of inorganic materials have also been prepared as monodisperse samples of nonspherical colloids. For example, it was recently demonstrated that silver and gold could be synthesized as uniform nanocubes and nanoboxes, respectively, by controlling the growth kinetics with an appropriate capping reagent. Figure 8.3D shows an SEM image of some of silver nanocubes that were synthesized by reducing silver nitrate with ethylene glycol in the presence of poly(vinyl pyrrolidone) (PVP).[86] The silver nanocubes could then be utilized as sacrificial templates for the production of gold-silver alloyed nanoboxes that were characterized by a truncated, cubic shape. Another unique class of nonspherical colloids is based on iron oxides. Figure 8.3E shows an SEM image of α-Fe_2O_3 peanuts that were synthesized by reacting ferric chloride with sodium hydroxide in the presence of sodium sulfate, and aging for eight days at 100 °C.[87–89] The size, shape, and growth rate of the particles were strongly dependent on the rate at which the base was added to the ferric chloride solution. Nanorods represent another class of nonspherical building block. Currently there are many methods for generating monodisperse nanorods from a variety of materials. An example of these is given by the SEM image in Figure 8.3F. This image shows nanorods of trigonal tellurium that were grown by reducing orthotelluric acid with hydrazine in a mixture of ethylene glycol and water at 178 °C.[90] They were about 100 nm in diameter and ~1.8 μm in length. While the shape of these particles was rather interesting, it has been very difficult to organize them into an ordered, three-dimensional structure. As a matter of fact, most work related to colloidal assembly has concentrated on spherical building blocks, rather than nonspherical ones.

8.3. ASSEMBLY OF SPHERICAL COLLOIDS INTO WELL-DEFINED AGGREGATES

Spherical colloids suspended in a liquid medium may form discrete aggregates of relatively small sizes, albeit the shapes and dimensions of the aggregates are often polydispersed as caused by random collisions. To solve this problem, physical templates have to be introduced to control the aggregation of spherical colloids into well-defined structures. As demonstrated by a number of studies, the excluded volume interactions between colloids and the walls of templates, and those among colloids themselves—a physical constraint that excludes spatial overlapping between any two physical objects—can always lead to the formation of colloidal aggregates with complex and controllable sizes, shapes, and structures. Physical templates can be used in a number of different ways to both direct and control the self-assembly of spherical colloids into well-defined aggregates, although this is not an entirely new concept. In a paper published in 1969, Stöber *et al.* demonstrated the use of liquid droplets as templates to assemble polymer latexes into polygonal and polyhedral aggregates with a range of sizes and shapes.[91] The assembly was accomplished by nebulizing a concentrated suspension of polystyrene beads into a specially designed centrifuge, and they obtained dimeric, trimeric, tetrahedral, and octahedral clusters as well-separated aggregates. The idea of using emulsion droplets as templates to confine and organize spherical colloids into large aggregates was further exploited by Velev and coworkers.[92–94] Whitesides *et al.* also explored this method to generate porous spherical structures by using hexagonal rings as the building blocks.[95] Weitz *et al.* were able to generate spherical, hollow aggregates using this method with polystyrene beads as the building block.[96] Most recently, Pine *et al.* employed a similar method to assemble polystyrene latexes into polyhedral objects with well-controlled sizes and shapes.[97,98] In a different approach, Whitesides *et al.* pioneered the use of patterned monolayers as physical templates to direct the deposition and then self-organization of charged objects in the designated regions of a solid substrate.[99] Following this work, Aizenberg *et al.*,[100] Hammond *et al.*,[101–103] and several other groups[104] modified the procedure to organize spherical colloids into controllable aggregates supported on flat substrates. Most recently, Mallouk and coworkers further extended this method to demonstrate the organization of metal nanorods into patterned structures.[105]

This section will only concentrate on the use of physical templates that are patterned as relief structures on the surfaces of solid substrates. Relief structures patterned on solid substrates have been exploited as templates by a number of research groups to dictate the nucleation and growth of colloidal crystals with specific planes oriented parallel to the substrates.[106–116] This templating process can be considered as a mesoscopic analogue to epitaxial growth, which has been used on the atomic and molecular level to control the crystallographic orientation of a deposited film. Here we only discuss how such templates can be used to organize spherical colloids into discrete aggregates with controllable sizes, shapes, and structures.[117–118] This approach is now referred to as TASA, for template-assisted self-assembly.

8.3.1. Templated-Assisted Self-Assembly (TASA)

Figure 8.4A gives a schematic illustration of the TASA process. The key component is a fluidic cell fabricated by sandwiching a square gasket (~20 μm thick, cut from a piece of Mylar™ film, not shown in the drawing) between two glass substrates.[117–122] The template (*e.g.*, a two-dimensional array of cylindrical holes or trenches) can be lithographically patterned either in a thin film of photoresist spin-coated on the surface of bottom substrate or in the surface of a Si(100) wafer via anisotropic wet etching. The spherical colloids used in these demonstrations were either polymer beads or silica spheres. As the liquid slug slowly dewetts through the confined space, spherical colloids are pushed into each physical template to form a densely packed structure. The maximum number of spherical colloids that each template hole can host and the structural arrangement among these particles can all be determined geometrically, and can be controlled by changing the ratios between the dimensions of template and the size of particles. As a result, the use of templates assures the quantitative formation of colloidal aggregates with pre-specified sizes, shapes, and geometric structures.

The efficiency of a TASA process is dependent upon the balanced interplay of a number of forces. As illustrated in Figure 8.4A, there are three major forces exerted on each colloidal particle as the liquid dewetts across the cell: the capillary force ($\boldsymbol{F}_c$) associated with the meniscus of the liquid slug; the gravitational force ($\boldsymbol{F}_g$) due to the difference in density between the particle and the dispersion medium; and the electrostatic force ($\boldsymbol{F}_e$) caused by charges resting on the surface of the particle and the bottom substrate. When colloidal

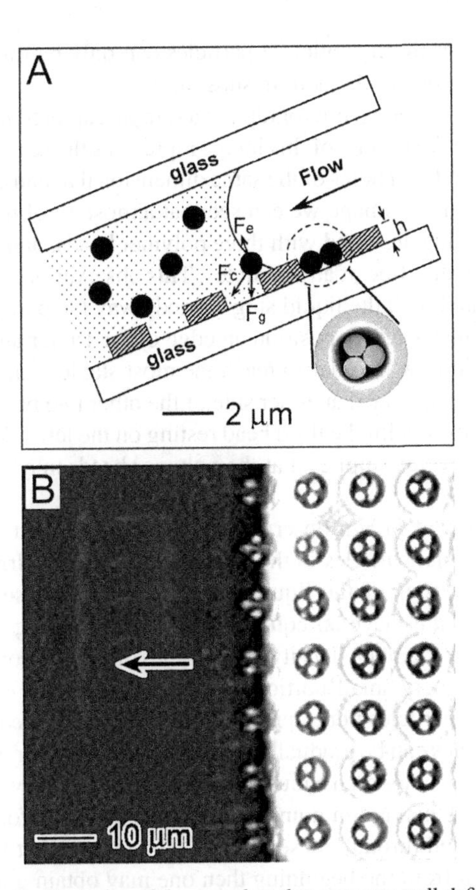

FIGURE 8.4. (A) A schematic illustration of the procedure that generates well-defined aggregates of spherical colloids under the physical confinement of templates. A trigonal cluster of polymer beads is shown to demonstrate the concept. For a colloidal particle next to the rear edge of the liquid slug, there are three possible forces exerting on it: the capillary force (F_c), the gravitational force (F_g), and the electrostatic force (F_e). (B) Optical micrograph of trigonal clusters assembled from 2.5-μm polystyrene beads in an array of cylindrical holes that were 6 μm in diameter and 2.3 μm in depth. The direction of liquid flow is indicated by an arrow. Note the preferential positioning and/or spatial orientation of the colloidal aggregates in the plane of the substrate.

particles smaller than ~1 μm are used in this process, the effect of Brownian motion may be greater than negligible. For polymeric latexes where water is usually used as the suspension media, the gravitational force is negligible because the densities of organic polymers (*e.g.*, 1.05 g/cm^3 for polystyrene) are very close to that of water (1.00 g/cm^3). A slight repulsive electrostatic force between the particles and the bottom substrate is required to levitate the colloids and prevent random sticking of particles to the bottom substrate. To accomplish this it may be necessary to modify the surface of the colloids and/or the bottom substrate, as well as to adjust the pH of the suspension.[120,121] Overall, the most important force in determining the yield of a TASA process seems to be the capillary force originating from the liquid meniscus. As shown in Figure 8.4A, this force points partially along the moving direction of the liquid slug and partially towards the surface of bottom substrate, and thus is capable of pushing particles into template holes and taking excess ones along with the dewetting liquid. As a result, the colloidal particles can only be trapped in the templates, and not on the raised regions of the bottom substrate.

Figure 8.4B shows the optical micrograph of a sample captured *in situ* during the TASA process. The dark line in the middle of this image represents the rear edge of the liquid slug. The templates were cylindrical holes of the same dimension that had been patterned in a thin film of photoresist. From this image we can see that almost all of the cylindrical template holes behind the water slug are filled with three polymer beads, and that the raised portion of the template is essentially free of any particles. This observation indicates that the strong capillary force associated with the liquid slug carried away the particles not trapped in the template holes. It was necessary to position an edge of the trimer aggregate perpendicular to the direction of liquid flow in order to reach the most stable configuration. In principle, the third particle can be positioned at either side of the other two beads. However, only one of these two configurations (with the third bead resting on the left side) was observed in our experiments. This observation implies that the polymer beads may be sequentially pushed into the template hole.

Based on the results of *in situ* observation, it is believed that the self-assembly and spatial ordering of colloidal particles in the templates are mainly driven by capillary force associated with the meniscus of the liquid slug. As a result of sedimentation, particles suspended in the liquid slug have an equal probability of settling in the holes or on the raised portion. Since the number density of particles in the colloidal suspension is not sufficiently high, only a very small portion of the template holes can be filled through a settling mechanism. As particles not trapped by the holes are moved along with the liquid slug, the colloidal particles can be gradually concentrated at the rear edge of the slug. When the density of particles becomes high enough, most templates in the vicinity of the dewetting front will be filled with a maximum number of particles; even before the rear edge of the liquid slug has ever passed through these regions. This means that if the concentration of particles is high enough from the beginning then one may obtain a nearly perfect array of aggregates over the entire region of the template. If the starting concentration is not high enough then areas first uncovered during dewetting process will be often incompletely filled until the concentration at the dewetting front reaches a sufficiently high concentration.

8.3.2. Homo-Aggregates of Spherical Colloids

The size, shape, and structure of aggregates self-assembled from spherical colloids can all be controlled by varying the key components of a TASA process. Even with a simple array of cylindrical holes, it was possible to obtain a broad range of different polygonal aggregates by keeping the depth (H) of template holes on the same order as the size of spherical colloids. In this case, the maximum number of spherical colloids retained in each cylindrical hole and thereby the structural arrangement among the colloids were determined by the ratio between the diameter (D) of holes and the diameter (d) of spherical colloids. By simply varying this ratio, one could easily generate many types of polygonal aggregates as displayed in Figure 8.5. Figures 8.5A and 8.5B show the SEM images of two typical examples–triangular and pentagonal aggregates–that were fabricated by templating polystyrene beads (0.9 μm for Figures 8.5A and 0.7 μm for Figure 8.5B) against 2D arrays of cylindrical holes. The holes were patterned in thin films of photoresist with dimensions of $D = 2$ μm and $H = 1$ μm. In general, it was possible to fabricate uniform aggregates in the form of dimer, trimer, square,

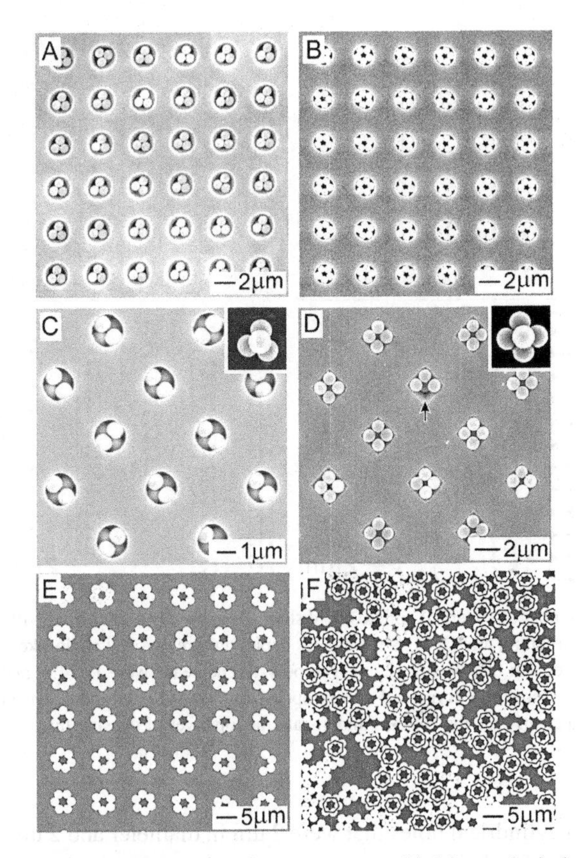

FIGURE 8.5. SEM images of typical examples of aggregates assembled from spherical colloids by templating against arrays of holes having different dimensions and/or shapes. (A) trigonal planar clusters formed with 0.9-μm polymer beads in cylindrical holes of 2 μm in diameter and 1 μm in depth; (B) pentagonal rings of 0.7-μm polymer beads formed in an array of cylindrical holes of 2 μm in diameter and 1 μm in depth; (C) tetrahedra that were assembled from 1.0-μm polymer beads in cylindrical holes of 2 μm in diameter and 2 μm in depth; (D) square pyramidal clusters assembled from 1.0-μm silica colloids in pyramidal cavities; (E) hexagonal rings assembled from 2-μm polymer beads in cylindrical holes of 6 μm in diameter (with 2-μm posts in their centers); and (F) hexagonal rings after they had been released from the original support and re-deposited onto a silicon substrate. Hexagons are sketched to assist the visualization of rings.

pentagon, and hexagon simply by using spherical colloids with diameters ranging from 0.6 to 1.0 μm. When the samples were dried via solvent evaporation, another type of capillary force came into play, and this force could drive the spherical colloids in each template hole into a closely packed structure.[123]

The aggregates of spherical colloids within each template hole could be welded together into a permanent entity by heating the entire system at a temperature slightly above the glass transition temperature of the colloidal material. When this was done, the surfaces of colloids in physical contact were fused together as a result of viscoelastic deformation of the spherical colloids.[124–126] Once welded together, the photoresist film could be dissolved with isopropanol to release the colloidal aggregates from the substrate by sonicating in a water bath. Figure 8.5C shows the SEM image of an array of tetrahedra assembled from

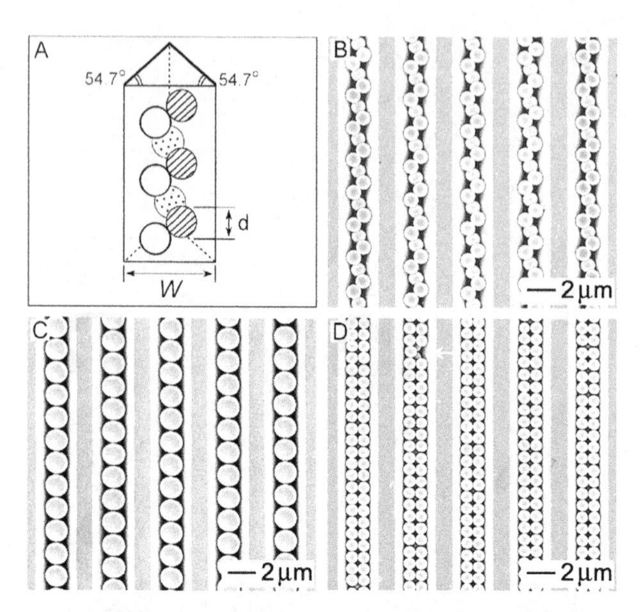

FIGURE 8.6. Self-assembly of spherical colloids in V-shaped grooves: (A) a schematic illustration showing the formation of a helical structure; and (B-D) SEM images depicting three typical chain-like aggregates assembled in 2D arrays of V-grooves that had W = 2.72 μm. The polystyrene beads were 1.0, 1.6, and 0.8 μm in diameter, respectively. Helical structures only formed at an appropriate ratio between *W* and d. The arrow in (D) indicates a defect, where one can clearly see the colloids underneath the top layer of the structure.

1.0 μm PS beads in cylindrical holes that were 2 μm in diameter and 2 μm in depth, and the inset shows a cluster after being released.

Templates with shapes other than cylindrical holes can also be used to fabricate aggregates of spherical colloids with more complex structures. Figure 8.5D shows the SEM image of an array of square pyramidal clusters of 1.0-μm silica beads that were prepared by templating against square pyramidal cavities. The templates measured 2.2 μm at the base, and were generated in the surface of a Si(100) wafer by anisotropic etching. The arrow indicates a defect, from which the particle underneath the top layer can also be clearly seen. The inset shows a cluster after being released from the template.

Figure 8.5E shows the SEM image of a 2D array of six-member rings that were assembled using a template containing an array of cylindrical holes, with a cylindrical post in the center of each hole. After fusion, the photoresist template was dissolved, leaving the colloidal aggregates adhered to the surface of the glass substrate as a result of the thermal treatment. Figure 8.5F shows an SEM image of these aggregates after they have been released from the glass substrate and then deposited onto a silicon wafer. Some of these aggregates have been sketched with hexagons to help view individual rings. Although some of these aggregates have been broken into smaller fragments during the sonication process, the percentage of six-member rings calculated from this SEM image was still as high as 75%. These experimental results demonstrate the power of the TASA process to assemble spherical colloids into complex aggregates arrays on solid supports that are characterized by both positional and orientational orders.

When the TASA template is changed from an array of pits to a 2D array of parallel V-shaped grooves, helical and (100) cubic packed chains were obtained.[127] V-shaped grooves can also be readily generated by anisotropic wet etching of a Si (100) wafer. The packing of beads in the final chain was determined by the ratio between the width of the channel (W) and the diameter of beads (d) being packed. Figure 8.6A schematically illustrates the helical packing in a V-shaped groove. As seen in Figure 8.6B, double-layered structures with a helical morphology were formed when W/d was in the range of 2.70–2.85. The chirality of the chains could also be controlled by varying the direction of the liquid flow relative to the channel in the dewetting process. As $W/d$was reduced less than 2.70, linear chains were observed (Figure 8.7C). Chains with a (100) cubic packed morphology were generated when W/d was greater than 2.85 (Figure 8.7D).

In a typical TASA experiment, the overall yield of the pre-specified aggregates (calculated from the geometric parameters of the templates and spherical colloids) is usually in the order of ~90%. The most commonly observed defects are aggregates containing particles fewer than the calculated number (as marked in Figure 8.5D), and they can be eliminated by altering the concentration of colloidal dispersion and/or by flowing a moderately dilute (<0.1%) colloidal dispersion through the fluidic cell more than once. By optimizing

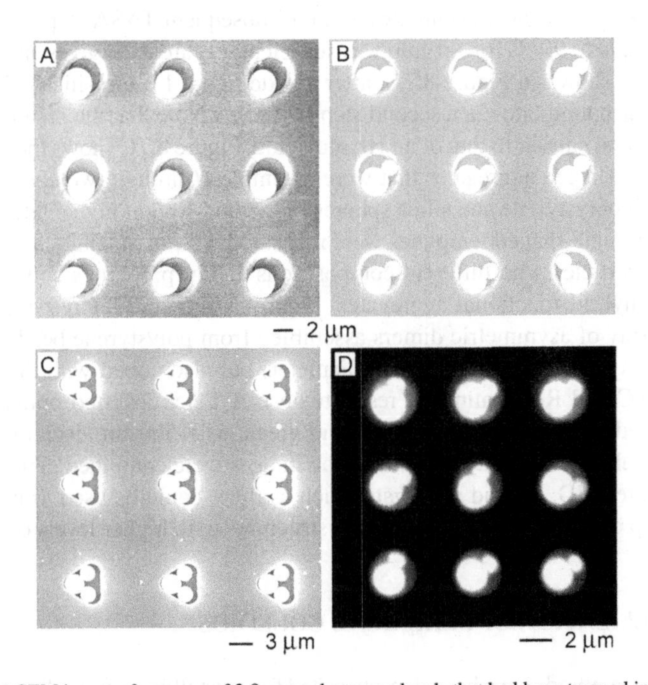

FIGURE 8.7. (A) SEM image of an array of 2.8-μm polystyrene beads that had been trapped in cylindrical holes of 5 μm in diameter and 2.5 μm in depth. (B) An SEM image of the same sample after a 1.6-μm silica sphere had been introduced into each hole through another step of dewetting. (C) SEM image of an array of H_2O-shaped aggregates that were assembled (through two dewetting steps) from polystyrene and silica spheres of 2.5 and 1.8 μm in diameter, respectively. The prism-shaped templates were ~5 μm in length at the edge. (D) Fluorescence microscopy image of an array of asymmetric dimers consisting of polystyrene beads that were different in both size and color: 3.0-μm beads labeled with a green dye (FITC) and 1.7-μm beads labeled with a red dye (Rhodamine 6G). These dyes were separately excited, imaged, and then superimposed to form a single picture.

these parameters, it is possible to generate defect-free arrays of colloidal aggregates as large as several mm^2 in area. This limit is mainly placed by the dimensional uniformity and areas of test patterns fabricated by photolithography and etching. It is believed that the self-assembly process, itself, should be extendible to substrates as large as several cm^2 in area.

8.3.3. Hetero-Aggregates of Spherical Colloids

The TASA method is not limited to the assembly of particles with the same size or composition. Consecutive TASA depositions can be used to create more complex, heterogeneous aggregates. As with other applications of TASA, the dimensional ratio between the template and spherical colloids has to be precisely controlled so that only a well-controlled number of particles are introduced into the template in each step.[120,121] Figure 8.7A shows the SEM image of a 2D array of 2.8-μm polystyrene beads that were trapped in cylindrical holes ($D = 5.0$ μm, $H = 2.5$ μm) patterned in a thin film of photoresist. Each bead was in physical contact with the wall of the template as a result of the shear force (as caused by liquid flow) and the attractive capillary force between these surfaces (due to solvent evaporation).[123] The position of the bead could be fixed in place using a quick heat treatment to prevent it from washing away in the subsequent TASA step. For polystyrene, heating the sample to 90 °C for ~1 min was sufficient. Figure 8.7B gives an SEM image of the same sample shown in Figure 8.7A, after an additional 1.6-μm silica colloid had been introduced to each template via a second step of TASA. Note that the resultant aggregates had a configuration similar to that of an HF molecule. Figure 8.7C shows the SEM image of an array of H_2O-shaped aggregates that were assembled (through two consecutive dewetting steps) from polystyrene and silica spheres of 2.5 and 1.8 μm in diameter, respectively. The rounded-triangle shaped templates were measured ~5 μm in length at the edge.

Colloidal particles with different compositions and/or properties could also be combined to form hybrid functional aggregates. Figure 8.7D is the fluorescence microscopy image of an array of asymmetric dimers assembled from polystyrene beads that were not only different in size (3.0 and 1.7 μm in diameter) but also doped with different fluorescent dyes (FITC and Rhodamin 6G, respectively). A Leica inverted optical microscope (DMIRBE, fitted with blue, 480 ± 40 nm; and green, 515–560 nm, excitation cubes) was used to selectively excite the dyes and then the images were combined generate the image as seen in Figure 8.7D. Hybrid aggregates such as these could be used as building blocks for another step of self-assembly to generate structures with higher levels of complexity.

8.4. CRYSTALLIZATION OF SPHERICAL COLLOIDS

A variety of methods have been demonstrated for crystallizing monodispersed spherical colloids (such as polymer beads and silica spheres) into long-range ordered lattices. Some of the commonly used ones include sedimentation, self-assembly via repulsive electrostatic interaction, ordering via attractive capillary forces, and crystallization under physical confinement.

While sedimentation may sound like the most straightforward method of crystallization, it is actually a balancing act between many complex processes that are often difficult to control. Sedimentation usually results in a solid mass accumulating on the bottom of a vessel, which is sometimes crystalline or amorphous.[128] The key to the success of this process is to optimize the rate of sedimentation by tightly controlling the uniformity and size of the colloidal particles. Crystallization by sedimentation of a colloidal suspension requires the particles to be in a spherical shape and a sufficiently slow sedimentation rate in order for the concentrate to undergo a hard-sphere, disorder-order phase transition.[129] In addition, the difference between the densities of the suspension medium and the colloidal particles needs to be sufficiently high for the particles to actually settle out rather than remain as a stable suspension. Opals represent an example of colloidal crystals formed via slow sedimentation.[130]

The use of electrostatic interactions in colloidal crystallization can be a powerful means of self-assembly. For colloidal suspensions, the pair-wise interaction can be accounted for by a curve known as the Sogami-Ise potential, which is actually a sum of three components: a short-range steric repulsive force, a moderate-range attractive van der Waals force, and a long-range Coulombic repulsive force.[131] If the concentration of stray electrolytes is high, the van der Waals interaction dominates and particles will begin to aggregate. When the concentration of electrolytes is sufficiently low ($<10^{-5}$ M), the intensity of electrostatic repulsion becomes strong enough to stabilize an ordered array for these particles at separations greater than the diameter of an individual particle. As a result, when highly charged spherical colloids are suspended in a liquid medium containing very few stray ions, they can spontaneously organize themselves into a variety of crystalline structures as driven by the minimization of electrostatic repulsive interaction.[132–137]

The method of crystallization via attractive capillary forces was originally developed by Nagayama and coworkers to form hexagonally packed 2D arrays of spherical colloids supported on various substrates.[138,139] They have also followed this self-assembly process experimentally by using an optical microscope.[140,141] It was found that nucleation usually occurred when the thickness of the liquid layer approached the diameter of the colloids. As the liquid evaporated more and more, colloids were driven towards the nucleus through convective transport, and eventually crystallized into a hexagonal lattice due to the attractive capillary forces. In order to generate long-range ordered lattices over relatively large areas, flat, clean, and chemically homogeneous surfaces are often required.

Monodispersed spherical colloids can also spontaneously organize into long-range ordered lattices when they are subjected to a physical confinement (provided, for example, by a pair of parallel substrates).[142] On the basis of this observation, our group has demonstrated an effective approach that allows for the fabrication of colloidal crystals with domain sizes as large as several square centimeters.[143–145] In a typical procedure, spherical colloids (including both PS beads and silica spheres, with diameters ranging from ~50 nm to ~5 μm) are injected into a specially designed fluidic cell, and crystallized into a cubic-close-packed lattice under constant agitation supplied by sonication or mechanical vibration. Figure 8.8 shows the schematic design of a fluidic cell, which can be fabricated using common microlithographic techniques. In the early work,[143–145] the fluidic cell was constructed by sandwiching a photoresist frame of uniform thickness (~10 μm) between two glass slides and then held together using binder clips. The surface of the photoresist frame had been patterned with parallel arrays of shallow channels, with dimensions less than those of the particles being crystallized, using a double-exposure procedure that involved overlapping of two photomasks. Recently, newer designs of the fluidic cell forego the requirement of

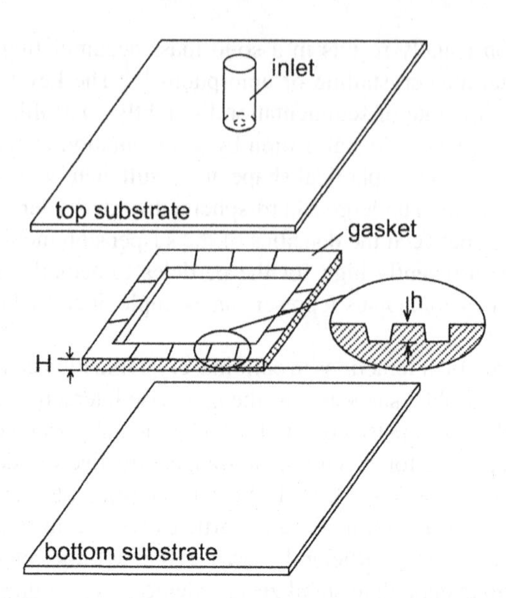

FIGURE 8.8. A schematic illustration of the flow cell used to crystallize spherical colloids into 3D crystalline lattices. The gasket was cut from a plastic film whose thickness could vary in the range of a few to several hundred microns. A glass tube was attached to the hole drilled into the top glass substrate whose surface had been made hydrophilic by treating with oxygen plasma. The cell was assembled together by sandwiching the gasket between the top and bottom glass substrates, and tightening with binder clips. The aqueous dispersion of monodispersed colloids was injected into the cell through the inlet.

clean-room facilities.[119] This was accomplished by replacing the photoresist frame with gasket cut from a sheet of uniform Mylar^{TM} film, which is available with thickness in the range of ~20 to ~100 μm from Fralock (Canoga, CA) or Dupont. Channels can also be quickly generated using the Mylar film: for example, the film can be rubbed with soft paper on both sides; colloids smaller than those to be crystallized can be sandwiched between the gasket and one of the substrates; or the glass substrate(s) can be patterned with arrays of channels in thin films of gold through the use of microcontact printing (μCP) and selective wet etching.[146] Although these methods are not as reproducible as the photolithographic technique, a yield of higher than ~90% can still be routinely accomplished.

The aqueous dispersion of spherical colloids is added to the cell through a glass tube glued to the top substrate over a small hole (1–3 mm in diameter) that had been drilled with a diamond bit. As shown in Figure 8.8, colloids with a diameter (d) larger than the depth (h) of channels in the gasket surfaces will be concentrated at the bottom of the fluidic cell and forced to crystallize into a long-range ordered lattice. The cell is constantly agitated using a Branson 1510 sonicator (Danbury, CT), and this agitation helps to ensure that the particles only settle in the energetically favorable positions. This procedure allows one to routinely assemble monodispersed spherical colloids into a crystalline lattice with well-controlled thickness that is often homogeneous over several square centimeters in area. Both scanning electron microscopy and optical diffraction studies indicate that the (111) planes of such a crystalline lattice are oriented parallel to the substrate.[144] For any crystalline lattice fabricated using the fluidic cell shown in Figure 8.8, the number of crystalline planes

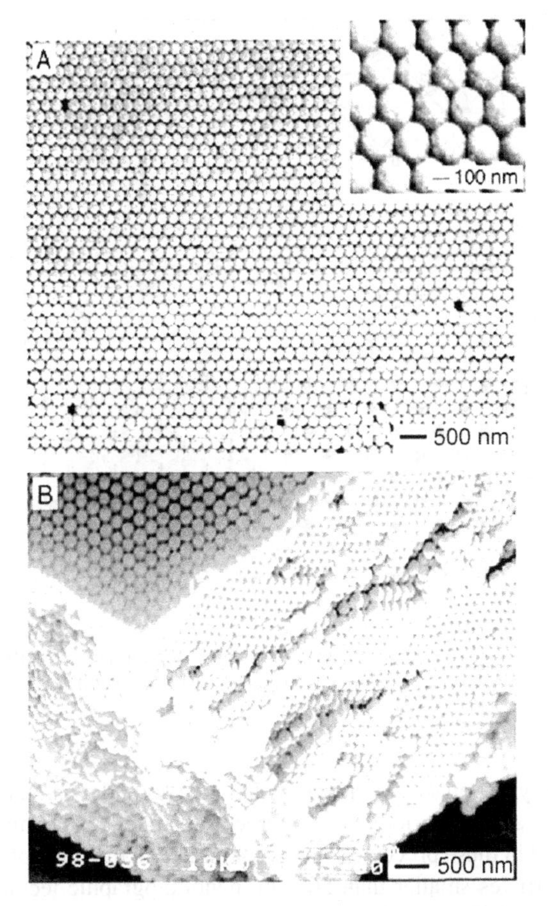

FIGURE 8.9. SEM images of 3D colloidal crystals grown using the packing cell shown in Figure 8: A) the top view, which shows long-range ordering; packing; and (B) the cross-sectional view of another crystal grown in a similar packing cell. The inset shows a close-up of the hexagonal packing of (111) planes.

is largely determined by the ratio between the thickness (H) of the cell and the diameter (d) of the spherical colloids, and this ratio can be conveniently tuned from 1 to more than 200 by using gaskets with different thicknesses.[147] When spherical colloids are crystallized on flat substrates, the resultant 3D lattice usually exhibit a face-centered cubic structure with its (111) planes parallel to the substrate. Figure 8.9 shows SEM images of a typical colloidal crystal grown in the packing cell illustrated in Figure 8.8.

A template surface is usually needed in order to obtain colloidal crystals with orientations other than (111). When a substrate with the appropriate arrays of relief structures is used, the pits can serve as physical templates to confine and control the nucleation and growth of colloidal crystals that will exhibit a pre-specified spatial orientation. van Blaaderen and Wiltzius first demonstrated the use of lithographically defined surfaces as templates to grow specifically oriented colloidal crystals. They accomplished this by utilizing a slow sedimentation process and were able to produce crystals with either (100) or (110) planes oriented parallel to the supporting substrates.[148] This work was followed by Yodh and coworkers

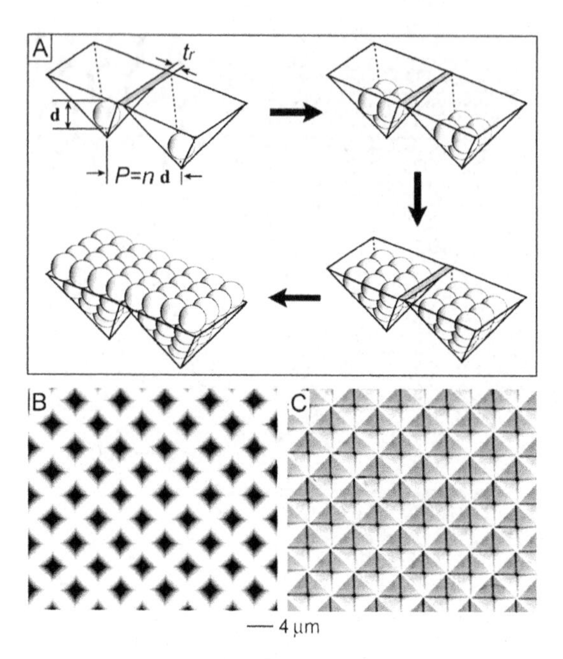

FIGURE 8.10. (A) A schematic illustration showing the step-by-step formation of (100)-oriented 3D opaline lattice by templating against a 2D array of square pyramidal pits. (B, C) Two SEM images of square pyramidal templates with different spacing between pits.

who used entropically driven self-organization to make colloidal crystals with (100) orientations by templating against periodically patterned substrate.[149] The minimum feature size in the templates used in these processes was often limited to the scale of the particles being crystallized. For particles smaller than 500 nm, nanolithographic techniques like e-beam lithography were required to fabricate the templates. Ozin and coworkers demonstrated that this limitation could be overcome by using tapered templates that were anisotropically etched in the surface of a Si (100) substrate. They have also grown (100)-oriented colloidal crystals by templating against such relief structures.[150–152] Our group further extended the potential of these templates to assemble spherical colloids into discrete clusters characterized by a square pyramidal shape.[153] We further demonstrated that similar templates could be adopted to form (100)-oriented colloidal crystals as large as several square centimeters in area.[154] As illustrated schematically in Figure 8.10A, the polymer beads could be directed by the square pyramidal pits in Si(100) surface to grow into a 3D crystalline lattice with its (100) planes oriented parallel to the surface of supporting substrate. Figures 8.10B and 10C show SEM images of two such templates having different widths for the ridges. In this template-directed process, the atomic scale symmetry of the underlying Si(100) substrate was faithfully transferred into the mesoscale colloidal crystal. Figure 8.11 shows the SEM images of two samples fabricated by crystallizing 1.0-μm PS beads against 2D arrays of pyramidal pits etched in the surfaces of Si(100) wafers. Figure 8.11A shows the top view of a portion of the 3D lattice, clearly indicating the square symmetry of (100) planes parallel to the surface of supporting substrate. Figure 11B shows an SEM of the backside of this crystal, whose surface is decorated by an array of nuclei formed

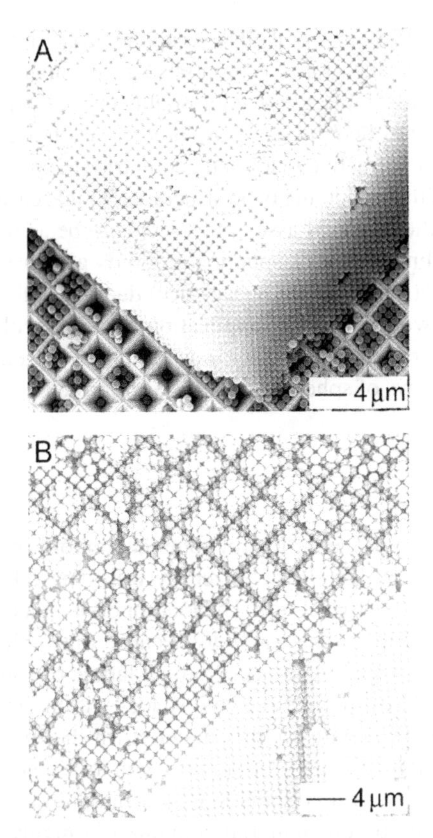

FIGURE 8.11. (A, B) SEM images of (100)-oriented colloidal crystals generated by templating 1.0-μm polystyrene beads against a 2D array of 4-μm wide square pyramidal pits. The crystals belong to the same sample that happened to stick to the bottom (A) and top (B) substrate when the top glass substrate was separated from the cell.

in the pyramidal pits. It is worth noting that spherical colloidal as small as 0.25 μm have been successfully demonstrated for use with this procedure.

8.5. APPLICATIONS OF COLLOIDAL ASSEMBLIES

Colloidal assemblies have been explored for use in a number of applications. For example, 2D assemblies of spherical colloids have been successfully demonstrated as arrayed microlenses in image processing[155,156] and photolithography;[157] as physical masks for evaporation or reactive ion etching to generate patterned arrays of micro- or nanostructures;[158,159] and as regular arrays of relief structures to cast elastomeric stamps for use in soft lithographic techniques.[160] Three-dimensional assemblies have been actively exploited as sacrificial templates to fabricate inverse opals having highly ordered macroporous structures[161–163] that could serve as photonic crystals, diffractive optical sensors, and band filters.[164,166] They have also been employed as a directly observable (in the real space) model system to study a range of fundamental phenomena such as crystallization, phase transition, diffusion, and fracture mechanics.[167,168] The success of all these applications and studies strongly depends on the availability of spherical colloids as monodisperse samples, and on the ability to organize them into uniform assemblies with controllable crystal structures, sufficiently large domain sizes, and well-controlled thickness or volume. When used as photonic crystals, it is also desirable to vary the refractive index and geometric shape of the colloidal particles in an effort to better control their photonic band structures. In this case, a tight control over the degree of perfection on the colloidal assemblies seems to be as necessary to the photonic exploitation of a crystalline lattice of spherical colloids as has been the situation in the microelectronic usage of a single crystalline semiconductor. As there is limited space, here we only briefly discuss two examples of application related to colloidal assemblies: *i*) the use of crystalline lattices of spherical colloids as templates to generate inverse opals, and *ii*) the use of patterned arrays of spherical colloids as microlenses in image production and size reduction.

8.5.1. Fabrication of Inverse Opals

Templating against crystalline lattices of spherical colloids represents a generic route to macroporous materials that are characterized by well-controlled pore sizes and highly ordered, interconnected porous structures. The macroporous materials prepared using this method have been referred to as *inverse opals* because they exhibit a periodic structure complementary to that of an opal. After a crystalline lattice of spherical colloids has been dried, there is at least 26% (by volume) of void spaces within the lattice that can be infiltrated (partially or completely) with another material (in the form of a gaseous or liquid precursor, as well as a colloidal suspension) to form a matrix around the spherical colloids. Selective removal of the spherical colloids (*e.g.*, via dissolution or calcination) will lead to the formation of a macroporous material containing a highly ordered architecture of uniform air balls interconnected to each other through a number of small "windows". This approach offers a number of advantages over other methods: *i*) It offers a tight control over the size, shape, structure, and surface density of the pores. Because the pores are generated by removal of colloidal templates, the size of pores in the resultant material can be readily controlled and tuned in the range that spans from tens of nm to hundreds of microns. *ii*) It provides a simple, versatile, and reliable route to porous materials with three-dimensionally periodic structures. One of the attractive features associated with these porous materials is their interconnected network of uniform air balls that inherits all ordering and symmetry features from the original template. *iii*) In principle, such a template-directed method could be extended to any particular material of interest. The only requirement seems to be the availability of a gaseous or liquid precursor that can fill the voids in the template lattice without significantly swelling or dissolving the colloidal particles (usually made of an organic polymer or silica). When a liquid precursor is involved, the fidelity of this procedure is mainly determined by the van der Waals interaction, wetting of the template surfaces by the precursor, kinetic factors such as capillary filling of the small channels within the colloidal template, and volume shrinkage of a precursor during the solidification process.

A rich variety of materials (including precursors) have been incorporated into the template-directed synthesis.[33,169,170] Curable liquid prepolymers are particularly attractive because they have been widely used in replica molding with resolutions better than a few nanometers. By judicially selecting a monomer, the volume shrinkage involved in

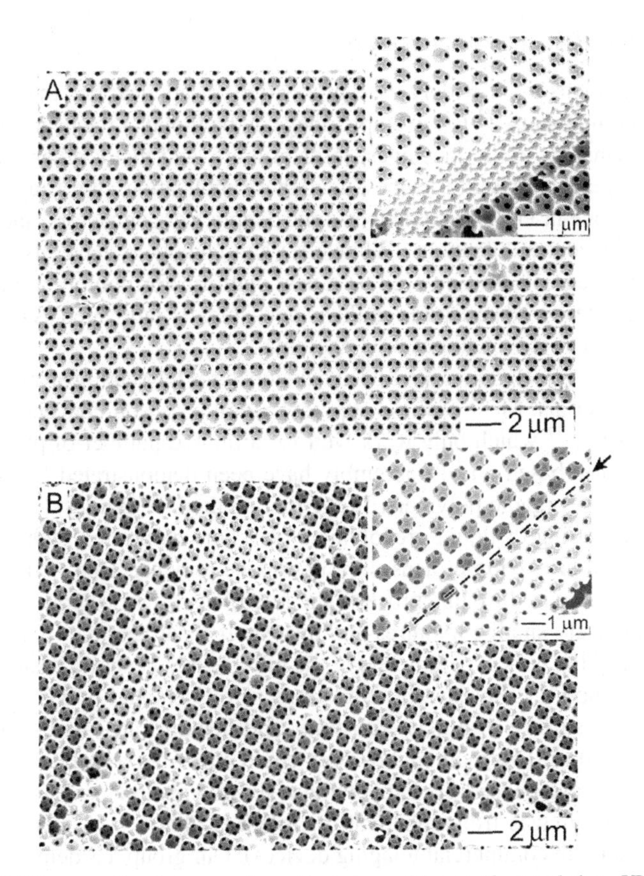

FIGURE 8.12. SEM images of two inverse opal samples that were fabricated by templating a UV-curable prepolymer to the poly(acrylate methacrylate) copolymer against the opaline lattices of polystyrene beads with different orientations.

a templating process could be controlled well below 1%.[171] When a liquid prepolymer is applied along the edges of a colloidal lattice, the voids within the lattice can be spontaneously filled as driven by capillary action. After curing, the polymer beads can be selectively dissolved by immersing the sample in an organic solvent for a few hours. For silica beads, they can be selectively removed using hydrofluoric acid (49% in water), followed by rinsing with water. We have demonstrated the fabrication of inverse opals using liquid precursors to polymers such as polyurethanes (PU) and poly(acrylate methacrylate) copolymers (PAMC). Figure 8.12 shows SEM images of two typical examples – inverse opals of PAMC that were fabricated by templating against (111)- and (100)-oriented crystalline lattices of polystyrene beads. These images also clearly reveal the long range ordering and interconnection between adjacent spherical pores. Using this approach, inverse opals with a uniform pore size and structure could be readily fabricated with areas $\geq$1 cm^2. Such a polymeric inverse opal was sufficiently flexible to withstand mechanical deformations and thus could be easily obtained as a freestanding film. In addition to their use as photonic crystals, such porous membranes could be employed to measure and study the permeabilities of various kinds of gases and liquids.

8.5.2. *Fabrication of Arrayed Microlenses*

Arrayed microlenses are widely used in a variety of applications that involve miniaturized optical components.[172] For example, they can be found at the heart of optical communication systems, facsimile machines, laser printers, and many other kinds of digital information storage or processing devices. In all these applications, the arrayed microlenses simply serve as diode laser correctors, fiber-optic couplers or connectors, and optical scanners. In a set of recent publications, Whitesides and coworkers have also demonstrated that arrayed microlenses could be used as a new platform for photolithography, through which submicrometer-sized structures could be conveniently fabricated as patterned arrays by reducing mm to cm scale features on a photomask.[157]

A large number of techniques have been developed for fabricating arrayed microlenses in a variety of materials. Most of these approaches are based on conventional microfabrication methods,[173–177] which only work well for a limited number of photoresists. Two chemical approaches based on self-assembly have been demonstrated for fabricating arrayed microlenses of organic polymers. In the first method,[178] Whitesides and coworkers used selective dewetting of liquid prepolymers on a surface printed with self-assembled monolayers to form arrayed microlenses of an organic polymer. Due to the low viscosity of the prepolymer required by this process, the microlenses fabricated using this method used to have relatively small curvatures and thus short focal lengths. It is also nontrivial to precisely control the optical parameters of the microlenses generated using this method. In the second method,[179] Hayashi and coworkers self-assembled polystyrene beads into a 2D lattice on the surface of a glass substrate, and demonstrated its use as an array of microlenses in imaging. This method, however, could only offer a specific patterned array: that is, hexagonal-close-packed 2D lattice with a six-fold symmetry. It is very difficult to obtain high quality images from spherical microlenses (that is why spherical lenses are seldom used directly in commercial imaging devices). Our group has demonstrated the use of TASA in generating arrayed microlenses with controllable optical parameters.

Figure 8.13A schematically outlines the procedure. An aqueous dispersion of monodisperse polystyrene beads were deposited via the TASA method into a 2D array. A convenient template could be a 2D array of cylindrical holes patterned in a photoresist film spin-coated on a solid substrate. Once the 2D array had been assembled into the template and dried, the sample could be heated above the glass transition temperature of polystyrene to deform the spheres into mushroom-shaped structures. The photoresist was then dissolved with ethanol, leaving behind a 2D array of mushroom-shaped lenses on the glass substrate as demonstrated by the SEM image in Figure 8.13B. If a hemispherical shape is desired, it can be easily accomplished by another step of heat treatment. Figure 8.13C shows the SEM image of an array of hemispherical microlenses that was fabricated in this fashion. It is worth noting that the filling factors and patterned designs associated with these microlenses could both be varied independently by controlling the templates and the sizes of spherical colloids.

Figure 8.14 compares optical images obtained using the mushroom-shaped and hemispherical microlenses, respectively. The images were recorded on a Leica optical microscope in the transmission mode, with an F-shaped object (cut out of a thick piece of black paper) placed in the path of the light source at the field lens position. Figure 8.14A shows schematically how this object was projected through an array of microlenses to form a patterned array of micrometer-sized features. Figures 8.14B and 8.14C show the optical micrographs

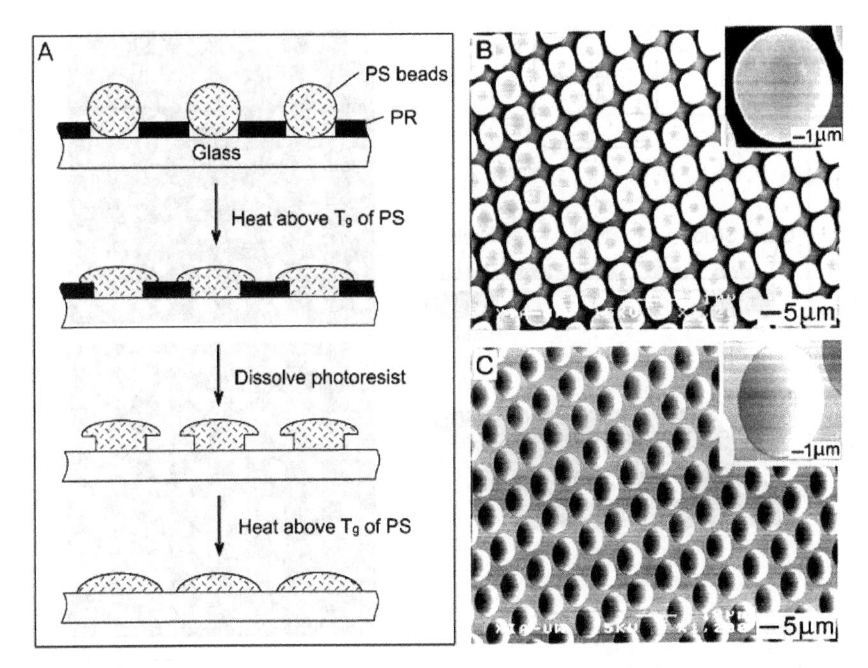

FIGURE 8.13. (A) Fabrication of microlenses with hemispherical and mushroom-shaped profiles by annealing the sample at a temperature higher than the glass transition temperature of polystyrene (~93 °C). The formation of a hemispherical shape was driven by the minimization of the surface free energy. (B) SEM image of a 2D array of polystyrene beads after the sample had been heated at 96 °C for ~10 min, followed by dissolution of the photoresist with ethanol. (C) An SEM image of the same sample after it had been heated again at 96 °C for ~10 min. In this second step of thermal annealing, the mushroom-shaped microlenses were converted to the most commonly used, hemispherical ones.

of images obtained through the mushroom-shaped and hemispherical microlenses, respectively. Both of them were able to generate well-resolved images with slightly different sizes, indicating their difference in optical focal length. In principle, these images could also be recorded using a thin film of photoresist to transfer the patterned features into the resist film via optical exposure and chemical developing.[157] It should provide a simple and convenient method for generating submicrometer-sized structures with certain types of regular designs.

8.6. CONCLUDING REMARKS

Monodisperse spherical colloids have emerged as the material of choice for a broad range of fundamental studies and niche applications. They also represent a class of simple and versatile building blocks that can be readily synthesized in large scales and be organized into long-range ordered crystals. By templating against colloidal crystals, new types of 3D periodic structures such as inverse opals can also be conveniently fabricated. The self-assembly approach based on spherical colloids offers a flexible and practical route to three-dimensionally periodic structures with remarkably little investment as compared to the conventional microfabrication techniques. More importantly, the feature sizes of these

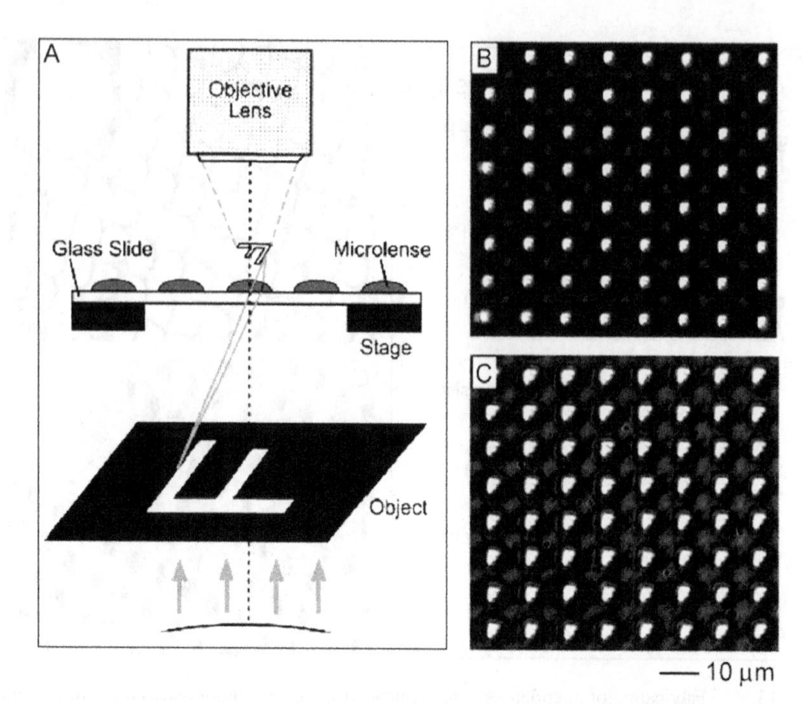

FIGURE 8.14. (A) A schematic illustration of the setup used to project an F-shaped object onto the focal plane through an array of microlenses. (B, C) images formed at the focal plane of an array of mushroom-shaped (B) and hemispherical (C) microlenses. The microlenses used in all of these demonstrations were fabricated by templating 5.7-μm polystyrene beads against 2D arrays of cylindrical holes that were 5 μm in height and diameter.

mesoscale structures can be varied in a controllable fashion to cover a broad range of scales that span from tens of nanometers to hundreds of micrometers. The ability to generate such periodic structures with well-controlled feature sizes has already enabled the quest of useful functionality not only from the constituent materials but also from the long-range order that characterizes these lattices. A good example can be found in the so-called photonic band gap crystals that can be used to control the emission and/or propagation of photons with wavelengths in the entire spectrum of light.

Spherical colloids also provide the simplest building blocks to generate complex structures (such as polygonal or polyhedral aggregates) through self-assembly. The dewetting of colloidal dispersions from a contoured surface, for example, could be combined with capillary forces and geometric confinement to provide an effective route to assemble spherical colloids into discrete aggregates with well-controlled sizes, shapes, and structures. The success of this self-assembly approach depends on a number of parameters such as the meniscus curvature of the rear edge of dewetting liquid, charges on the surfaces of colloidal particles and templates, and concentrations of the colloidal suspensions. The smallest building blocks that have been demonstrated for use with this process were ~50 nm in diameter, and this low limit was mainly restricted by our ability to fabricate templates with smaller feature sizes. By employing nanolithographic techniques such as e-beam writing, we should be able to further push the size down to tens of nanometers. The self-assembly strategy demonstrated here should be extendable to many other colloidal systems, including nonspherical building blocks that are truly monodispersed in size and shape. In addition, it will be possible to apply this method to the deposition and organization of biofunctional particles such as cells and lipid vesicles.[180]

It is worth emphasizing that the experiments described here always led to the quantitative formation of colloidal aggregates with uniform and controllable structures. The availability of well-controlled aggregates of spherical colloids will provide an opportunity to experimentally probe the hydrodynamic, aerodynamic, and optical properties associated with colloidal particles that are highly complex in morphology or topology. For example, the aggregates can serve as a good model to investigate the deviation of Stokes-Einstein formula as colloidal particles become nonspherical in shape. It will also be interesting to study the interaction potentials between these aggregates to achieve a better understanding on this matter. In addition, the ability to generate hybrid colloidal particles should make it possible to design some experiments that cannot be performed using colloidal particles that are usually homogeneous in composition and surface functionality. Some of the colloidal aggregates (*e.g.*, the dimers) can even be explored as a class of new building blocks for self-assembly to generate mesostructured systems with paramount complexities that will find use in areas such as photonics, electronics, and condensed matter physics. On the other hand, the self-assembly scheme described here may find use as a means to study the crystallization of monodisperse colloids under physical confinement.

The ability to assemble monodisperse, spherical colloids into regular arrays (both 2D and 3D) is immediately useful in a range of applications. For example, it has been demonstrated that the 2D lattice of polystyrene beads could serve as arrayed microlenses with controllable focal lengths. The spherical beads could also be transformed into mushroom-shaped or hemispherical microlenses to improve the imaging quality by annealing the samples at temperatures slightly higher than the glass transition temperature of the polymer. When imaged through such an array of microlenses, objects of ~1 cm in lateral dimensions could be reduced and projected into arrays of submicrometer-sized structures with well-resolved features. In other demonstrations, the 3D crystalline lattices of spherical colloids have been actively exploited as templates to generate inverse opals that are potentially useful as photonic band gap crystals.

Monodisperse spherical colloids and most of the applications derived from these materials are still in an early stage of technical development. Many issues still need to be addressed before these materials can reach their potential in industrial applications. For example, the diversity of materials must be greatly expanded to include every major class of functional materials. At the moment, only silica and a few organic polymers (*e.g.*, polystyrene and polymethylmethacrylate) can be prepared as *truly* monodispersed spherical colloids. These materials, unfortunately, do not exhibit any particularly interesting optical, nonlinear optical or electro-optical functionality. In this regard, it is necessary to develop new methods to either dope currently existing spherical colloids with functional components or to directly deal with the synthesis of other functional materials. Second, formation of complex crystal structures other than closely packed lattices has been met with limited success. As a major limitation to the self-assembly procedures described in this chapter, all of them seem to lack the ability to form 3D lattices with arbitrary structures. Recent demonstrations based on optical trapping method may provide a potential solution to this problem, albeit this approach seems to be too slow to be useful in practice.[181–184] Third, the density of defects in the crystalline lattices of spherical colloids must be well-characterized and kept below the level tolerated by a particular application. Some recent work has started to address this issue by studying the crystallization process of spherical colloids with mixed sizes. In fact, the use of polydispersed spherical colloids as building blocks in self-assembly may provide another route to the fabrication of structures with higher levels of complexity.

ACKNOWLEDGMENT

This work was supported in part by the Office of Naval Research (N00014-01-1-0976), a Career Award from the National Science Foundation (DMR-9983893), and a Fellowship from the David and Lucile Packard Foundation. Y. X. is an Alfred P. Sloan Research Fellow and a Camille Dreyfus Teacher Scholar. J. M. thanks the Center for Nanotechnology at the UW for the Nanotech Early Bird Award funded by the IGERT program of the National Science Foundation (DGE-9987620). Y.L. thanks the Center for Nanotechnology at the UW for the Graduate Fellowship Award.

REFERENCES

1. D. H. Everett, *Basic Principles of Colloid Science*. (Royal Society of Chemistry, UK, 1988).
2. W. B. Russel, D. A. Saville, and W. R. Schowalter, *Colloidal Dispersions* (Cambridge University Press, USA, 1989).
3. R. J. Hunter, *Introduction to Modern Colloid Science* (Oxford University Press, UK, 1993).
4. A. Einstein, *Brownian Motion Investigations: On the Theory of the Brownian Movement*, (Dover Publications, USA, 1956).
5. A. Henglein, *Chem. Rev.* **89**, 1861 (1989).
6. G. Schmid, *Chem. Rev.* **92**, 1709 (1992).
7. G. Schon, U. Simon, *Colloid Polym. Sci.* **273**, 101 (1995).
8. G. Schmid, L. F. Chi, *Adv. Mater.* **10**, 515 (1998).
9. M. L. Steigerwald and L. E. Brus, *Acc. Chem. Res.* **23**, 183 (1990).
10. M. G. Bawendi, M. L. Steigerwald, and L. E. Brus, *Annu. Rev. Phys. Chem.* **41**, 477 (1990).
11. H. Weller, *Angew. Chem. Int. Ed. Engl.* **32**, 41 (1993).
12. A. P. Alivisatos, *Science* **271**, 933 (1996).
13. J. H. Fendler, *Chem. Rev.* **87**, 877 (1987).
14. G. Ozin, A. *Adv. Mater.* **4**, 612 (1992).
15. D. A. Tomalia, *Adv. Mater.* **6**, 529 (1994).
16. T. Sugimoto, *Fine Particles: Synthesis, Characterization, and Mechanisms of Growth* (Marcel Dekker, USA, 2000).
17. G. Mie, *Ann. Phys.* **25**, 377 (1908).
18. B. V. Derjaguin and L. Landau, *Acta Physicochimica (USSR)* **14**, 633 (1941).
19. E. J. Verwey and J. T. G. Overbeek, *Theory of the Stability of Lyophobic Colloids* (Elsevier, Netherlands, 1948).
20. E. Matijestié, *Chem. Mater.* **5**, 412 (1993).
21. I. Piirma, *Emulsion Polymerization*, (Academic Press, USA, 1982).
22. G. W. Poehlein, R. H. Ottewill, and J. W. Goodwin, *Science and Technology of Polymer Colloids* (Martinus Nijhoff Publishers, USA, 1983) Vol. II.
23. Y. Lu, Y. Yin, and Y., Xia, *Adv. Mater.* **13**, 415 (2001).
24. M. I. Mishchenko, J. W. Hovenier, and L. D. Travis, *Light Scattering by Nonspherical Particles: Theory, Measurements, and Applications*, (Academic Press, USA, 2000).
25. J. Visser, in *Surface and Colloid Science*, edited by E. Matijevié (John Wiley & Sons, USA, 1976), Vol. 8, pp. 3–84.
26. R. L. Rowell, *NATO ASI Ser., Ser. C* **303**, 187 (1990).

27. T. Radeva, *Curr. Top. Coll. Interf. Sci.* **2**, 131 (1997).
28. P. Pieranski, *Contemp. Phys.* **24**, 25 (1983).
29. W. van Megan and I. Snook, *Adv. Coll. Interf. Sci.* **21**, 119 (1984).
30. A. P. Gast and W. B. Russel, *Physics Today* **December**, 24 (1998).
31. D. G. Grier, *From Dynamics to Devices: Directed Self-Assembly of Colloidal Materials*, a special issue in *MRS Bull.* **23**, 21 (1998).
32. A. K. Arora and B. V. R. Tata, *Ordering and Phase Transitions in Colloidal Systems* (VCH, USA, 1996).
33. Y. Xia, B. Gates, Y. Yin, and Y. Lu, *Adv. Mater.* **12**, 693 (2000).
34. V. L. Colvin, *MRS Bull.* **26**, 637 (2001).
35. A. D. Dinsmore, J. C. Crocker and A. G. Yodh, *Curr. Opin. Coll. Interf.* **3**, 5 (1998).
36. J. V. Sanders, *Acta Crystallogra.* **24**, 427 (1968).
37. See, for example the special issue in *Adv. Mater.* **13**, 369 (2001).
38. Polman and P. Wiltzius, Materials science aspects of photonic crystals, a special issue in *MRS Bull.* **26**, 608 (2001).
39. E. C. C. Goh and H. D. H. Stover, *Macromolecules* **35**(27), 9983 (2002).
40. W. Stöber and A. Fink, *J. Coll. Interf. Sci.* **26**, 62 (1968).
41. E. Matijevié, *Acc. Chem. Res.* **14**, 22 (1981).
42. E. Matijestic, Fine particles, a special issue in *MRS Bull.* **14**, 18 (1989).
43. E. Matijevié, *Langmuir* **10**, 8 (1994).
44. F. Caruso, M. Spasova, V. Saigueirino-Maceira, and L. M. Liz-Marzan, *Adv. Mater.* **13**, 11 (2001).
45. A. L. Aden, *J. Appl. Phys.* **22**, 1242 (1951).
46. T. Sugimoto, *MRS Bull.* **14**, 23 (1989).
47. M. Ohmori and E. Matijevié, *J. Coll. Interf. Sci.* **150**, 594 (1992).
48. W. P. Hsu, R. Yu, and E. Matijevié, *J. Coll. Interf. Sci.* **156**, 56 (1993).
49. A. P. Philipse, M. P. B. van Bruggen, and C. Pathmamanoharan, *Langmuir* **10**, 92 (1994).
50. R. D. Averitt, D. Sarkar, and N. J. Halas, *Phys. Rev. Lett.* **78**, 4217 (1997).
51. M. Giersig, L. M. Liz-Marzan, T. Ung, D. Su, and P. Mulvaney, *Bunsenges. Phys. Chem.* **101**, 1617 (1997).
52. S. J. Olderburg, R. D. Averitt, S. L. Westcott, and N. J. Halas, *Chem. Phys. Lett.* **288**, 243 (1998).
53. A. Garg and E. Matijevié, *J. Coll. Interf. Sci.* **126**, 243 (1988).
54. N. Kawahashi and E. Matijevic, *J. Coll. Interf. Sci.* **138**, 534 (1990).
55. X. C. Guo and P. Dong, *Langmuir* **15**, 5535 (1999).
56. Y. Yin, Y. Lu, Y. Sun, and Y. Xia, *Nano Lett.* **2**, 427 (2002).
57. Y. Lu, Y. Yin, and Y. Xia, *Nano Lett.* **2**, 183 (2002).
58. Y. Lu, Y. Yin, and Y. Xia, *Nano Lett.* **2**, 785 (2002).
59. T. E. Patten and K. Matyjaszewski, *Adv. Mater.* **10**, 901 (1998).
60. J. Xia, X. Zhang, and K. Matyjaszewski, *Macromolecules* **30**, 2249 (1999).
61. T. Werne and T. E. Patten, *J. Am. Chem. Soc.* **121**, 7409 (1999).
62. C. Perruchot, M. A. Khan, S. P. Armes, T. Werne, and T. E. Patten, *Langmuir* **17**, 4479 (2001).
63. X. Huang and M. J. Wirth, *Macromolecules* **32**, 1694 (1999).
64. X.-X. Kong, T. Kawai, J. Abe, and T. Iyoda, *Macromolecules* **34**, 1837 (2001).
65. K. Kamata, Y. Lu, and Y. Xia, *J. Am. Chem. Soc.* **125**(9), 2384 (2003).
66. X. Jiang, T. Herricks, and Y. Xia, *Adv. Mater.* **15**, 1205 (2003).
67. W. P. Hsu, R. C. Yu, and E. Matijevié, *J. Coll. Interf. Sci.* **156**, 56 (1993).
68. M. Anpo, T. Shima, S. Kodama, and Y. Kubokawa, *J. Phys. Chem.* **91**, 4305 (1987).
69. G. A. Battiston, R. Gerbasi, M. Porchia, and L. Rizzo, *Chem. Vapor Depos.* **5**, 73 (1999).
70. Z.-Y. Li, J. Wang, and B.-Y. Gu, *J. Phys. Soc. Jpn.* **67**, 3288 (1998).
71. S. P. Sutera and C. W. Boylan, *J. Coll. Interf. Sci.* **73**, 29 (1980).
72. M. Nagy and A. Keller, *Polymer Comm.* **30**, 130 (1989).
73. K. M. Keville, E. I. Franses, and J. M. Caruthers, *J. Coll. Interf. Sci.* **144**, 103 (1991).
74. P. Jiang, J. F. Bertone, and V. L. Colvin, *Science* **291**, 453 (2001).
75. Y. Lu, Yin Y. And Y. Xia, *Adv. Mater.* **13**, 271 (2001).
76. A. T. Skjeltorp, J. Ugelstad, and T. Ellingsen, *J. Coll. Interf. Sci.* **113**, 577 (1996).
77. H. R. Sheu, M. S. El-Aasser, and J. W. Vanderhoff, *Polym. Mater. Sci. Eng.* **59**, 1185 (1988).

78. M. Okubo, T. Yamashita, T. Suzuki, and T. Shimizu, *Coll. Polym. Sci.* **275**, 288 (1997).
79. H. Takei and N. Shimizu, *Langmuir* **13**, 1865 (1997).
80. L. Petit, E. Sellier, E. Duguet, S. Ravaine, and C. Mingotaud, *J. Mater. Chem.* **10**, 253 (2000).
81. C. Alexander and E. N. Vulfson, *Adv. Mater.* **9**, 751 (1997).
82. Bao Z. L. Chen, M. Weldon, E. Chandross, O. Cherniavskaya, Y. Dai, and J. B.-H. Tok, *Chem. Mater.* **14**, 24 (2002).
83. O. D. Velev, A. M. Lenhoff, and E. W. Kaler, *Science* **287**, 2240 (2000).
84. J. P. Novak, C. Nickerson, S. Franzen, and D. L. Feldheim, *Anal. Chem.* **73**, 5758 (2001).
85. J. C. Love, B. D. Gates, D. B. Wolfe, K. E. Paul, and G. M. Whitesides, *Nano Lett.* **2**, 891 (2002).
86. Y. Sun, Y. Xia, *Science* **298**, 2176 (2002).
87. E. Matijevié and P. Scheiner, *J. Coll. Sci.* **63**, 509 (1978).
88. T. Sugimoto, M. Khan, and A. Muramatsu, *Coll. Surf. A* **70**, 167 (1993).
89. T. Sugimoto, M. Khan, A. Muramatsu, and H. Itoh, *Coll. Surf. A* **79**, 233 (1993).
90. B. Mayers and Y. Xia, *J. Mater. Chem.* **12**, 1875 (2002).
91. W. Stöber, A. Berner, and R. Blaschke, *J. Coll. Interf. Sci.* **29**, 710 (1969).
92. O. D. Velev, K. Furusawa, and K. Nagayama, *Langmuir* **12**, 2374 (1996).
93. O. D. Velev, K. Furusawa, and K. Nagayama, *Langmuir* **12**, 2385 (1996).
94. O. D. Velev, A. M. Lenhoff, and E. W. Kaler, *Science* **287**, 2240 (2000).
95. W. T. S. Huck, J. Tien, and G. M. Whitesides, *J. Am. Chem. Soc.* **120**, 8267 (1998).
96. A. D. Dinsmore, M. F. Hsu, M. G. Nikolaides, M. Marquez, A. R. Bausch, and D. A. Weitz, *Science* **298**, 1006 (2002).
97. G. R. Yi, V. N. Manoharan, S. Klein, K. R. Brzezinska, D. J. Pine, F. F. Lange, and S.-M. Yang, *Adv. Mater.* **14**, 1137 (2002).
98. V. N. Manoharan, M. T. Elsesser, and D. J. Pine, *Science*, **301**, 483 (2003).
99. J. Tien, A. Terfort, and G. M. Whitesides, *Langmuir* **13**, 5349 (1997).
100. J. Aizenberg, P. V. Braun, and P. Wiltzius, *Phys. Rev. Lett.* **84**, 2997 (2000).
101. K. M. Chen, X. Jiang, L. C. Kimerling, and P. T. Hammond, *Langmuir* **16**, 7825 (2000).
102. H. Zheng, I. Lee, M. F. Rubner, and P. T. Hammond, *Adv. Mater.* **14**, 569 (2002).
103. I. Lee, H. Zheng, M. F. Rubner, and P. T. Hammond, *Adv. Mater.* **14**, 572 (2002).
104. Y. Masudo, M. Itoh, T. Yonezawa, and K. Koumoto, *Langmuir* **18**, 4155 (2002).
105. B. R. Martin, S. K. St. Angelo, and T. E. Mallouk, *Adv. Func. Mater.* **12**, 759 (2002).
106. H. W. Deckman, J. H. Dunsmuir, S. Garoff, J. A. McHenry, D. G. Peiffer, *J. Vac. Sci. Technol. B* **6**, 333 (1988).
107. A. van Blaaderen, R. Ruel, and P. Wiltzius, *Nature* 385, 321 (1997).
108. A. van Blaaderen and P. Wiltzius, *Adv. Mater.* **9**, 833 (1997).
109. K.-H. Lin, J. C. Crocker, V. Prasad, A. Schofield, D., A. Weitz, T. C. Lubensky, and A. G. Yodh, *Phys. Rev. Lett.* **85**, 1770 (2000).
110. S. M. Yang and G. A. Ozin, *Chem. Commun.* 2507 (2000).
111. G. A. Ozin and M. Y. Yang, *Adv. Funct. Mater.* **11**, 95 (2001).
112. F. Burmeister, C. Schafle, B. Keilhofer, C. Bechinge, J. Boneberg, and P. Leiderer, *Adv. Mater.* **10**, 495 (1998).
113. Y.-H. Ye, S. Badilescu, V.-V. Truong, P. Rochon, and A. Natansohn, *Appl. Phys. Lett.* **79**, 872 (2001).
114. D. K. Yi, E.-M. Seo, and D.-Y. Kim, *Appl. Phys. Lett.* **80**, 225 (2002).
115. J. Y. Chen, J. F. Klemic, and M. Elimelech, *Nano Lett.* **2**, 393 (2002).
116. J. Zhang, A. Alsayed, K. H. Lin, S. Sanyal, Zhang F. Pao W.-J. and V. S. K. Balagurusamy, *Appl. Phys. Lett.* **81**, 3176 (2002).
117. Y. Yin and Y. Xia, *Adv. Mater.* **13**, 267 (2001).
118. Y. Yin, Y. Lu, B. Gates, and Y. Xia, *J. Am. Chem. Soc.* **123**, 8718 (2001).
119. Y. Lu, Y. Yin, B. Gates, and Y. Xia, *Langmuir* **17**, 6344 (2001).
120. Y. Yin, Y. Lu, and Y. Xia, *J. Am. Chem. Soc.* **123**, 771 (2000).
121. Y. Xia, B. Gates, and Y. Yin, *Aust. J. Chem.* **54**, 287 (2001).
122. Y. Yin, Y. Lu, and Y. Xia, *J. Mater. Chem.* **11**, 987 (2001).
123. P. A. Kralchevsky and K. Nagayama, *Langmuir* **10**, 23 (1994).
124. B. Gates, S. H. Park and Y. Xia, *Adv. Mater.* **12**, 653 (2000).
125. S. Mazur, R. Beckerbauer, and J. Buckholz, *Langmuir* **13**, 4287 (1997).

126. Y. Xia, Y. Yin, Y. Lu, and J. McLellan, *Adv. Func. Mater.* **13**, 907 (2003).
127. Y. Yin, and Y. Xia, *J. Am. Chem. Soc.* **125**, 2048 (2003).
128. K. E. Davis, W. B. Russel, and W. J. Glantschnig, *Science* **245**, 507 (1989).
129. P. N. Pusey and W. Vanmegen, *Nature* **320**, 340 (1986).
130. J. V. Sanders, *Nature* **204**, 1151 (1964).
131. I. Sogami and N. Ise, *J. Chem. Phys.* **81**, 6320 (1984).
132. A. Kose and S. Hachisu, *J. Colloid Interface Sci.* **46**, 460 (1974).
133. N. A. Clark and B. J. Ackerson, *Nature* **281**, 57 (1979).
134. D. H. Vanwinkle and C. A. Murray, *Phys. Rev. A* **34**, 562 (1986).
135. Harland J. L. S. I. Henderson, S. M. Underwood, and W. Vanmegen, *Phys. Rev. Lett.* **75**, 3572 (1995).
136. T. Okubo, *Langmuir* **10**, 1695 (1994).
137. T. Palberg, W. Monch, J. Schwarz, and P. Leiderer, *J. Chem. Phys.* **102**, 5082 (1995).
138. N. D. Denkov, O. D. Velev, P. A. Kralchevski, I. B. Ivanov, H. Yoshimura, and K. Nagayama, *Langmuir* **8**, 3183 (1992).
139. P. Kralchevski, V. Paunov, I. Ivanov, and K. Nagayama, *J. Coll. Interf. Sci.* **151**, 79 (1992).
140. N. D. Denkov, O. D. Velev, P. A. Kralchevsky, I. B. Ivanov, H. Yoshimura, and K. Nagayama, *Nature* **361**, 26 (1993).
141. P. A. Kralchevsky, V. N. Paunov, N. D. Denkov, I. B. Ivanov, and K. Nagayama, *J. Coll. Interf. Sci.* **155**, 420 (1993).
142. P. Pieranski, L. Strzelecki, and B. Pansu, *Phys. Rev. Lett.* **50**, 900 (1983).
143. S. H. Park, D. Qin, and Xia Y. *Adv. Mater.* **10**, 1028 (1998).
144. S. H. Park and Y. Xia, *Langmuir* **15**, 266 (1999).
145. B. Gates, D. Qin, and Y. Xia, *Adv. Mater.* **11**, 466 (1999).
146. Y. Xia and G. M. Whitesides, *Angew. Chem. Int. Ed.* **37**, 550 (1998).
147. B. Gates, Y. Lu, Z.-Y. Li, and Y. Xia, *Appl. Phys. A* **76**, 509 (2003).
148. A. van Blaaderen, R. Ruel, and P. Wiltzius, *Nature* **385**, 321 (1997).
149. K. H. Lin, J. C. Crocker, V. Prasad, A. Schofield, D. A. Weitz, T. C. Lubensky, and A. G. Yodh, *Phys. Rev. Lett.* **85**, 1770 (2000).
150. S. M. Yang and G. A. Ozin, *Chem. Commun.* 2507 (2000).
151. G. A. Ozin and S. M. Yang, *Adv. Funct. Mater.* **11**, 95 (2001).
152. V. Kitaev, S. Fournier-Bidoz, S. M. Yang, and G. A. Ozin, *J. Mater. Chem.* **12**, 966 (2002).
153. Y. Yin and Y. Xia, *Adv. Mater.* **14**, 605 (2002).
154. Y. Yin, Z.-Y. Li, and Y. Xia, *Langmuir* **19**, 622 (2003).
155. S. Hayashi, Y. Kumamoto, T. Suzuki, and T. Hirai, *J. Coll. Interf. Sci.* **144**, 538 (1991).
156. Y. Lu, Y. Yin, and Y. Xia, *Adv. Mater.* **13**, 34 (2001).
157. H. Wu, T. W. Odom, and G. M. Whitesides, *J. Am. Chem. Soc.* **124**, 7288 (2002).
158. H. W. Deckman and J. H. Dunsmuir, *Appl. Phys. Lett.* **41**, 377 (1982).
159. J. C. Hulteen, D. A. Treichel, M. T. Smith, M. L. Duval, T. R. Jensen, and Van R. P. Duyne, *J. Phys. Chem. B* **103**, 3854 (1999).
160. Y. Xia, J. Tien, D. Qin, and G. M. Whitesides, *Langmuir*, **12**, 4033 (1996).
161. M. E. Turner, T. J. Trentler, and V. L. Colvin, *Adv. Mater.* **13**, 180 (2001).
162. A. Stein and R. C. Schroden, *Curr. Opin. Solid State Mater. Sci.* **5**, 553 (2001).
163. P. V. Braun and P. Wiltzius, *Curr. Opin. Coll. Interf.* **7**, 116 (2002).
164. R. C. Schroden, M. Al-Daous, C. F. Blanford, and A. Stein, *Chem. Mater.* **14**, 3305 (2002).
165. M. Weissman, and H. B. Sunkara, A. S. Tse, and S. A. Asher, *Science* **274**, 959 (1996).
166. P. L. Flaugh, S. E. O'Donnell, and S. A. Asher, *Appl. Spect.* **38**, 847 (1984).
167. C. A. Murray, *MRS Bull.* **23**, 33 (1998).
168. C. A. Murray, and D. G. Grier, *Am. Sci.* **83**, 238 (1995).
169. K. M. Kulinowski, P. Jiang, H. Vaswani, and V. L. Colvin, *Adv. Mater.* **12**, 833 (2000).
170. D. V. Orlin and M. L. Abraham, *Current Opin. Colloid Interf. Sci.* **5**, 56 (2000).
171. Y. Xia, E. Kim, X.-M. Zhao, J. A. Rogers, M. Prentiss, and G. M. Whitesides, *Science* **273**, 347 (1996).
172. M. C. Hutley, *Microlens Arrays* (Institution of Physics, Teddington, 1991).
173. J. Bähr and K.-H. Brenner, *Appl. Opt.* **35**, 5102 (1996).
174. J. Hahns and S. J. Walker, *Appl. Opt.* **29**, 931 (1990).
175. Z. L. Liau, and D. E. Mull', C. L. Dennis, and R. C. Williamson, *Appl. Opt. Lett.* **64**, 1484 (1994).

176. H. Chase, M. A. Handschy, M. J. O'Callaghan, and F. W. Supon, *Opt. Lett.* **20**, 1444 (1995).
177. S. Lazare, J. Lopez, J.-M. Turlet, M. Kufner, S. Kufner, and P. Chavel, *Appl. Opt.* **35**, 4471 (1996).
178. H. A. Biebuyck, and G. M.Whitesides, *Langmuir* **10**, 2790 (1994).
179. S. Hayashi, Y. Kumamoto, T. Suzuki, and T. Hirai, *J. Coll. Interf. Sci.* **144**, 538 (1991).
180. H. Wu, V. R. Thalladi, S. Whitesides, and G. M. Whitesides, *J. Am. Chem. Soc.* **124**, 14495 (2002).
181. M. M. Burns, J. M. Fournier, and J. A. Golovchenko, *Science* **249**, 749–754 (1990).
182. H. Misawa, K. Sasaki, M. Koshioka, N. Kitamura, and H. Masuhara, *Appl. Phys. Lett.* **60**, 310–312 (1992).
183. W. Hu, H. Li, B. Chang, J. Yang, Z. Li, J. Xu, and D. Zhang, *Opt. Lett.* **20**, 964–966 (1995).
184. C. Mio, M. D. W. Mar *Langmuir*, **15**, 8565–8568 (1999).

9

Self-Assembly and Nanostructured Materials

George M. Whitesides, Jennah K. Kriebel, and Brian T. Mayers

"Nanostructured materials" are those having properties defined by features smaller than 100 nm. This class of materials is interesting for the reasons: i) They include *most* materials, since a broad range of properties—from fracture strength to electrical conductivity—depend on nanometer-scale features. ii) They may offer *new* properties: The conductivity and stiffness of buckytubes, and the broad range of fluorescent emission of CdSe quantum dots are examples. iii) They can mix classical and quantum behaviors. iv) They offer a bridge between classical and biological branches of materials science. v) They suggest approaches to "materials-by-design". Nanomaterials can, in principle, be made using both top-down and bottom-up techniques. Self-assembly bridges these two techniques and allows materials to be designed with hierarchical order and complexity that mimics those seen in biological systems. Self-assembly of nanostructured materials holds promise as a low-cost, high-yield technique with a wide range of scientific and technological applications.

9.1. INTRODUCTION

9.1.1. *Materials*

Materials are what the world is made of. They are hugely important, and hugely interesting. They are also intrinsically complicated. Materials comprise, in general, large numbers of atoms, and have properties determined by complex, heterogeneous structures. Historically, the heterogeneity of materials—regions of different structure, composition,

Department of Chemistry and Chemical Biology, Harvard University, Cambridge MA 02138. Email: Gwhitesides@gmwgroup.harvard.edu and properties, separated by interfaces that themselves have nanometer-scale dimensions and that may play crucial roles in determining properties[1]—have been determined largely empirically, and manipulated through choice of the compositions of starting materials and the conditions of processing.

The phrase "nanostructured materials," implies two important ideas: i) that at least some of the property-determining heterogeneity in materials occurs in the size range of nanostructures (~1–100 nm), and ii) that these nanostructures might be synthesized and distributed (or organized), at least in part, by design. The idea of "nanostructured materials" thus focuses on four key questions: i) What nanostructures are interesting? ii) How can they be synthesized? iii) How can they be introduced into materials? iv) How can the relationships between their structures and compositions, their matrices, and their interfaces control the properties of the materials that incorporate them? The last question is an old one: "materials by design" has been a goal of materials science since its inception[2]. It remains, however, a difficult one—sufficiently difficult, in fact, that the majority of research still focuses on the first three, where progress is easier to achieve and recognize.

Two broad strategies are commonly employed for generating nanostructures. The first is "bottom-up[3,4]": that is, to use the techniques of molecular synthesis[4], colloid chemistry[5], polymer science[6], and related areas to make structures with nanometer dimensions. These nanostructures are formed in parallel and can sometimes be nearly identical, but usually have no long-range order when incorporated into extended materials. The second strategy is "top-down[7]": that is, to use the various methods of lithography to pattern materials. Currently, the maximum resolution of these patterns is significantly coarser than the dimensions of structures formed using bottom-up methods. Materials science needs an accessible strategy to bridge these two methods of formation, and to enable the fabrication of materials with the fine resolution of bottom-up methods and the longer-range and arbitrary structure of top-down processes. This bridging strategy is "self-assembly[7–9]": that is, to allow structures (in principle, structures of any size, but especially nanostructures) synthesized bottom-up to organize themselves into regular patterns or structures by using local forces to find the lowest-energy configuration, and to guide this self assembly using templates fabricated top-down.

The literature contains many examples of self-assembly bridging top-down and bottom-up structures[10]: photolithography can be used to direct the phase separation of block co-polymers into patterns[11]; (an example of top-down control constraining a self-assembling system); or alkanethiols can form self-assembled monolayers (SAMs) on gold colloids[5,12] (an example of self-assembly increasing the complexity of a bottom-up structure). Self-assembly is particularly useful because it allows the aggregation of structures too small to be manipulated individually (or manipulated *conveniently*) into the ordered arrays or patterns that often give function to materials. The development of hierarchically ordered structures—structures in which self-assembly has been at work at different scales, each bringing a different property—is one that permeates biology, but is just beginning to be exploited consciously and rationally in synthetic materials[13–15] (although, of course, the concept that underlies composite materials is hierarchical structure). Self-assembly can also both generate structures with true three-dimensional order, and do so in bulk and inexpensively. An example is the self-assembly of atoms and molecules into stable crystalline CdSe nanoparticles, and the subsequent self-assembly of the nanoparticles into three-dimensional photonic crystals[16].

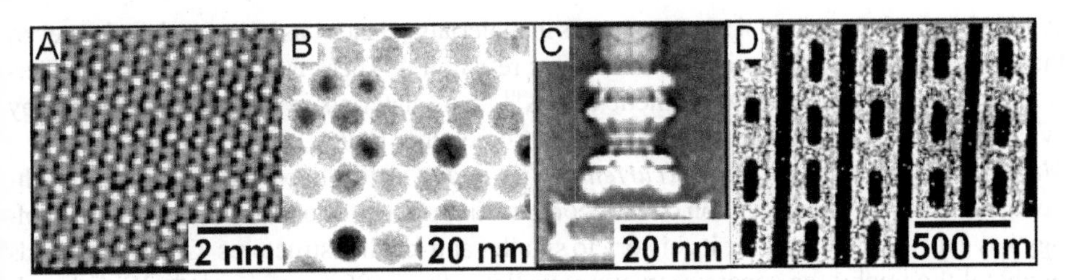

FIGURE 9.1. Examples of different nanostructures. A) Molecular: an STM image of the surface of a SAM of alkanethiolates[17]. B) Colloidal: TEM image of self-assembled iron oxide nanoparticles. C) Biological: rotationally averaged electron microscope image of a flagellar motor[60]. D) Microfabricated: SEM image of a pattern generated lithographically[44].

Whatever the method of synthesis or fabrication, there is general agreement for the moment that the most inclusive definition of "nanostructures" is that they are structures with smallest dimensions less than 100 nm. Further differentiation of the field is, however, useful in considering strategies for integrating nanoscience and materials science. The types of nanostructures can be organized into groups based on their size, function, and structure; this organization will help to define some of the potential of the field.

9.1.2. Nanostructured Materials by Size

We introduce the question of "size" by examining several classes of nanostructures, grouped according to the methods by which they were formed (Figure 9.1).

Molecules (1–6 nm) The most sophisticated and structurally complex nanostructures are molecules[17]. Chemists have, albeit unwittingly, been doing nanoscience since the beginning of chemistry. One important task and competence of chemistry is to place atoms into molecular or extended structures with atomic-level precision. The art of chemical synthesis—especially synthesis of organic and organometallic molecules—is one of the most sophisticated in all of science. It has been most highly developed in making functional materials—especially drugs[18], but also a wide variety of polymers[19], adhesives[20,21], dyes[22], detergents[23], explosives[24], and other materials.

Because most organic molecules are not electrically conducting, chemistry has historically been connected with information technology (IT)—the area of nanotechnology that is commercially most important—only insofar as it has contributed passive components[25–27] such as insulators, adhesives, structural materials, and bulk starting materials, such as single-crystal silicon and doping gases. Since it is now clear that organic compounds can be conductors[28] and semiconductors[29,30] in addition to insulators, one of the opportunities in nanomaterials science is to use organic synthesis and molecular design to make electronically useful structures starting with organic molecules.

Colloids, small crystals, and aggregates (1–100 nm) The chemistry of colloids and small aggregates (nanocrystals[31], micelles[32], small particles of synthetic polymers[33–36], and phase-separated polymers[37,38]) also has a rich background and history, but only recently an association with nanoscience. Nanoscale objects have always been difficult to work with: they cannot be characterized by molecular techniques; they are too small to see optically; and they are usually heterogeneous in size and properties (at least compared to molecular synthesis). As their potential as components in materials science has become clear, however, they are increasingly attractive objectives for research[39]. There is great interest in understanding and developing new synthetic routes[40–42] to materials in this regime whether by synthesis, phase-separation, self-assembly, or some other route.

Micro/nanofabricated structures (currently, 70 nm and up) Microelectronics is the technology that has focused attention on the economic importance of nanoscience. This technology has evolved steadily over its history to smaller and smaller structures. Current methods of photolithography can generate structures well into the <100 nm regime[43,44]. With the addition of difficult but well-understood extensions based on phase-shifting methods[45] and immersion optics[46,47], that limit can certainly be extended to 70 nm in practical processes[47,48], and probably to still smaller dimensions. By moving to shorter wavelengths (deep- or extreme-UV photolithography[49]) and front-surface optics[50], it may be possible to generate features as small as 10 nm[51–53]. Electron beam lithography[52] and perhaps other methods (x-ray lithography[54] and scanning probe lithography[55]) can also generate structures in this size range. There is, thus, a substantial overlap in sizes that the methods of lithography and of colloid and molecular synthesis can access. An important difference is, however, that lithographically fabricated structures are ordered, but are expensive and largely limited to small, planar patterns, whereas bottom-up methods often generate nanostructures with little or no long-range order, but can do so in large quantities and inexpensively.

Biological structures (2 nm and up) One of the inspirations for nanoscience is biology[56]. There are many descriptions of a cell, one of which is, "an ensemble of functional nanostructures, enclosed in a molecule- and ion-selective, semi-permeable membrane, that replicates itself." These functional structures—proteins and nucleic acids, often associated into aggregates, modified with oligosaccharides, and associated with lipids, are the most sophisticated nanostructures known[57]. These structures, from the simplest small RNA molecules[58], to functional organelles such as ribosomes[59] or flagellar nanomotors[60,61], to viruses[62] and complete cells, are all, of course, self-assembled: no hand—robotic or human—places their components together. The myriad examples of functional, self-assembled nanostructures found in biology provide an encyclopedia of demonstrations and strategies for those wishing to learn how to use self-assembly of molecules at the nanoscale to generate complex function.

Although much of the interest in biological nanostructures has focused on relatively complex functionality, cells and organisms themselves can be considered as a collection of self-assembled *materials*: lipid bilayers, the extracellular matrix, tendon and connective tissue, skin, spider silk, cotton fiber, wood, and bone are all self-assembled biological materials, with an internal structure hierarchically ordered from the molecular to the macroscopic scale.

9.1.3. By Function

Nanostructured materials can also be classed according to their function.

Electronic "Micro" electronics is, without any ground-breaking new technology, on the verge of becoming "nano" electronics. Some structures already have one nanoscale dimension—for example, gate dielectrics in CMOS technology[63] and vertical structures in giant magnetoresistive devices[64]. Lateral dimensions of other types of structures—especially those used for information storage—are also shrinking rapidly. IT is highly developed and technologically sophisticated, and its immediate evolutionary extension well beyond 100 nm-scale structures is inevitable, albeit technically difficult.

There is also a possibility that nanoscience may bring *revolutionary* new technology to IT. Many possibilities have been suggested: buckytubes[65–67] or silicon nanowires[68] as transistors; arrays of gold quantum dots as components of cellular automata[69–71]; random tangles ("spaghetti") of nanowires as the basis for hyper-defect-tolerant computers; and computers using devices based on single molecules[72] or single colloid particles[73]. Most of these possibilities are at the level of "suggestion" and it is not clear if any will ever see commercial reality. Regardless, their exploration is a source of stimulation for the field. In particular, they require an emphasis on the design of properties—electrical and magnetic—that have not conventionally been effectively generated by bottom-up methods.

Optical The importance of optical materials in IT is comparable to that of electronic materials, since most information is shipped optically; there is probably a comparable opportunity for nanostructured materials in optical applications. Photonic bandgap materials—although typically structured with repeat distances larger than 100 nm[74–76]—are properly considered part of nanotechnology, because the structure is modified with defects that create the desired band-gap properties, and the placement of these defects must be precise on the nanoscale[77]. Self-assembly is an obvious strategy for the fabrication of photonic band-gap materials[78,79]. Quantum dots[80]—most commonly CdSe nanoparticles having diameters of tens of nm, and with protective (and usually self-assembled) surface coatings[16]—are remarkable for their ability to fluoresce over the complete visible spectral range, with the controlling parameter being their size[81]. Substrates for surface-enhanced Raman spectroscopy (SERS, a technique with very high sensitivity)—typically silver particles—give the best performance when these particles are engineered into shapes that maximize the electric field gradients that underlie the spectroscopic phenomena[82]. Sub-wavelength optical structures—for example, polarizers[83], filters[84], and superluminescent antennae based on nm-sized holes[85]—all require manipulating materials with optical functionality into structures having nanoscale periodicity.

Magnetic The magnetic properties of materials depend on the structure and interaction of magnetic domains; these domains typically have <100 nm dimensions[86]. Currently these systems are structurally uncomplicated, although even they are often challenging to fabricate. Some examples of these systems include: 2D crystals formed from monodisperse superparamagnetic alloy colloids, with potential use in very dense magnetic storage media[87]; magnetite colloids stabilized against aggregation in a magnetic field by self-assembled dextran coatings and used to increase contrast, *in vivo*, in magnetic resonance imaging[88,89]; metal nanorods stabilized and functionalized with alkanethiolate SAMs and used to characterize the mechanical properties of living mammalian cells[90–92].

9.1.4. By Structure

Certain structural classes are especially relevant to nanoscience.

Surfaces and interfaces As the dimensions of structures become smaller, their ratio of surface-to-volume increases. Molecules are essentially "all surface," as are the smallest nanostructures. The ability to make materials with high ratios of surface-to-volume by building them from nanostructures is an important opportunity for surface science and technology. The surface is an extremely interesting state of matter: it determines many properties crucial for materials[93]—wettability, adhesion, friction, susceptibility to corrosion, some aspects of biocompatibility, and many others. Self-assembly has already made a large contribution to surface science through the introduction of self-assembled monolayers[94] (SAMs), which allow significant control and tunability of the surface properties. The field of engineered, nanostructured surfaces is still in its infancy, and only a few of the opportunities in this area have been exploited: the great majority of research has focused on SAMs of alkanethiolates on gold and silver[95]. The complexity of self-assembling systems can be increased by exploring new components (e.g. complex organic and organometallic molecules) or supramolecular structures (e.g. monodisperse colloids[96], or proteins[97]). Surfaces and interfaces are intrinsically nanoscale structures that are key in determining the properties and behavior of many important systems. They are, also, uniquely amenable to investigation using the tools already available in nanoscience, especially scanning probe[98] and particle (electron and ion) beam[99] devices.

Mechanical properties The mechanical properties of materials are strongly influenced by nanoscale structure. Fracture strength and character, ductility, and various mechanical moduli all depend on the substructure of the materials over a range of scales. This dependence has been extensively exploited throughout the field of structural materials: formation of grains in metallic alloys[100], phase-separation in polymers[37] and ceramics[101], and toughening of materials with nanoparticle additives[102–104] (carbon black and silica) are examples. Since much of the development of nanostructured materials has been carried out empirically, the opportunity to redevelop a science of materials that are nanostructured by design is largely open.

9.2. WHY BUILD NANOSTRUCTURES?

"Nanotechnology" has become a word around which a remarkable range of science and engineering is being organized. Why has it emerged into the limelight, while other areas of technology that might have as much potential ("intelligent machines", the biological/computational interface, sustainable development, and others) have not? The answer to this question is complicated, with components of economic necessity, scientific opportunity, and public engagement. Some of the more technical aspirations of nanotechnology are these:

9.2.1. Unique Properties

A key aspiration of nanotechnology is to demonstrate the proposition that as things become small, they become different. "Different" often translates into "interesting", and sometimes into "useful and valuable". There are a number of demonstrations of the emergence of differences in nanostructures. Although buckyballs[105] (among the first of the synthetic nanostructures to catch the interest of both the technical community and the public) have so far fallen in the category of "interesting but not especially useful", bucky*tubes* (or carbon nanotubes) have genuinely remarkable properties: especially their high electrical conductivity and unique mechanical strength[106,107]. The properties of surfactant-stabilized colloids are the basis for many bioanalytical systems[108,109]. Fluorescent CdSe quantum dots, unlike fluorescent organic dyes, photobleach only slowly (or not at all)[110], and show interesting (if annoying) optical phenomena such as "blinking"[111]. SAMs provide an unequaled ability to tailor the properties of surfaces[112].

9.2.2. *Quantum Behavior*

Quantum behavior becomes increasingly prominent as structures become smaller. Many of the behaviors of atoms and molecules are, of course, only explicable on the basis of quantum mechanics. The properties of objects and structures larger than a few microns are usually classical. In the intermediate region—the region of nanometer-scale structures—quantum and classical behaviors mix. This mixture offers the promise of new phenomena and/or new technologies. The fluorescent behavior of semiconductor quantum dots can only be explained quantum mechanically[110]; as can the tunneling currents that characterize scanning tunneling microscopes, and electron emission from the tips of buckytubes[107]. The response of electrical resistance to magnetic field in GMR materials is already useful in magnetic information storage[64,113,114], and the behavior of spin-polarized electrons in magnetic semiconductors forms one foundation for the emerging field of spintronics[115]. The ability to make structures in the region where quantum behavior emerges, or where classical and quantum behaviors merge in new ways, is one with enormous opportunity for discovery. And because quantum behavior is fundamentally counterintuitive, there is the optimistic expectation that nanostructures and nanostructured materials will found fundamentally new technologies.

9.2.3. *Microelectronics*

The argument for the development of nanotechnology for use in the microelectronics industry is clear. Information technology (IT) has been the technology that has most changed society in the last 50 years. Its development has not yet subsided, although the progression of dimensions to ever-smaller sizes—described by Moore's Law[116]—must inevitably come to an end when these dimensions reach the size of molecules and individual atoms. In between the current structures ($\sim$100 nm) and the minimum size limit ($\sim$1 nm), developments of immense economic importance are inevitable.

Beyond evolutionary developments in silicon-based technology, there are a host of possibilities in *new*, but not necessarily fundamentally different, technologies for IT. Will there be important technologies built around organic semiconductors? Will it be possible and practical to take advantage of the high mobility of electrons in semiconducting buckytubes to make new electronic devices? Will some combination of bottom-up synthesis of monodisperse, magnetic colloids, surfactant-assisted crystallization, and materials fabrication generate practical, ultradense magnetic information storage media? Is there a way to use self-assembly of small circuit elements as the basis for a new strategy for fabricating microprocessors, mass storage devices, or displays? It is too early to judge the practical importance of these areas, although they are showing technical feasibility in demonstrations in research laboratories.

In the longer term loom the potentially *revolutionary* technologies, to which nanostructures and nanomaterials may make a contribution: quantum computing[117–119], computing using cellular automata[120,121], photonic computing[122,123], semiconductor/biological hybrid computing[124,125], molecular electronics[95], and others. The history of predicting revolutions is poor, and it is likely that any revolution will emerge from an unexpected direction. Since, however, nanostructures constrain electrons and photons in new ways, and since nanostructures have been difficult to fabricate, and hence are still relatively unexplored, the possibility that a revolution in IT will appear, unexpectedly, from the exploration of some area of nanoscience is higher than it might be in more familiar areas.

9.2.4. *Manufacturing*

The relation between manufacturing and nanotechnology is less explored than that in many apparently higher-technology areas. Manufacturing is a field that permeates technology: any successful technology must be transferred from its developer to its users, and most technologies (even software is ultimately housed in hardware) require manufacturing something. There are complex but important relations between nanoscale features of manufacturing systems, the cost of manufacturing processes, and the performance of manufactured objects. Would the performance of ball-bearings in a heavy-duty transmission improve if there were no defects larger than a few nanometers? What would robotic assembly systems be like if every part were identical to within a few nanometers? It is not possible, at present, to answer these and related questions, since nanoscience is just beginning to generate the tools and metrologies necessary to explore them. Nanoscience does, however, have the potential to make important contributions to future manufacturing systems.

9.2.5. *Fundamental Science*

At the foundation, underlying the technologies, is the fundamental science of phenomena at the nanoscale. Nanoscience, in its broad sense, is a new area. We do not know what will develop, but we do know that in order for it to develop, it must have materials, procedures, and tools. Developing new ways of manipulating matter at the smallest scales—scales that bridge between atoms and molecules (chemistry) and mesoscopic matter, (materials science)—is a centrally important part of fundamental scientific inquiry.

9.3. WHY USE SELF-ASSEMBLY TO BUILD NANOSTRUCTURES?

Self-assembly has a special place in nanoscience. Top-down methods of fabrication provide the ability to build patterns, but are capital intensive, two dimensional, and limited in their ability to provide materials in quantity. Bottom-up methods can make large quantities of nanostructures (including nanostructures too small to be made by any top-down method: e.g., molecules), but without pattern or regularity to their arrangement. Self-assembly bridges the two: it provides a strategy that makes possible the patterning (in a broad sense) of nanostructures made by bottom-up synthesis; it can also use patterns generated by top-down fabrication to guide the ordering of nanostructures made by bottom-up methods.

This capability of self-assembly to make ordered arrays of nanostructures is, in essence, nothing new. Crystallization of molecular or atomic species (whether it is the phase transition of liquid water into solid ice, or of liquid silicon into semiconductor-grade silicon crystal) is an example of self-assembly, as are the formation of surfactants in soap bubbles[126], the crystallization of viruses for x-ray structure determination[127], and the ordering of liquid crystals in displays[128]. The novelty of self-assembly is in the focus on the formation of matter *structured rationally* at scales less than 100 nm, and the realization that the only practical method of achieving these structure is to have the components assemble themselves spontaneously.

We examine several general areas where self-assembly seems to be the best method (by whatever metric: practicality, cost, order, dimensionality) for building materials with nanodimensional structural regularities.

9.3.1. Components too Small for Top-Down

Self-assembly provides the only approach to nanostructured materials that simply cannot be fabricated by current top-down methods. Although top-down methods are versatile[53], and can fabricate astonishingly small structures[52], there are many types of structures that they can not fabricate, the most important and obvious of which are molecules.

The ability of scanning probe devices to arrange atoms on surfaces is a demonstration of the power of these devices, but it is a methodology that is limited in its practicality, and in the complexity of the systems that it can make. A circle of xenon atoms on a surface is a practical target[129,130]; cholesterol is not. Self-assembly will, thus, be an essential part of the generation of materials starting from their atomic or molecular components. Top-down lithographic methods will, of course continue to be the best for the important, specialized task of making planar structures with arbitrary patterns (e.g. circuits).

9.3.2. Too Many Components for Conventional Placement

Self-assembly is a massively parallel process, and can normally involve very large numbers of components (a large crystallization might involve 10^{27}molecules). Robotic pick-and-place methods for placement are limited by the fact that they are serial. Although they can be accelerated by using a number of robotic devices in parallel (for example, the multiple scanning probe heads of the IBM "millipede"[131]), they cannot approach the number of molecules in a test tube, for example.

There is no clear understanding at this point of the parameters that dictate the cases for which "self-assembly" will be superior to "externally directed assembly". For numerous, small components, self-assembly will probably always be superior; for unsymmetrical patterns with relatively large components (e.g., microelectronic circuits) top-down fabrication will probably remain superior. In problems involving intermediate sizes and intermediate numbers of components, it remains to be seen which strategy is best. Pick-and-place assembly with 100-nm components will never be straightforward practically, although it may find uses in research.

9.3.3. Too Many Dimensions

The technologies for making nm-scale structures by current lithographic methods are largely restricted to planar or quasi-planar geometries. (An array of trenches carved into silicon by reactive ion etching is formally a 3D structure, but its method of production precludes more complex 3D structures without elaborate stacking and registration.) Self-assembly is not limited to planar surfaces, and may work better in 3D than in 2D in some cases. Crystallization is a good example of 3D self-assembly. In crystallization each atom benefits from bonding interactions with all of its neighboring atoms. While there are some materials that form stable 2D crystals (e.g. SAMs), the majority of materials form solids in three dimensions because of energy minimization through interactions with neighbors.

9.3.4. Fragility of Biological Systems

Manipulating biological nanostructures is already a technologically important area. Fabricating DNA and protein arrays[132,133], analyzing small quantities of biological materials[134,135], and manipulating or examining the cell with nanometer scale probes[136–138] are examples of problems in which the combination of the small sizes of components, and their sensitivity to damage outside of a narrow range of environmental parameters, requires mild conditions for fabrication and assembly—conditions where self-assembly can work uniquely well.

9.3.5. Cost

Because self-assembly is a parallel process, and because it does not involve robotic or other devices to impose order on nanostructures, it will probably, when applicable, ultimately always have an advantage in cost over other methods of fabrication.

9.4. WHAT ARE TARGETS FOR THE FIELD OF NANOSTRUCTURED MATERIALS BY SELF-ASSEMBLY?

The field of nanotechnology is in its infancy, and it is too early to identify targets with any certainty. Some systems are, however, obvious candidates for research and development, either because there is a clear economic imperative requiring their exploration in order to have options on future technologies as they emerge, or because they hold promise for exploratory research (Figure 9.2).

9.4.1. Information Technology (IT): Electronics and Electronic Components; Photonic Materials—Magnetic Materials

IT is an area in which it is already clear that emerging nanotechnologies will be centrally important; the only question is "Which ones?"

Electron transport in organic molecules The ultimate electronic nanotechnology—single molecule electronics—has had a checkered start, with the concern that a number of the initial experiments that stimulated interest in the field are not reproducible, or over-interpreted, or in one unfortunate set of experiments, fraudulent. This field has yet to sort itself out–one *goal* for the field of self-assembled nanomaterials in electronics is to build a set of experiments that provides a believable basis for estimating its potential[139,140], either as the object of fundamental science, or, perhaps, for commercialization.

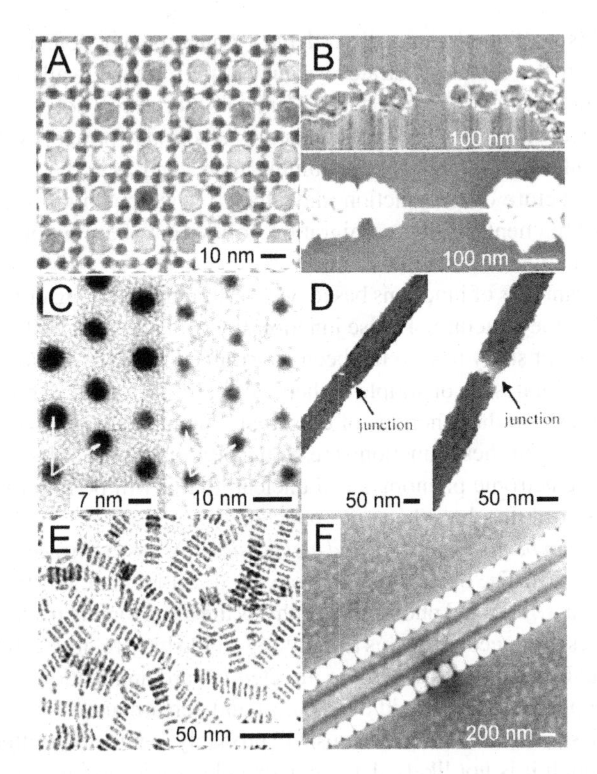

FIGURE 9.2. Examples of self-assembled nanostructures: A) Close packed binary nanoparticle superlattice[185]. B) DNA templated carbon nanotube transistor, (top) single nanotube, (bottom) carbon nanotube bundle[170]. C) Nanoparticles on S-layer protein lattices[171]. D) Molecular junctions in nanowires[140]. E) Self-assembled chains of prismatic BaCrO4 nanoparticles on a TEM grid[186]. F) Template-assisted self-assembly of colloids on a photoresist pattern[183].

The most reliable current method for examining the mechanism of electron transport through *single* molecules is probably to incorporate the molecule of interest in a SAM, and probe it with an STM[141]. This system has two advantages: i) it involves making one chemical contact—at the end where the thiol-metal bond forms; ii) it allows examination of a large number of molecules, and hence the accumulation of useful statistics on the variation in measured properties. It has the disadvantage that the molecule of interest is in a matrix of other molecules in the SAM, but the one through which electron transport occurs is often located at a "special" site (a site where the exchange of molecules with the solution is relatively rapid and the surface structure is distinct and less well characterized)[95]. The results from STM-based examinations are in general agreement with results from experiments in which the junction consists of a SAM in contact with a mercury drop covered with a second SAM[142], or even a SAM onto which a second electrode is formed by evaporation of metal[143,144]. Similar results are also obtained from electrochemical studies of SAMs[142].

Although these SAM-based systems have been studied extensively in terms of electron transport, and although self-assembly plays a crucial role in determining the degree of order in these systems, there remain a number of uncertainties. i) It is now clear that the structures of SAMs—although generally ordered—contain many defects, due to steps and grain boundaries in the metal substrate, vacancies and domain walls in the SAM, and other details of structure[17,95,145]. ii) The SAM-metal interface is not well defined, and in circumstances where there is a second, contacting electrode formed by evaporation, the structure of *this* interface is only just beginning to be understood[146,147]. iii) There is a concern that the structure of the junction may change when potential is applied across it: formation of metal filaments by electromigration, followed by burnout of these filaments at high currents, is a matter of particular concern[148].

One of the advantages of junctions based on SAMs and containing organic molecules is the ability to design the structure of these junctions by using organic molecules of different structures. This type of study has, so far, been constrained to relatively simple systems (for example, n-alkanethiolates[139], or simple biphenyls[149]). More complex structures have been incorporated as SAMs in the junctions of cross-bar structures using nanowires[150], but the structure of the SAMs in these junctions are so uncertain, and the values of the processes that determine the electronic phenomena so unclear, that it will be impossible to interpret these experiments until they have been reproduced.

Very low cost, medium performance printed electronics A less controversial area is printed (or organic) electronics. The objective in this rapidly developing field is to develop alternatives to silicon and conventional photolithography as the basis for electronic systems[151–153]. Initially, the devices produced using this technology would have relatively low performance, but very low cost. These devices would be directed toward applications (for example, RF ID tags[154]) where one-time use would dictate cost and performance.

The economics of these kinds of systems will probably require that patterns be produced by printing. Although it is not likely that the active electronic systems produced by printing methods would have nm-scale features, microcontact printing of SAMs—ultimately, probably, by reel-to-reel printing—is being explored for forming metal features[155,156]

Low-cost, high-performance electronics A more ambitious challenge is that of developing a technology for fabricating high-performance (>GHz clock rate) electronics. There are several methods being considered for this type of technology. One method would use high-mobility materials such as carbon nanotubes[65] or silicon nanowires[68], and orient these materials in the gate region by self-assembly[157,158]. A second would use lower-mobility materials (e.g., organic semiconductors[29] or amorphous silicon) and fabricate very narrow gates by self-assembly[159].

9.4.2. Sensors and Analytical Systems

The ability to fabricate small cantilevers, tips, and wires[160] (and to cross the wires, in some circumstances), opens the possibility of making nanoscale sensors[161]. A number of these systems have been demonstrated in laboratory experiments; in most cases, SAMs have provided the functionality that gives the systems their selectivity for particular analytes[162,163].

This area is clearly an interesting one for exploratory research. An important question is: "For what applications does one *need* nanoscale sensors?" Maximization of surface area is one consideration, and nanoscale devices have the advantage of being largely surface. The list of potential applications that exploit this advantage is, however, not currently long. One potential use may exist in the study of individual cells (either intact, or after lysis); in this application, the quantity of sample is very limited, and sensing regions should ideally be sufficiently small that they do not to deplete materials present on the cell membrane, or in solution[164].

9.4.3. *Structural and Multifunctional Materials: Controlled Heterogeneity*

The development of composite structural materials based on ordered nanostructures in a matrix material has been surprisingly difficult to accomplish, and has only proved successful in a few cases[165]. Part of the difficulty is that the surface chemistry of the nanostructures must be controlled carefully so that the structure truly *is* a composite and that the nanostructured phase in the matrix is ordered. Given the importance of nano-scale heterogeneity in determining the mechanical properties of materials, this area is one of great theoretical and practical interest.

9.4.4. *Surfaces*

Surface science has been one of the first beneficiaries of self-assembled nanostructures (in the form of SAMs). Self-assembly is a very general strategy for forming molecularly tailored interfaces, and, other than the few systems that have formed the basis for the majority of work in SAMs, almost none of the obvious opportunities to use self-assembly to build ordered, nanostructured interfaces have been examined. The preparation of more sophisticated structures based on molecules with complex structures, on self-assembled colloids, on multilayered polymers formed by electrostatic interactions between charged groups[166,167], or on biologically derived structures is just beginning[168–171].

A number of other types of processes that can be considered a form of self-assembly at surfaces are just beginning to appear. The selective oxidation of silicon, followed by etching of the silicon dioxide, as a route to silicon nanowires is an example[172]; the galvanic deposition of platinum on selenium nanostructures, followed by removal of the selenium, to make nanowalls with complex shapes is a second[173,174].

9.4.5. *Structured Colloids and Other Mesoscale Systems*

One of the most successful areas of early-stage nanotechnology is the preparation of structured colloids, and the modification of the structures and properties of these systems. Examples include: i) the formation of nanorods by electrodeposition, followed by selective formation of SAMs on regions of these nanorods[140,175]; ii) the formation of liquid crystal structures from these nanorods[176]; iii) the formation of metallic nanopores, and the modification of the interior surface of these pores with SAMs (and the remarkable demonstration of ion selectivity of the resulting nanochannels[177,178]); iv) the formation of a wide variety of colloids, some with remarkable structures[179–181]; v) the self-assembly of small spheres into wells, and the use of this templating to make ordered aggregates of these spheres[182–184]; and vi) the formation of colloids, and their use as the basis for the formation of more complex structures by self-assembly[185,186].

9.4.6. *Manufacturing Processes*

There has been relatively little work on the relationship between nanoscience and non-microelectronic (or MEMS) manufacturing. The development of the scanning probe microscope as a relatively inexpensive instrument with which to measure surface properties is beginning to be used in industrial research laboratories,[187,188] but not yet in manufacturing process control.

9.5. OPPORTUNITIES, CHALLENGES, AND PROBLEMS

Self-Assembly Works! An important conclusion from the early phase of nanoscience and nanotechnology is that self-assembly provides a useful and general strategy for organizing simple nanostructures—molecules or colloid particles (or other particles that are colloidal in size, such as viruses)—into more ordered structures. The final measure of success in nanoscience will be *function*—regardless of whether as a tool or substrate in fundamental science. Since simple nanoparticles usually have relatively little independent function, the formation of more complex aggregates and structured aggregates will be a constitutive part of nanoscience and nanotechnology. Self-assembly is the only general strategy at this time for accomplishing this type of organization.

9.5.1. Biology and Biomimetics

Biologically derived materials carry with them a special set of problems. They are often thermally unstable, scarce, impure, and difficult to manipulate. Certain of the problems of biological materials are being addressed: developing strategies for designing surfaces that will not adsorb proteins is one example[189]. Others—for example, designing nanostructures that are compatible with the interior of the cell, and capable of reporting information about this interior—are still at early stages of investigation[5,109].

Biomimetic systems also hold much promise. Self-assembly offers, in principle, the opportunity to make materials showing multiple levels of structural organization, and, in principle, multiple contributions to properties. This type of hierarchical self-assembly and self-organization is used throughout biology, but to a relatively low degree in nanoscience[13–15,190–192]. Another interesting characteristics of self-assembly in biology is that it is usually *dynamic*[193]: that is, self-assembly proceeds only while there is a flux of energy through the system. Understanding dynamic self-assembly (as opposed to the equilibrium or steady-state self-assembly normally practiced in chemistry and materials science) will, in the long term, lead to new concepts[194,195].

Biology, on the whole, offers an incredible range of functional nanoscale structures, and processes for forming these structures. Man-made processes almost never resemble the biological ones, in strategy (or often in quality of product). Exploration, understanding, and exploitation of processes mimicking those used in biology are exciting areas of fundamental science, and may eventually contribute to technology.

9.5.2. Materials and Engineering

The key to self-assembly is the components that are involved: the information required for efficient self-assembly must, in general, be embedded in these components. The first step toward the capability to design and synthesize nanostructures for self-assembly is simply to expand the range of syntheses and methods of fabrication that lead to nanostructures. For the most part, the requirements of self-assembly will necessitate that these syntheses generate highly homogeneous (both structurally and functionally) nanostructures. The polydispersity that characterizes most syntheses of colloids, for example, is sufficient to make uniform crystallization impossible without purification[78,196–198].

Regardless of the homogeneity of the components, self-assembled structures will always have some level of defects. In a three-dimensional colloidal crystal, for example, a single vacancy will not largely affect the gross optical properties of the structure[196,198]. Multiple vacancies or cracks can, however, affect the performance of such a crystal. Defects in SAMs can be equally deleterious, for example, to electronic properties. Characterizing these types of defects, and then designing the systems in which the nanostructured materials are to be used, may be a minor task (inspecting the assembled structure to find the occasional defective one, if defects are rare) or a major one (if defects are relatively common). The repair of defective structures is also important, and can be difficult.

For certain types of functional structures—especially for those to be used in information processing and storage—the interface between the nanoscopic world and the macroscopic world of pins, solder pads, and optical fibers may be as important as the nanostructures themselves. Understanding, in a particular case, if this problem of interfacing nanostructures with the macroscopic world can be solved by some application of conventional technology, or if it will require some new solution, is a key part of the problem.

As with any system in materials science, self-assembled nanostructures are unlikely to be useful on a larger scale until they can be generated reproducibly and on a large scale. Developing a *strategy* for applying nanomaterials, nanofabrication, and nano-scale metrology to manufacturing—especially in precision manufacturing outside the area of microelectronics—remains to be done.

9.5.3. The Fundamentals of Self-Assembly

If one accepts the premise that self-assembly will be an important component of the formation of nanomaterials, it is clearly important to understand it as a process (or, better, class of processes). The fundamental thermodynamics, kinetics, and mechanisms of self-assembly are surprisingly poorly understood. The basic thermodynamic principles derived for molecules may be significantly different for those that apply (or do not apply) to nanostructures: the numbers of particles involved may be small; the relative influence of thermal motion, gravity, and capillary interactions may be different; the time required to reach equilibrium may be sufficiently long that equilibrium is not easily achieved (or never reached); the processes that determine the rates of processes influencing many nanosystems are not defined.

Putting together a fundamental science of nanostructures and self-assembly will probably take many years. We should not assume that because there is a rich base of information about molecules and their reactions, that there is a closely analogous or equally detailed body of information about nanostructures and self-assembly. Along the same vein, although the combination of top-down patterning and self-assembly promises to be a powerful one, the development of general strategies, based on a sound understanding of the relationship between the template and the self-assembled nanoscale components, remains to be worked out.

REFERENCES

1. Nemat-Nasser, S. & Hori, M. (eds.) Micromechanics, Part 1: Overall Properties of Heterogeneous Materials, Second Revised Edition (Elsevier: Amsterdam, 1998).
2. Jones, W., Rao, C. N. R. & Editors. Supramolecular Organization and Materials Design (Cambridge University Press: Cambridge, UK, 2002).
3. Lieber, C. M. Nanoscale science and technology: building a big future from small things. MRS Bull. **28**, 486–491 (2003).
4. Hecht, S. Welding, organizing, and planting organic molecules on substrate surfaces-promising approaches towards nanoarchitectonics from the bottom up. Angew. Chem., Int. Ed. Engl. **42**, 24–26 (2003).
5. Daniel, M.-C. & Astruc, D. Gold Nanoparticles: Assembly, Supramolecular Chemistry, Quantum-Size-Related Properties, and Applications toward Biology, Catalysis, and Nanotechnology. Chem. Rev. **104**, 293–346 (2004).
6. Park, C., Yoon, J. & Thomas, E. L. Enabling nanotechnology with self assembled block copolymer patterns. Polymer **44**, 6725–6760 (2003).
7. Whitesides, G. M. & Grzybowski, B. Self-assembly at all scales. Science (Washington, D. C.) **295**, 2418–2421 (2002).
8. de Wild, M., Berner, S., Suzuki, H., Ramoino, L., Baratoff, A. & Jung, T. A. Molecular assembly and self-assembly: Molecular nanoscience for future technologies. Chimia **56**, 500–505 (2003).
9. Zhang, S. Building from the bottom up. Mater. Today **6**, 20–27 (2003).
10. Walt, D. R. Nanomaterials: Top-to-bottom functional design. Nature Mater. **1**, 17–18 (2002).
11. Kim, S. O., Solak, H. H., Stoykovich, M. P., Ferrier, N. J., de Pablo, J. J. & Nealey, P. F. Epitaxial self-assembly of block copolymers on lithographically defined nanopatterned substrates. Nature (London) **424**, 411–414 (2003).
12. Shenhar, R. & Rotello Vincent, M. Nanoparticles: scaffolds and building blocks. Acc. Chem. Res. **36**, 549–61 (2003).
13. Park, J.-W. & Thomas, E. L. Multiple ordering transitions: Hierarchical self-assembly of rod-coil block copolymers. Adv. Mater. (Weinheim, Ger.) **15**, 585–588 (2003).
14. Coelfen, H. & Mann, S. Higher-order organization by mesoscale self-assembly and transformation of hybrid nanostructures. Angew. Chem., Int. Ed. Engl. **42**, 2350–2365 (2003).
15. Wong, G. C. L., Tang, J. X., Lin, A., Li, Y., Janmey, P. A. & Safinya, C. R. Hierarchical self-assembly of F-actin and cationic lipid complexes: Stacked three-layer tubule networks. Science (Washington, D. C.) **288**, 2035–2039 (2000).
16. Seifert, G. Nanomaterials: Nanocluster magic. Nature Mater. **3**, 77–78 (2004).
17. Liu, G.-Y., Xu, S. & Qian, Y. Nanofabrication of Self-Assembled Monolayers Using Scanning Probe Lithography. Acc. Chem. Res. **33**, 457–466 (2000).
18. Lednicer, D. & Mitscher, A. The Organic Chemistry of Drug Synthesis, Volume 6 (Wiley: New York, 1998).
19. Sawamoto, M., Percec, V., Hawker, C. J. & Editors. Polymer Chemistry. [In: J. Polym. Sci., Part A: Polym. Chem., 2002; 40(6)] (Wiley: New York, 2002).
20. Bauer, M. & Schneider, J. Adhesives in the electronics industry: Handbook of Adhesive Technology (2nd Edition, Revised and Expanded) (Marcel Dekker: New York, 2003).
21. Chaudhury, M., Pocius, A. V. & Editors. Adhesion Science and Engineering, Volume 2: Surfaces, Chemistry and Applications (Elsevier Science: Amsterdam, 2002).
22. Peters, A. T., Freeman, H. S. & Editors. Analytical Chemistry of Synthetic Colorants (Blackie: Glasgow, UK, 1995).
23. Hargreaves, T. Roast beef and ashes to vegetarian shampoos. Chem. Rev. **12**, 6–9 (2002).
24. Oxley, J. C. The chemistry of explosives. Explosive Effects Appl., 137–172 (1998).
25. Swalen, J. D. Some emerging organic-thin-films technologies. ACS Symp. Ser. **695**, 2–8 (1998).
26. Malik, J. & Clarson, S. J. The chemistry and technology of reworkable polymeric materials for electronic applications. Surf. Coat. Int., Part B: Coat. Trans. **86**, 9–20 (2003).
27. Koehler, M. & Biaggio, I. Influence of diffusion, trapping, and state filling on charge injection and transport in organic insulators. Phys. Rev. B: Condens. Matter **68**, 075205/1-075205/8 (2003).

28. Nalwa, H. S. & Editor. Handbook of Advanced Electronic and Photonic Materials and Devices, Volume 3: High Tc Superconductors and Organic Conductors (Academic: San Diego, 2001).
29. Horowitz, G. Charge transport in polycrystalline organic field-effect transistors. Diffus. Defect Data, Pt. B **80–81**, 3–14 (2001).
30. Farchioni, R., Grosso, G. & Editors (eds.) Organic Electronic Materials: Conjugated Polymers and Low Molecular Weight Organic Solids. [In: Springer Ser. Mater. Sci., 2001; 41] (Springer: Berlin, 2001).
31. Rao, C. N. R., Kulkarni, G. U., Thomas, P. J. & Edwards, P. P. Size-dependent chemistry: properties of nanocrystals. Chem.–Eur. J. **8**, 28–35 (2002).
32. Chattopadhyay, A. Special Issue—Dynamics of Organized Molecular Assemblies: From Micelles to Cells. [In: J. Fluoresc., 2001; 11(4)] (Kluwer Academic/Plenum Publishers: New York, 2001).
33. Lansalot, M., Elaissari, A. & Mondain-Monval, O. Polymer colloids: widespread and novel techniques of characterization. Surf. Sci. Ser. **115**, 381–418 (2003).
34. Soula, R., Claverie, J., Spitz, R. & Guyot, A. Catalytic polymerization of olefins in emulsion: a breakthrough in polymer colloids. Surf. Sci. Ser. **115**, 77–92 (2003).
35. Rieger, J. Polymer crystallization viewed in the general context of particle formation and crystallization. Lect. Notes Phys. **606**, 7–16 (2003).
36. Ikkala, O. & ten Brinke, G. Functional materials based on self-assembly of polymeric supramolecules. Science (Washington, D. C.) **295**, 2407–2409 (2002).
37. Hamley, I. The Physics of Block Copolymers (Oxford University Press: Oxford, UK, 1998).
38. Elaissari, A. Colloidal Polymers: Synthesis and Characterization. [In: Surfactant Sci. Ser., 2003; 115] (Marcel Dekker: New York, 2003).
39. Radloff, C., Moran, C. E., Jackson, J. B. & Halas, N. J. Nanoparticles: building blocks for functional nanostructures. Molec. Nanoelectr., 229–262 (2003).
40. Landfester, K. & Antonietti, M. Miniemulsions for the convenient synthesis of organic and inorganic nanoparticles and "single molecule" applications in materials chemistry. Colloids Colloid. Assembl., 175–215 (2004).
41. Johnston, K. P. & Shah, P. S. Materials science: Making nanoscale materials with supercritical fluids. Science (Washington, D. C.) **303**, 482–483 (2004).
42. Yonezawa, T. Well-dispersed bimetallic nanoparticles. Springer Ser. Mater. Sci. **64**, 85–112 (2004).
43. Boerger, B. E., McLeod, S., Forber, R. A., Turcu, I. C. E., Gaeta, C. J., Bailey, D. K. & Ben-Jacob, J. Advances in CPL, collimated plasma source and full field exposure for sub-100-nm lithography. Proc. SPIE-Int. Soc. Opt. Eng. **5037**, 1112–1122 (2003).
44. Michel, B., Bernard, A., Bietsch, A., Delamarche, E., Geissler, M., Juncker, D., Kind, H., Renault, J. P., Rothuizen, H., Schmid, H., Schmidt-Winkel, P., Stutz, R. & Wolf, H. Printing meets lithography: Soft approaches to high-resolution printing. IBM J. Res. Dev. **45**, 697–719 (2001).
45. Fritze, M., Tyrrell, B., Mallen, R. D., Wheeler, B., Rhyins, P. D. & Martin, P. M. Dense only phase-shift template lithography. Proc. SPIE-Int. Soc. Opt. Eng. **5042**, 15–29 (2003).
46. Owa, S. & Nagasaka, H. Advantage and feasibility of immersion lithography. J. Microlith. Microfab. Microsys. **3**, 97–103 (2004).
47. Switkes, M., Kunz, R. R., Rothschild, M., Sinta, R. F., Yeung, M. & Baek, S. Y. Extending optics to 50 nm and beyond with immersion lithography. J. Vac. Sci. Technol., B **21**, 2794–2799 (2003).
48. Mulkens, J., McClay, J. A., Tirri, B. A., Brunotte, M., Mecking, B. & Jasper, H. Optical lithography solutions for sub-65-nm semiconductor devices. Proc. SPIE-Int. Soc. Opt. Eng. **5040**, 753–762 (2003).
49. Brainard, R. L., Cobb, J. & Cutler, C. A. Current status of EUV photoresists. J. Photopolym. Sci. Tech. **16**, 401–410 (2003).
50. Mandelis, A., Batista, J. & Shaughnessy, D. Infrared photocarrier radiometry of semiconductors: Physical principles, quantitative depth profilometry, and scanning imaging of deep subsurface electronic defects. Phys. Rev. B: Condens. Matter **67**, 205208/1-205208/18 (2003).
51. Brueck, S. R. J. There are no fundamental limits to optical lithography. Int. Trends Appl. Opt., 85–109 (2002).
52. Wallraff, G. M. & Hinsberg, W. D. Lithographic Imaging Techniques for the Formation of Nanoscopic Features. Chem. Rev. **99**, 1801–1821 (1999).
53. Brunner, T. A. Why optical lithography will live forever. J. Vac. Sci. Technol., B **21**, 2632–2637 (2003).
54. Sumitani, H. X-ray lithography. Optronics **256**, 128–133 (2003).

55. Zhou, H., Li, Z., Wu, A., Zheng, J., Zhang, J. & Wu, S. The influence of tip performance on scanning probe lithography. Appl. Surf. Sci. **221**, 402–407 (2004).
56. Hamley, I. W. Nanotechnology with soft materials. Angew. Chem., Int. Ed. Engl. **42**, 1692–1712 (2003).
57. Whitesides, G. M. The 'right' size in nanobiotechnology. Nature Biotech. **21**, 1161–1165 (2003).
58. Perumal, K. & Reddy, R. The 3′ end formation in small RNAs. Gene Exp. **10**, 59–78 (2002).
59. Dundr, M. & Misteli, T. Functional architecture in the cell nucleus. Biochem. J. **356**, 297–310 (2001).
60. Berg, H. C. Constraints on models for the flagellar rotary motor. Philos. Trans. R. Soc. London, Ser. B **355**, 491–501 (2000).
61. Oster, G. & Wang, H. Rotary protein motors. Trends Cell Bio. **13**, 114-121 (2003).
62. Chiu, W. & Johnson, J. E. Virus Structure. [In: Adv. Protein Chem., 2003; 64] (Elsevier Science: San Diego, CA, 2003).
63. Stathis, J. H. Reliability limits for the gate insulator in CMOS technology. IBM J. Res. Dev. **46**, 265–286 (2002).
64. Tsymbal, E. Y. & Pettifor, D. G. Perspectives of giant magnetoresistance. Solid State Phys. **56**, 113–237 (2001).
65. Martel, R. Nanotube electronics. High-performance transistors. Nature Mater. **1**, 203–204 (2002).
66. Avouris, P. Molecular Electronics with Carbon Nanotubes. Acc. Chem. Res. **35**, 1026–1034 (2002).
67. Baughman, R. H., Zakhidov, A. A. & de Heer, W. A. Carbon nanotubes-the route toward applications. Science (Washington, D. C.) **297**, 787–792 (2002).
68. Zhang, Z., Chen, L. S. & Zhou, G. W. Low dimensional materials and their microstructures studied by high resolution electron microscopy. Springer Ser. Surf. Sci. **39**, 105–169 (2001).
69. Snider, G. L., Orlov, A. O., Kummamuru, R. K., Bernstein, G. H., Lent, C. S., Lieberman, M., Felhner, T. P. & Ramasubramaniam, R. Experimental progress in quantum-dot cellular automata. Proc. SPIE-Int. Soc. Opt. Eng. **5023**, 436–440 (2003).
70. Snider, G. L., Orlov, A. O., Kummamuru, R. K., Ramasubramaniam, R., Amlani, I., Bernstein, G. H. & Lent, C. S. Quantum-dot cellular automata. Mater. Res. Soc. Symp. Proc. **696**, 221–231 (2002).
71. Lieberman, M., Chellamma, S., Varughese, B., Wang, Y., Lent, C., Bernstein, G. H., Snider, G. & Peiris, F. C. Quantum-dot cellular automata at a molecular scale. Ann. N. Y. Acad. Sci. **960**, 225–239 (2002).
72. Wada, Y. Prospects for single-molecule information-processing devices for the next paradigm. Ann. N. Y. Acad. Sci. **960**, 39–61 (2002).
73. Simon, U. & Schon, G. Electrical properties of chemically tailored nanoparticles and their application in microelectronics (ed. Nalwa, H. S.) (Academic Press: San Diego, CA, 2000).
74. Lopez, C. Materials aspects of photonic crystals. Adv. Mater. (Weinheim, Ger.) **15**, 1679–1704 (2003).
75. Knight, J. C. Photonic crystal fibers. Nature (London) **424**, 847–851 (2003).
76. Downer, M. C. Optics. A new low for nonlinear optics. Science (Washington, D. C.) **298**, 373–375 (2002).
77. De La Rue, R. Photonic crystals: microassembly in 3D. Nature Mater. **2**, 74–76 (2003).
78. Landon, P., Glosser, R. & Zakhidov, A. Self-assembly methods for photonic crystals. Trends Opt. Photon. **91**, 52–54 (2003).
79. Xia, Y., Gates, B. & Li, Z.-Y. Self-assembly approaches to three-dimensional photonic crystals. Adv. Mater. (Weinheim, Ger.) **13**, 409–413 (2001).
80. Zrenner, A. A close look on single quantum dots. J. Chem. Phys. **112**, 7790–7798 (2000).
81. Asahi, T. & Masuhara, H. Microspectroscopy of single nanoparticles (ed. Masuhara, H. N., Hachiro; Sasaki, Keiji) (Springer: Berlin, 2003).
82. Kneipp, K., Kneipp, H., Itzkan, I., Dasari, R. R. & Feld, M. S. Single molecule detection using near infrared surface-enhanced Raman scattering. Springer Ser. Chem. Phys. **67**, 144–160 (2001).
83. Bomzon, Z. e., Kleiner, V. & Hasman, E. Computer-generated space-variant polarization elements with sub-wavelength metal stripes. Opt. Lett. **26**, 33–35 (2001).
84. Deguzman, P. C. & Nordin, G. P. Stacked subwavelength gratings as circular polarization filters. Appl. Opt. **40**, 5731–5737 (2001).
85. Rajic, S., Corbeil, J. L. & Datskos, P. G. Feasibility of tunable MEMS photonic crystal devices. Ultramicroscopy **97**, 473–9 (2003).

86. Dennis, C. L., Borges, R. P., Buda, L. D., Ebels, U., Gregg, J. F., Hehn, M., Jouguelet, E., Ounadjela, K., Petej, I., Prejbeanu, I. L. & Thornton, M. J. The defining length scales of mesomagnetism: a review. J. Phys.: Condens. Matter **14**, R1175–R1262 (2002).
87. Wirth, S. & von Molnar, S. Magnetic interactions in nanometer-scale particle arrays grown onto permalloy films. J. Appl. Phys. **87**, 7010–7012 (2000).
88. Hong, X., Guo, W., Yuan, H., Li, J., Liu, Y., Ma, L., Bai, Y. & Li, T. Periodate oxidation of nanoscaled magnetic dextran composites. J. Magn. Magn. Mater. **269**, 95–100 (2004).
89. Lacava, L. M., Lacava, Z. G. M., Azevedo, R. B., Chaves, S. B., Garcia, V. A. P., Silva, O., Pelegrini, F., Buske, N., Gansau, C., Da Silva, M. F. & Morais, P. C. Use of magnetic resonance to study biodistribution of dextran-coated magnetic fluid intravenously administered in mice. J. Magn. Magn. Mater. **252**, 367–369 (2002).
90. Chen, J., Fabry, B., Schiffrin, E. L. & Wang, N. Twisting integrin receptors increases endothelin-1 gene expression in endothelial cells. Am. J. Physiol. **280**, C1475–C1484 (2001).
91. Hu, S., Chen, J., Fabry, B., Numaguchi, Y., Gouldstone, A., Ingber, D. E., Fredberg, J. J., Butler, J. P. & Wang, N. Intracellular stress tomography reveals stress focusing and structural anisotropy in cytoskeleton of living cells. Am. J. Physiol. **285**, C1082–C1090 (2003).
92. Wang, N., Butler, J. P. & Ingber, D. E. Mechanotransduction across the cell surface and through the cytoskeleton. Science (Washington, D. C.) **260**, 1124–1127 (1993).
93. Kolasinski, K. Surface science: foundations of catalysis and nanoscience (Wiley: Chichester, NY, 2002).
94. Nuzzo, R. G. & Allara, D. L. Adsorption of bifunctional organic disulfides on gold surfaces. J. Am. Chem. Soc. **105**, 4481–4483 (1983).
95. Schreiber, F. Structure and growth of self-assembling monolayers. Prog. Surf. Sci. **65**, 151–256 (2000).
96. Xia, Y., Gates, B., Yin, Y. & Sun, Y. in Handbook of Surface and Colloid Chemistry (2nd Edition) (ed. Birdi, K. S.) 555–579 (CRC Press: Boca Raton, FL, 2003).
97. Pum, D., Neubauer, A., Gyorvary, E., Sara, M. & Sleytr, U. B. S-layer proteins as basic building blocks in a biomolecular construction kit. Nanotechnology **11**, 100–107 (2000).
98. Hofer, W. A., Foster, A. S. & Shluger, A. L. Theories of scanning probe microscopes at the atomic scale. Rev. Mod. Phys. **75**, 1287–1331 (2003).
99. Kogel, G. Microscopes/microprobes. Appl. Surf. Sci. **194**, 200–209 (2002).
100. Morris, D. G. & Munoz-Morris, M. A. Relationships between mechanical properties, grain size, and grain boundary parameters in nanomaterials prepared by severe plastic deformation, by electrodeposition and by powder metallurgy methods. J. Metastable Nanocrystal. Mater. **15–16**, 585–590 (2003).
101. Interrante, L. V., Moraes, K., Liu, Q., Lu, N., Puerta, A. & Sneddon, L. G. Silicon-based ceramics from polymer precursors. Pure Appl. Chem. **74**, 2111–2117 (2002).
102. Egerton, T. A. The modification of fine powders by inorganic coatings. Kona **16**, 46–59 (1998).
103. Duan, R.-G., Roebben, G., Vleugels, J. & Van der Biest, O. TiO2 additives for in situ formation of toughened silicon nitride-based composites. Mater. Lett. **57**, 4156–4161 (2003).
104. Chiu, H.-T. & Chiu, W.-M. The toughening behavior in propylene-ethylene block copolymer filled with carbon black and styrene-ethylene butylene-styrene triblock copolymer. Mater. Chem. Phys. **56**, 108–115 (1998).
105. Crane, J. Buckyballs bounce into action. Chem. Rev. **4**, 2–11 (1995).
106. Bernholc, J., Brenner, D., Nardelli, M. B., Meunier, V. & Roland, C. Mechanical and electrical properties of nanotubes. Ann. Rev. Mater. Res. **32**, 347–375 (2002).
107. Zhou, O., Shimoda, H., Gao, B., Oh, S., Fleming, L. & Yue, G. Materials Science of Carbon Nanotubes: Fabrication, Integration, and Properties of Macroscopic Structures of Carbon Nanotubes. Acc. Chem. Res. **35**, 1045–1053 (2002).
108. Willard, D. M. Nanoparticles in bioanalytics. Anal. Bioanal. Chem. **376**, 284–286 (2003).
109. Berry, C. C. & Curtis, A. S. G. Functionalisation of magnetic nanoparticles for applications in biomedicine. J. Phys. D: Appl. Phys. **36**, R198–R206 (2003).
110. Woo, W.-K., Shimizu, K. T., Jarosz, M. V., Neuhauser, R. G., Leatherdale, C. A., Rubner, M. A. & Bawendi, M. G. Reversible charging of CdSe nanocrystals in a simple solid-state device. Adv. Mater. (Weinheim, Ger.) **14**, 1068–1071 (2002).
111. Koberling, F., Mews, A. & Basche, T. Oxygen-induced blinking of single CdSe nanocrystals. Adv. Mater. (Weinheim, Ger.) **13**, 672–676 (2001).

112. Whitesides, G. M., Jiang, X., Ostuni, E., Chapman, R. G. & Grunze, M. SAMS and biofunctional surfaces. The "inert surface" problem. Polym. Prepr. (Am. Chem. Soc., Div. Polym. Chem.) **45**, 90–91 (2004).
113. Dai, J., Tang, J., Hsu, S. T. & Pan, W. Magnetic nanostructures and materials in magnetic random access memory. J. Nanosci. Nanotech. **2**, 281–291 (2002).
114. Butler, W. H., Heinonen, O. & Zhang, X. Theory of magnetotransport for magnetic recording. Springer Ser. Surf. Sci. **41**, 277–313 (2001).
115. Sarma, S. D. Ferromagnetic semiconductors: A giant appears in spintronics. Nature Mater. **2**, 292–294 (2003).
116. Marsh, G. Moore's law at the extremes. Mater. Today **6**, 28–33 (2003).
117. Herz, L. M. & Phillips, R. T. Quantum computing. Fine lines from dots. Nature Mater. **1**, 212–213 (2002).
118. Leggett, A. J. Superconducting qubits—a major roadblock dissolved? Science (Washington, D. C.) **296**, 861–862 (2002).
119. Lloyd, S. Perspectives: Quantum computing: Computation from geometry. Science (Washington, D. C.) **292**, 1669 (2001).
120. Hush, N. Molecular electronics. Cool computing. Nature Mater. **2**, 134–135 (2003).
121. Wang Kang, L. Issues of nanoelectronics: a possible roadmap. J. Nanosci. Nanotech. **2**, 235–66 (2002).
122. Takeuchi, S. Nanodevices for quantum computing using photons. Springer Ser. Chem. Phys. **70**, 183–193 (2003).
123. Monroe, C. Quantum information processing with atoms and photons. Nature (London) **416**, 238–246 (2002).
124. Kari, L. & Landweber, L. F. Computing with DNA. Meth. Molec. Bio. **132**, 413–430 (2000).
125. Kampfner Roberto, R. Digital and biological computing in organizations. Bio Sys. **64**, 179–88 (2002).
126. Mileva, E. & Exerowa, D. Foam films as instrumentation in the study of amphiphile self-assembly. Adv. Colloid Interface Sci. **100–102**, 547–562 (2003).
127. Bamford, D. H. Virus structures: Those magnificent molecular machines. Curr. Bio. **10**, R558–R561 (2000).
128. Chigrinov, V. G. Liquid crystal devices: physics and applications (Artech House: Boston, MA, 1999).
129. Stroscio, J. A. & Eigler, D. M. Atomic and molecular manipulation with the scanning tunneling microscope. Science (Washington, D. C.) **254**, 1319–26 (1991).
130. Eigler, D. M. & Schweizer, E. K. Positioning single atoms with scanning tunnelling microscope. Nature (London) **344**, 524–6 (1990).
131. Vettiger, P. & Binnig, G. The nanodrive project. Sci. Am. **288**, 47–48, 50–53 (2003).
132. Huang, R.-P. Protein arrays, an excellent tool in biomedical research. Frontiers Biosci. **8**, D559–D570 (2003).
133. Reimer, U., Reineke, U. & Schneider-Mergener, J. Peptide arrays: from macro to micro. Curr. Opin. Biotechnol. **13**, 315–320 (2002).
134. Beebe, D. J., Mensing, G. A. & Walker, G. M. Physics and applications of microfluidics in biology. Ann. Rev. Biomed. Eng. **4**, 261–286 (2002).
135. Andersson, H. & van den Berg, A. Microfluidic devices for cellomics: a review. Sens. Actuators, B **B92**, 315–325 (2003).
136. Hood John, D., Bednarski, M., Frausto, R., Guccione, S., Reisfeld Ralph, A., Xiang, R. & Cheresh David, A. Tumor regression by targeted gene delivery to the neovasculature. Science (Washington, D. C.) **296**, 2404–7 (2002).
137. Salem, A. K., Searson, P. C. & Leong, K. W. Multifunctional nanorods for gene delivery. Nature Mater. **2**, 668–671 (2003).
138. Dragnea, B., Chen, C., Kwak, E.-S., Stein, B. & Kao, C. C. Gold nanoparticles as spectroscopic enhancers for in vitro studies on single viruses. J. Am. Chem. Soc. **125**, 6374–5 (2003).
139. Salomon, A., Cahen, D., Lindsay, S., Tomfohr, J., Engelkes, V. B. & Frisbie, C. D. Comparison of electronic transport measurements on organic molecules. Adv. Mater. (Weinheim, Ger.) **15**, 1881–1890 (2003).
140. Mbindyo Jeremiah, K. N., Mallouk Thomas, E., Mattzela James, B., Kratochvilova, I., Razavi, B., Jackson Thomas, N. & Mayer Theresa, S. Template synthesis of metal nanowires containing monolayer molecular junctions. J. Am. Chem. Soc. **124**, 4020–6 (2002).

141. Ishida, T., Mizutani, W., Choi, N., Akiba, U., Fujihira, M. & Tokumoto, H. Structural Effects on Electrical Conduction of Conjugated Molecules Studied by Scanning Tunneling Microscopy. J. Phys. Chem. B **104**, 11680–11688 (2000).
142. Rampi, M. A. & Whitesides, G. M. A versatile experimental approach for understanding electron transport through organic materials. Chem. Phys. **281**, 373–391 (2002).
143. Wang, W., Lee, T. & Reed, M. A. Mechanism of electron conduction in self-assembled alkanethiol monolayer devices. Phys. Rev. B: Condens. Matter **68**, 035416/1-035416/7 (2003).
144. Zhou, C., Deshpande, M. R., Reed, M. A., Jones, K., II & Tour, J. M. Nanoscale metal/self-assembled monolayer/metal heterostructures. Appl. Phys. Lett. **71**, 611–613 (1997).
145. Poirier, G. E. Characterization of Organosulfur Molecular Monolayers on Au(111) using Scanning Tunneling Microscopy. Chem. Rev. **97**, 1117–1127 (1997).
146. Fisher, G. L., Walker, A. V., Hooper, A. E., Tighe, T. B., Bahnck, K. B., Skriba, H. T., Reinard, M. D., Haynie, B. C., Opila, R. L., Winograd, N. & Allara, D. L. Bond Insertion, Complexation, and Penetration Pathways of Vapor-Deposited Aluminum Atoms with HO- and CH3O-Terminated Organic Monolayers. J. Am. Chem. Soc. **124**, 5528–5541 (2002).
147. Chang, S.-C., Li, Z., Lau, C. N., Larade, B. & Williams, R. S. Investigation of a model molecular-electronic rectifier with an evaporated Ti-metal top contact. Appl. Phys. Lett. **83**, 3198–3200 (2003).
148. Walker, A. V., Tighe, T. B., Cabarcos, O. M., Reinard, M. D., Haynie, B. C., Uppili, S., Winograd, N. & Allara, D. L. The Dynamics of Noble Metal Atom Penetration through Methoxy-Terminated Alkanethiolate Monolayers. J. Am. Chem. Soc. **126**, 3954–3963 (2004).
149. Ulman, A. (ed.) Self-assembled monolayers of thiols (Academic Press: San Diego, CA, 1998).
150. Luo, Y., Collier, C. P., Jeppesen, J. O., Nielsen, K. A., Delonno, E., Ho, G., Perkins, J., Tseng, H.-R., Yamamoto, T., Stoddart, J. F. & Heath, J. R. Two-dimensional molecular electronics circuits. ChemPhysChem **3**, 519–525 (2002).
151. Crone, B., Dodabalapur, A., Lin, Y. Y., Filas, R. W., Bao, Z., LaDuca, A., Sarpeshkar, R., Katz, H. E. & Li, W. Large-scale complementary integrated circuits based on organic transistors. Nature (London) **403**, 521–523 (2000).
152. Blanchet, G. B., Loo, Y.-L., Rogers, J. A., Gao, F. & Fincher, C. R. Large area, high resolution, dry printing of conducting polymers for organic electronics. Appl. Phys. Lett. **82**, 463–465 (2003).
153. Sirringhaus, H., Kawase, T., Friend, R. H., Shimoda, T., Inbasekaran, M., Wu, W. & Woo, E. P. High-resolution inkjet printing of all-polymer transistor circuits. Science (Washington, D. C.) **290**, 2123–6 (2000).
154. Baude, P. F., Ender, D. A., Haase, M. A., Kelley, T. W., Muyres, D. V. & Theiss, S. D. Pentacene-based radio-frequency identification circuitry. Appl. Phys. Lett. **82**, 3964–3966 (2003).
155. Wilbur, J. L., Kumar, A., Biebuyck, H. A., Kim, E. & Whitesides, G. M. Microcontact printing of self-assembled monolayers: applications in microfabrication. Nanotechnology **7**, 452–457 (1996).
156. Wilbur, J. L., Kumar, A., Kim, E. & Whitesides, G. M. Microfabrication by microcontact printing of self-assembled monolayers. Adv. Mater. (Weinheim, Ger.) **6**, 600–4 (1994).
157. Diehl, M. R., Yaliraki, S. N., Beckman, R. A., Barahona, M. & Heath, J. R. Self-assembled, deterministic carbon nanotube wiring networks. Angew. Chem., Int. Ed. Engl. **41**, 353–356 (2002).
158. Marty, L., Bouchiat, V., Naud, C., Chaumont, M., Fournier, T. & Bonnot, A. M. Schottky barriers and Coulomb blockade in self-assembled carbon nanotube FETs. Nano Lett. **3**, 1115–1118 (2003).
159. Collet, J. & Vuillaume, D. Nano-field effect transistor with an organic self-assembled monolayer as gate insulator. Appl. Phys. Lett. **73**, 2681–2683 (1998).
160. Dresselhaus, M. S., Lin, Y. M., Rabin, O., Jorio, A., Souza Filho, A. G., Pimenta, M. A., Saito, R., Samsonidze, G. & Dresselhaus, G. Nanowires and nanotubes. Mater. Sci. Eng., C **C23**, 129–140 (2003).
161. Cui, Y., Wei, Q., Park, H. & Lieber, C. M. Nanowire nanosensors for highly sensitive and selective detection of biological and chemical species. Science (Washington, D. C.) **293**, 1289–1292 (2001).
162. Chaki, N. K. & Vijayamohanan, K. Self-assembled monolayers as a tunable platform for biosensor applications. Biosens. Bioelectron. **17**, 1–12 (2002).
163. Raj, C. R., Okajima, T. & Ohsaka, T. Gold nanoparticle arrays for the voltammetric sensing of dopamine. J. Electroanal. Chem. **543**, 127–133 (2003).
164. Yasukawa, T., Ikeya, T. & Matsuse, T. Fabrication and characterization of a microvial with a microdisk electrode for cellular measurements. Chem. Sens. **16**, 118–120 (2000).

165. Huynh, W. U., Dittmer, J. J. & Alivisatos, A. P. Hybrid nanorod-polymer solar cells. Science (Washington, D. C.) **295**, 2425–2427 (2002).
166. Schonhoff, M. Self-assembled polyelectrolyte multilayers. Curr. Opin. Colloid Interface Sci. **8**, 86–95 (2003).
167. Decher, G. Fuzzy nanoassemblies: toward layered polymeric multicomposites. Science (Washington, D. C.) **277**, 1232–1237 (1997).
168. Salditt, T. & Schubert, U. S. Layer-by-layer self-assembly of supramolecular and biomolecular films. Rev. Molec. Biotech. **90**, 55–70 (2002).
169. Li, X.-j. & Schick, M. Self-assembly of copolymers and lipids. Condens. Matter Phys. **26**, 325–333 (2001).
170. Keren, K., Berman Rotem, S., Buchstab, E., Sivan, U. & Braun, E. DNA-templated carbon nanotube field-effect transistor. Science (Washington, D. C.) **302**, 1380–2 (2003).
171. Hall, S. R., Shenton, W., Engelhardt, H. & Mann, S. Site-specific organization of gold nanoparticles by biomolecular templating. ChemPhysChem **2**, 184–186 (2001).
172. Yin, Y., Gates, B. & Xia, Y. A soft lithography approach to the fabrication of nanostructures of single crystalline silicon with well-defined dimensions and shapes. Adv. Mater. (Weinheim, Ger.) **12**, 1426–1430 (2000).
173. Mayers, B., Jiang, X., Sunderland, D., Cattle, B. & Xia, Y. Hollow nanostructures of platinum with controllable dimensions can be synthesized by templating against selenium nanowires and colloids. J. Am. Chem. Soc. **125**, 13364–13365 (2003).
174. Sun, Y., Mayers, B. & Xia, Y. Metal nanostructures with hollow interiors. Adv. Mater. (Weinheim, Ger.) **15**, 641–646 (2003).
175. Kovtyukhova, N. I. & Mallouk, T. E. Nanowires as building blocks for self-assembling logic and memory circuits. Chem.–Eur. J. **8**, 4354–4363 (2002).
176. Martin, B. R., St. Angelo, S. K. & Mallouk, T. E. Interactions between suspended nanowires and patterned surfaces. Adv. Funct. Mat. **12**, 759–765 (2002).
177. McCleskey, T. M., Ehler, D. S., Young, J. S., Pesiri, D. R., Jarvinen, G. D. & Sauer, N. N. Asymmetric membranes with modified gold films as selective gates for metal ion separations. J. Membr. Sci. **210**, 273–278 (2002).
178. Chun, K.-Y. & Stroeve, P. External Control of Ion Transport in Nanoporous Membranes with Surfaces Modified with Self-Assembled Monolayers. Langmuir **17**, 5271–5275 (2001).
179. Johnson, C. J., Dujardin, E., Davis, S. A., Murphy, C. J. & Mann, S. Growth and form of gold nanorods prepared by seed-mediated, surfactant-directed synthesis. J. Mater. Chem. **12**, 1765–1770 (2002).
180. Jana, N. R., Gearheart, L. & Murphy, C. J. Seed-mediated growth approach for shape-controlled synthesis of spheroidal and rod-like gold nanoparticles using a surfactant template. Adv. Mater. (Weinheim, Ger.) **13**, 1389–1393 (2001).
181. Kiraly, Z., Veisz, B., Mastalir, A. & Koefarago, G. Preparation of Ultrafine Palladium Particles on Cationic and Anionic Clays, Mediated by Oppositely Charged Surfactants: Catalytic Probes in Hydrogenations. Langmuir **17**, 5381–5387 (2001).
182. Xia, Y., Yin, Y., Lu, Y. & McLellan, J. Template-assisted self-assembly of spherical colloids into complex and controllable structures. Adv. Funct. Mat. **13**, 907–918 (2003).
183. Yin, Y., Lu, Y., Gates, B. & Xia, Y. Template-Assisted Self-Assembly: A Practical Route to Complex Aggregates of Monodispersed Colloids with Well-Defined Sizes, Shapes, and Structures. J. Am. Chem. Soc. **123**, 8718–8729 (2001).
184. Yin, Y. & Xia, Y. Self-Assembly of Spherical Colloids into Helical Chains with Well-Controlled Handedness. J. Am. Chem. Soc. **125**, 2048–2049 (2003).
185. Redl, F. X., Cho, K. S., Murray, C. B. & O'Brien, S. Three-dimensional binary superlattices of magnetic nanocrystals and semiconductor quantum dots. Nature (London) **423**, 968–971 (2003).
186. Li, M., Schnablegger, H. & Mann, S. Coupled synthesis and self-assembly of nanoparticles to give structures with controlled organization. Nature (London) **402**, 393–395 (1999).
187. De Stefanis, A. & Tomlinson, A. A. G. Scanning Probe Microscopies—From Surface Structure to Nano-Scale Engineering -. [In: Mater. Sci. Found., 2001; 14] (Trans Tech: Zurich, 2001).
188. Sakurai, T. & Watanabe, Y. (eds.) Advances in Scanning Probe Microscopy. [In: Adv. Mater. Res. (Berlin, Ger.), 2000; 2] (Springer: Berlin, 2000).

189. Ostuni, E., Chapman, R. G., Holmlin, R. E., Takayama, S. & Whitesides, G. M. A Survey of Structure-Property Relationships of Surfaces that Resist the Adsorption of Protein. Langmuir **17**, 5605–5620 (2001).
190. Lopes, W. A. & Jaeger, H. M. Hierarchical self-assembly of metal nanostructures on diblock copolymer scaffolds. Nature (London) **414**, 735–8 (2001).
191. Spillmann, H., Dmitriev, A., Lin, N., Messina, P., Barth, J. V. & Kern, K. Hierarchical Assembly of Two-Dimensional Homochiral Nanocavity Arrays. J. Am. Chem. Soc. **125**, 10725–10728 (2003).
192. Milic, T. N., Chi, N., Yablon, D. G., Flynn, G. W., Batteas, J. D. & Drain, C. M. Controlled hierarchical self-assembly and deposition of nanoscale photonic materials. Angew. Chem., Int. Ed. Engl. **41**, 2117–2119 (2002).
193. Cates, M. E. & Evans, M. R. (eds.) Soft and Fragile Matter, Nonequilibrium Dynamics, Metastability and Flow (IOP: Bristol, UK, 2000).
194. Ng, J. M. K., Fuerstman, M. J., Grzybowski, B. A., Stone, H. A. & Whitesides, G. M. Self-Assembly of Gears at a Fluid/Air Interface. J. Am. Chem. Soc. **125**, 7948–7958 (2003).
195. Grzybowski, B. A., Wiles, J. A. & Whitesides, G. M. Dynamic Self-Assembly of Rings of Charged Metallic Spheres. Phys. Rev. Lett. **90**, 083903/1-083903/4 (2003).
196. Pronk, S. & Frenkel, D. Large effect of polydispersity on defect concentrations in colloidal crystals. J. Chem. Phys. **120**, 6764–6768 (2004).
197. Sollich, P. Predicting phase equilibria in polydisperse systems. J. Phys.: Condens. Matter **14**, R79-R117 (2002).
198. Colvin, V. L. From opals to optics. Colloidal photonic crystals. MRS Bull. **26**, 637–641 (2001).

Index